AF820293

Modeling, Diagnostics and Process Control

Józef Korbicz and Jan Maciej Kościelny (Eds.)

Modeling, Diagnostics and Process Control

Implementation in the DiaSter System

Springer

Prof. Józef Korbicz
University of Zielona Góra,
Institute of Control & Computation Engineering
ul. Podgórna 50,
65-246 Zielona Góra
Poland
E-mail: J.Korbicz@issi.uz.zgora.pl

Prof. Jan Maciej Kościelny
Warsaw University of Technology,
Institute of Automatic Control and Robotics
Ul. Chodkiewicza 8
02-525 Warszawa
Poland
E-mail: jm.koscielny@mchtr.pw.edu.pl

ISBN 978-3-642-16652-5 e-ISBN 978-3-642-16653-2

DOI 10.1007/978-3-642-16653-2

Library of Congress Control Number: 2010938641

© 2010 Springer-Verlag Berlin Heidelberg

This work is subject to copyright. All rights are reserved, whether the whole or part of the material is concerned, specifically the rights of translation, reprinting, reuse of illustrations, recitation, broadcasting, reproduction on microfilm or in any other way, and storage in data banks. Duplication of this publication or parts thereof is permitted only under the provisions of the German Copyright Law of September 9, 1965, in its current version, and permission for use must always be obtained from Springer. Violations are liable to prosecution under the German Copyright Law.

The use of general descriptive names, registered names, trademarks, etc. in this publication does not imply, even in the absence of a specific statement, that such names are exempt from the relevant protective laws and regulations and therefore free for general use.

Typesetting: Data supplied by the authors

Cover Design: Scientific Publishing Services Pvt. Ltd., Chennai, India

Printed on acid-free paper

9 8 7 6 5 4 3 2 1

springer.com

Preface

The continuous increase in the complexity of modern industrial systems and objects as well as growing reliability demands regarding their operation and control quality are serious challenges for further development of the theory and practice of control and technical diagnostics. Thus modern control systems are complex in the sense of implementing numerous functions, such as process variable processing, digital control, process monitoring and alarm indication, graphic visualization of the course of a process, or data exchange with other systems or databases. Moreover, modern control systems are integrated with management systems, which very often cover production and corporate management problems. Hardware and software structures of control systems of complex processes are decentralized and space distributed. Decentralization consists in dividing system operation into many function units working simultaneously. In integrated systems, software for control and visualization creates one information system with a common database. The main driving force behind the development of modern control systems is rapid evolvement of computer techniques, which have forced the standardization of field networks and programming languages of control systems. The enormous possibilities of technical and program realization of control systems permit significant extension of their functions and tasks, including the introduction of advanced algorithms of process modeling, control and diagnostics.

The present book conveys a description of the developed *DiaSter* system as well as characteristics of advanced original methods of modeling, knowledge discovery, simulator construction, diagnostics, control, and supervision control applied in the system. The system gives the possibility of early recognition of abnormal states of industrial processes and faults or malfunctions of actuators as well as technological and measuring units. The universality of solutions assumed in *DiaSter* allows its broad application, for example, in power, chemical, pharmaceutical, metallurgical and food industries. The system is a world-scale unique solution, and due to its open architecture it can be connected practically with any other control systems.

In the main part of the book, analytical and artificial intelligence methods implemented in the *DiaSter* system are discussed. One of the first chapters is devoted to problems of physical process modeling, fundamental in modern control, diagnostics, or designing and searching for alternative solutions. The known analytical as well as neural, fuzzy and neuro-fuzzy models are presented. Particularly the latter, that is, artificial intelligence methods, are attractive for control systems as they give possibilities for describing nonlinear processes in quite an easy way. In turn, taking into account the fact that modern Distributed Control Systems (DCSs) as well as Supervision Control And Data Acquisition (SCADA) systems allow collecting a huge amount of process data coming from different sources, in another chapter methods and algorithms of knowledge discovery in databases are considered. In *DiaSter*, process data can be a source of diagnostic knowledge, which—obtained with the techniques of knowledge discovery—can be used for fault detection and localization of processes and systems.

An extensive chapter is devoted to describing diagnostics methods of processes and systems, which can be found useful in industry and were implemented in the *DiaSter* system. First of all, robust fault detection algorithms designed by employing the adaptive decision threshold approach are discussed. Such thresholds were assigned for fault detection systems with dynamical neural models such as multilayer perceptrons and the Group Method of Data Handling (GMDH). Methods of fault localization are considered mainly with the application of fuzzy logic. The proposed approach is based on inference rules robust with respect to structure changes of diagnosed processes or systems. It includes inference algorithms implemented in *DiaSter* for single and multiple faults, different ways of dealing with symptoms occurrence delays, or reference algorithms in a hierarchical structure. Moreover, taking into account advantages and disadvantages of methods of symptom diagnostics and the model based approach, in the book a belief-network-based model is presented as well. It is a heuristical model that permits sequential application of methods characteristic for both classes of diagnostic investigation.

In another chapter, methods of supervision control implemented in the system are discussed. Elementary structures and algorithms of predictive control, the so-called MPC (Model-based Predictive Control), including the DMC (Dynamic Model Control) algorithm for linear models, are presented. Also, fundamental methods of automatic adjustment of the PID control loop followed by precision adjustment and adaptation are considered. Describing a set point control system, the way of transmitting the set amount generated in the optimization layer of the process working point to the control loop in the direct control layer is presented.

The last chapter illustrates the operation of different functions of *DiaSter* with a simple system of three tanks. The chosen plain example has many educational advantages and gives an excellent possibility for exact studying of fundamental characteristics and possibilities of the *DiaSter* system. By using system tools, a simulator of the three-tank system, dynamical models of the GMDH and the multilayer perceptron type as well as the TSK (Takagi–Sugeno–Kang) model were built. The realization of diagnostic tasks was shown both for systematic diagnostics of abrupt faults as well as the detection and tracking of the development of slowly increasing faults. Here, the control systems used in the object, that is, traditional PID controllers with automatic adjustment of settings and predictive controllers, are presented as well.

The present monograph is in a sense a continuation of our earlier book entitled *Fault Diagnosis. Models, Artificial Intelligence, Applications* (Springer, 2004), and its first edition was published in Polish by Wydawnictwa Naukowo-Techniczne, WNT, in 2009, (Warsaw, Poland). This english language version is not merely a translation of the original—many chapters contain some significant improvements, such as new or extended parts and examples, found especially in Chapter 7. The book presents theoretical and practical results of research into fault diagnosis and control conducted over many years within cooperation between Polish research teams from the Warsaw University of Technology, the University of Zielona Góra, the Silesian University of Technology in Gliwice, and the Technical University of Rzeszów. In the years 2007–2009, the above-mentioned consortium of universities conducted a developmental project entitled *Intelligent diagnostic and control assistance system for industrial processes DiaSter.*

The editors wish to express their gratitude to all authors for preparing joint chapters and for very fruitful collaboration during the editorial process.

July 2010
Zielona Góra/Warsaw

Józef Korbicz
Jan Maciej Kościelny

Contents

List of Contributors

Paweł Chrzanowski
Department of Fundamentals of Machinery Design, Silesian University of Technology, ul. Konarskiego 18A, 44–100 Gliwice, Poland
pawel.chrzanowski@polsl.pl

Wojciech Cholewa
Department of Fundamentals of Machinery Design, Silesian University of Technology, ul. Konarskiego 18A, 44–100 Gliwice, Poland
wojciech.cholewa@polsl.pl

Bartłomiej Fajdek
Institute of Automatic Control and Robotics, Warsaw University of Technology, ul. Św. Andrzeja Boboli 8, 02–525 Warsaw, Poland
b.fajdek@mchtr.pw.edu.pl

Krzysztof Janiszowski
Institute of Automatic Control and Robotics, Warsaw University of Technology, ul. Św. Andrzeja Boboli 8, 02–525 Warsaw, Poland
kjanisz@mchtr.pw.edu.pl

Józef Korbicz
Institute of Control and Computation Engineering, University of Zielona Góra, ul. Podgórna 50, 65–246 Zielona Góra, Poland
j.korbicz@issi.uz.zgora.pl

Jan Maciej Kościelny
Institute of Automatic Control and Robotics, Warsaw University of Technology, ul. Św. Andrzeja Boboli 8, 02–525 Warsaw, Poland
jmk@mchtr.pw.edu.pl

Maciej Ławryńczuk
Institute of Control and Computation Engineering, Warsaw University of Technology, ul. Nowowiejska 15/19, 00–665 Warsaw, Poland
m.lawrynczuk@ia.pw.edu.pl

Piotr Marusak
Institute of Control and Computation Engineering, Warsaw University of Technology, ul. Nowowiejska 15/19, 00–665 Warsaw, Poland
p.marusak@ia.pw.edu.pl

Wojciech Moczulski
Department of Fundamentals of Machine Design, Silesian University of Technology, ul. Konarskiego 18A, 44–100 Gliwice, Poland
wojciech.moczulski@polsl.pl

Krzysztof Patan
Institute of Control and Computation Engineering, University of Zielona Góra, ul. Podgórna 50, 65–246 Zielona Góra, Poland
k.patan@issi.uz.zgora.pl

Tomasz Rogala
Department of Fundamentals of Machinery Design, Silesian University of Technology, ul. Konarskiego 18A, 44–100 Gliwice, Poland
tomasz.rogala@polsl.pl

Andrzej Stec
Department of Computer and Control Engineering, Rzeszów University of Technology, ul. W. Pola 2, 35–959 Rzeszów, Poland
astec@prz-rzeszow.pl

Michał Syfert
Institute of Automatic Control and Robotics, Warsaw University of Technology, ul. Św. Andrzeja Boboli 8, 02–525 Warsaw, Poland
msyfert@mchtr.pw.edu.pl

Robert Szulim
Institute of Electrical Metrology, University of Zielona Góra, ul. Podgórna 50, 65–246 Zielona Góra, Poland
r.szulim@ime.uz.zgora.pl

Zbigniew Świder
Department of Computer and Control Engineering, Rzeszów University of Technology, ul. W. Pola 2, 35–959 Rzeszów, Poland
swiderzb@prz-rzeszow.pl

Piotr Tatjewski
Institute of Control and Computation Engineering, Warsaw University of Technology, ul. Nowowiejska 15/19, 00–665 Warsaw, Poland
p.tatjewski@ia.pw.edu.pl

Piotr Tomasik
Department of Fundamentals of Machine Design, Silesian University of Technology, ul. Konarskiego 18A, 44–100 Gliwice, Poland
piotr.tomasik@polsl.pl

Leszek Trybus
Department of Computer and Control Engineering, Rzeszów University of Technology, ul. W. Pola 2, 35–959 Rzeszów, Poland
ltrybus@prz-rzeszow.pl

Dominik Wachla
Department of Fundamentals of Machine Design, Silesian University of Technology, ul. Konarskiego 18A, 44–100 Gliwice, Poland
dominik.wachla@polsl.pl

Marcin Witczak
Institute of Control and Computation Engineering, University of Zielona Góra, ul. Podgórna 50, 65–246 Zielona Góra, Poland
m.witczak@issi.uz.zgora.pl

Paweł Wnuk
Institute of Automatic Control and Robotics, Warsaw University of Technology, ul. Św. Andrzeja Boboli 8, 02–525 Warsaw, Poland
p.wnuk@mchtr.pw.edu.pl

Chapter 1
Introduction

Jan Maciej Kościelny

1.1 Control System Structures

Modern control systems are complex, i.e., they implement many different functions. The basic ones include process variables processing, binary control, direct digital control, supervisory control, batch control, balance calculations, statistic indices calculation, process protection, the sequence of event processing, alarm signaling, as well as data storing in archives, reporting, graphic visualization of process changes, hardware structures configuration, data exchange with other systems or databases, calculation sheets, etc., control with the use of recipes, statistic process control.

The functional structure defines the set of implemented tasks and relations that exist between them. Modern control systems and manufacturing management have a hierarchical functional structure (Isermann, 2006; Niederliński, 1984; Tatjewski, 2007; Korbicz and Kościelny, 2009). The hardware-software structure defines the way of technical implementation of tasks with the use of hardware and software. Many different hardware-software systems can correspond with the functional structure.

Within the automatic control system functional structure, one can distinguish the following:

- the process automatic control system, containing
  - the control subsystem,
  - the visual presentation and maintenance subsystem,
  - the safety subsystem,
  - the plant equipment maintenance subsystem;

Jan Maciej Kościelny
Institute of Automatic Control and Robotics,
Warsaw University of Technology,
ul. Św. Andrzeja Boboli 8, 02–525 Warsaw,
Poland
e-mail: jmk@mchtr.pw.edu.pl

- the management system, containing
  - the manufacturing management subsystem,
  - the company management subsystem.

The following levels are included in the control subsystem (Fig. 1.1):

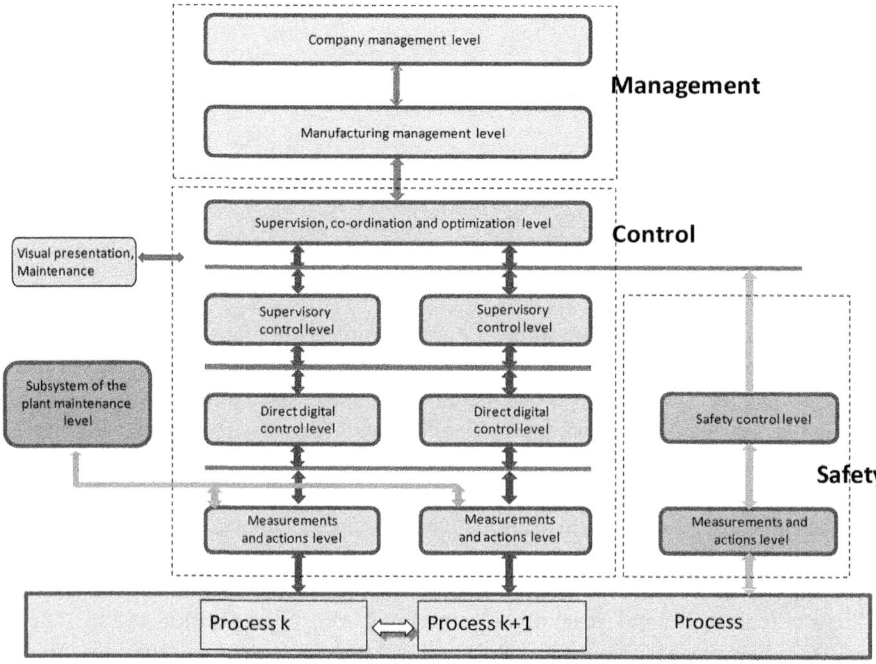

Fig. 1.1 Functional structure of the control and manufacturing management system

- *the measurement and controlling action level:* in the hardware structure, this is the set of instruments and actuators. More and more often, it consists of the so-called intelligent transducers and actuators equipped with microprocessor units that implement many functions of preprocessing signal conversion, actuator position control and communication with control units via Fieldbus;
- *the direct digital control level:* control and discrete automatic control are the main tasks of this level. Different kinds of controllers and control devices are placed within this level;
- *the supervisory control level:* it implements control algorithms, disturbance compensation algorithms, adaptation algorithms and optimum control of particular technology nodes algorithms. Optimal control signals worked out at this level do not act directly on actuators but are inputs assigned (lead) to algorithms at the direct control level. Algorithms at the supervisory control level may be implemented both by high capacity controllers (called "process" stations in Distributed Control Systems (DCSs)) as well as computers;

- *the supervisory, co-ordination and process optimization level:* the surveillance includes the tasks of detection, registration and the signaling of alarms. Control tasks implemented for continuous processes at this level include the co-ordination of flows of materials and energies that flow between different parts of the process (technological nodes), as well as the optimization of the process points of operation. For discrete processes, algorithms for the cooperation of groups of machines and devices are implemented. These functions are executed by computers.

The characteristic features of the hierarchical structure of the automatic control structure include the fact that lower levels of the automatic control structure are attributed to particular parts of the process, but higher levels of the automatic control structure may be common for several parts or even for the whole process. The functional structure is therefore characterized by vertical divisions, i.e., particular parts of the control system are attributed to the existing parts of the process (technological nodes, stations or manufacturing departments). Even smaller manufacturing subsystems are isolated. For instance, in a sugar factory, in the raw state subproduct station, the following substations can be distinguished: the sugar beet juice extraction substation, the juice refining substation, and the evaporator substation. The sugar factory power station consists of the boiler house substation, the turbine room substation, and the boiler water conditioning substation. For each one of the substations, most often separate subsystems are implemented which include process measurements and actions as well as direct control. However, the supervisory control and surveillance levels may be selected individually for each substation, but most often they are common for the whole raw state subproduct station or the thermal-electric power station.

In automatic control structures, higher-level algorithms calculate parameters for lower-level algorithms, but the latter are usually implemented more often than the former (Niederliński, 1984). For example, supervisory control algorithms calculate set-point values for control algorithms and they in turn direct control signals to the level of acting on the process where the actuators (often servo-mechanisms) follow control signals changes. Supervisory control algorithms, however, are usually implemented with much lower frequencies (e.g., several times per hour) than the control algorithm (several or several dozens of times per second).

The main task of the supervision subsystem is to ensure proper process operation. Stations can present adequately converted and prepared data from other levels. Process changes can be observed with the use of synoptic diagrams of the process, process variables diagrams, lists of events and alarms, etc. Not only current data can be observed but also past ones extracted from the archives. Reports give the operators data composed adequately to the users' needs. Operators' stations allow us to observe the process changes and to implement necessary actions (switching the devices on or off, set-point changes, manual control, etc.). Engineers' stations allow us to make hardware configuration changes, software (algorithm) changes of the system and to introduce modifications, but surveillance stations allow us to observe the visual presentation of the process and data changes only.

Each process has its safety subsystem that is independent of the automatic control system at the measurements and actions level, as well as at the control level. This means that the protection functions are implemented with the use of other devices (acting usually in the redundancy structure) such as the ones applied for control tasks. The visual presentation for control and process safety is usually integrated.

During the last couple of years, automatic control systems have begun to implement additional tasks related to the service of intelligent plant-based devices. They can be separated into individual subsystems that contain databases in which all configuration-related parameters of measuring devices and actuators are gathered, along with information on the implemented modifications of these parameters and device repairs, etc. Such a subsystem also permits remote configuration and calibration of plant-based devices, as well as off-line diagnostics (when a device does not work within the process) and—more and more often—current (on-line) diagnostics of the measuring devices and actuators that work within the process.

Modern automatic control systems are integrated with management systems. Such an integrated control and management system is presented in Fig. 1.1. It possesses two additional levels: one for manufacturing management and the other one for company management implemented in computer-based technology exclusively. The management tasks are implemented by Manufacturing Execution Systems (MESs) and Enterprise Resource Planning (ERP) systems.

In the hardware structure of automatic control systems one can distinguish the following units:

- measuring devices and actuators,
- controlling devices (Programmable Logic Controllers (PLCs), Programmable Automation Controllers (PACs), apparatus controllers, process stations),
- operators' panels and stations,
- engineers' and surveillance stations,
- servers,
- networks.

Networks unite all of the remaining devices into one system. In automatic control systems, both computer-based Local Area Networks (LANs) as well as field (industrial) networks usually called "Fieldbus" are applied. In the structure of the system, not just one network exists but usually several ones, adapted to different needs, conditions and technical requirements concerning data transmission in various places of the system structure. Separate networks are usually applied to connect of intelligent measuring devices and actuators with controlling devices, controlling units with observation units and devices for controlling system maintenance as well as the control and management system.

Hardware-software structures of systems controlling complex industrial processes are de-centralized and space-distributed. De-centralization consists in the distribution of system operation into many functional units working in parallel attributed to different parts of the process and (usually) placed apart one from the other. Most often, two kinds of hardware-software structures are applied:

- integrated systems (Distributed Control Systems (DCSs) class),
- systems implemented with the use of Supervisory Control And Data Acquisition (SCADA) monitoring systems as well as programmable controllers.

In integrated systems, software for control and visual presentation creates one system with a common database and is supplied by one manufacturer. The second group of solutions, however, is characterized by the fact that systems are composed from the SCADA monitoring system and programmable controllers. The controlling devices can be supplied by manufacturers different than the ones that supply us with the SCADA system. Therefore, the controlling system is composed of separate software systems for the monitoring and control of the process.

To the control of the largest technical installations in the power, chemical, iron and steel industries, DCSs are applied. They are characterized by the following:

- the possibility of controlling large processes—up to 50 to 200 thousand process variables;
- the possibility of the redundancy of all elements of the system structure;
- autodiagnostics;
- the scaling ability—the possibility to obtain economically rational solutions for large, medium and small plants, and structure expansion as the plant is enlarged;
- complexity—the possibility of the implementation of all tasks mentioned in Chapter 3;
- backward compatibility—compliance with the rule that new generations of the system should cooperate with the earlier ones so that system expansion is possible without additional costs of the modernization of parts installed earlier;
- the use of real-time operational systems (e.g., *VxWorks, VRTX, OS9*) in controlling devices;
- the possibility of inserting I/O modules in to the automation system be means of hot swap;
- the possibility of system software upgrading in the run mode without disturbing the controlled process itself or its control;
- the use of the time stamp for measuring signals, events and alarms;
- the availability of intrinsically safe I/O modules;
- the availability of safety controllers;
- the availability of software needed for the implementation of advanced modeling functions, diagnostics and automatic control.

The hardware structure does not always correspond with the hierarchy of the functional structure. Figure 1.2 presents an example of the hardware structure of a system (without transducers and actuators).

1.2 Trends in the Development of Modern Automatic Control Systems

The progress of information technology gave impetus towards the development of control systems: computational power of the new generation of processors,

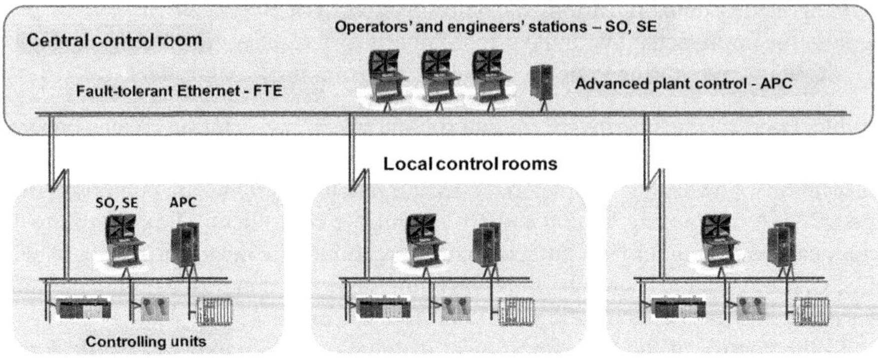

Fig. 1.2 Example of the hardware structure of a control system (based on the Honeywell company's materials)

memory capacity, the speed of data transmission in LANs and Fieldbus networks, the development of the Internet and system software. Another element that changed automatic control systems was the standardization of software languages of automatic control systems and field networks. One could mention the following directions of the development of automatic control systems observed in the last couple of years:

- the application of standard computers and operating systems for process monitoring and surveillance. The dominating position is occupied here by the *Windows* operating system. The *Unix* system is applied more rarely. The use of non-standard operation systems, so often applied a dozen or so years ago, has been completely abandoned;
- the application of the Intranet and Internet for interchanging information on process changes within the plant as well as long-range data transmission;
- the introduction of standard field networks (Profibus FMS/DP/PA, CAN, Fieldbus Foundation H1, H2);
- the development and gradual implementation of wireless technology for data transmission for measuring devices and actuators;
- broadening the system structure by the introduction of an additional level of information stations assigned for plant management, engineering supervision, etc;
- the introduction of intelligent measuring devices and actuators, i.e., devices supplied with microprocessor-based units connected with control devices with the use of field networks;
- the introduction of the IEC 61131 standard, which defines five languages for process automatic control (FBD, LD, SFC, DT, IL, CFC);
- the increase of the system openness degree;
- the ensurance of system reliability factors at acceptable levels as a compromise between system safety demands and costs. This is implemented by optional (and not obligatory) application of the redundancy of particular units of the system;

- the introduction of the diagnostics (autodiagnostics) of the automatic control system (process and operators' stations as well as network elements) in integrated systems;
- ensuring the scalability of systems, which allows us to apply it to automatic control of large, medium and small plants;
- the application of multi-media and management techniques to operators' and information stations;
- the integration of automatic control and management systems with the use of data warehouses;
- the introduction of new functions of automatic control systems, such as advanced modeling, control and process diagnostics algorithms.

The above-mentioned directions of automatic control systems development probably do not exhaust all of the tendencies that appear in new solutions. However, they do show that the new generation of automatic control systems is better than the previous ones. It has better openness, high comfort of service, the easiness of design, the possibility of application to technical installations of various dimensions. Integrated controlling software, as well as software for visual presentation and configuration, ensures also current update of the system documentation and easy access to it. New functions of automatic control systems are becoming—apart from the development and standardization of hardware and software—one of main areas of rivalry between manufacturers offering automatic control systems. These new tasks of automatic control systems are characterized in Section 1.3.

1.3 New Functions of Advanced Automatic Control Systems

Process control and supervision functions in automatic control and monitoring of industrial processes are still relatively limited. In control, mainly classic PID controllers are applied, and software for advanced control is available in few modern integrated systems or separate software packages. Supervision functions rely on the detection and signaling of events and alarms. Such a supervision functional range can be described as basic.

This situation, however, is rapidly changing. Recently, fast development of methods and software for advanced automatic control modeling and process diagnostics has been observed. Very many papers are written on the subject. The implementation of advanced calculation algorithms is possible thanks to modern computer technology, while modern teaching programs at universities allow us to break down the barrier related to the necessity of training specialists prepared for the implementation and use of advanced modeling, control and diagnostic techniques.

Models of processes are necessary for the implementation of advanced algorithms of process control optimization, fault detection, the implementation of virtual sensors, analyzers and simulators of processes. Problems of process identification and modeling are broadly discussed by Walter and Pronzato (1997), Ljung (1999), Janiszowski (2002), Duda (2003), Landau and Zito (2006), Isermann and Münchhof (2009). Apart from analytic methods of modeling, neural and fuzzy techniques are also widely applied (Ying, 2000; Piegat, 2001; Rutkowski, 2004b; Witczak, 2007; Patan, 2008b; Świątek, 2009; Talebi et al., 2010).

Software for physico-chemical modeling and modeling based on available measurement data is also being developed. With the use of process models, virtual sensors and analyzers are built that constitute information redundancy of measured signals. Measurements of some physical quantities are costly, so in many cases it is more profitable to use process models to calculate signal values using other measured process variables than to use hardware redundancy. When the measuring path is faulty, the variable value can be reconstructed based on the virtual sensor or analyzer. Thus, we obtain system robustness to faults of these measuring lines for which virtual substitutes exist. In the power industry, it is especially advisable to construct virtual analyzers of NOx and COx emissions since hardware analyzers cause many exploitation problems. Process simulators are more and more widely applied mainly to the training of operators. They can also be applied to the testing of new control strategies, process state changes predictions, etc.

In modern automatic control systems, advanced control algorithms are more and more widely applied. Advanced control techniques use algorithms which are more complex than classic PID algorithms (Tatjewski, 2007). Classic PID type algorithms and their expansions are still the dominating solution at the direct control level, which results from their robustness, the fact that operators are familiar with them, and their simple service. In distributed control systems, self-tuning and control loop adapting methods are applied (Bobál et al., 2005; Świder and Trybus, 2009).

The development of new automatic control techniques results from difficulties that arise during the control of difficult processes. Control difficulties are related to strong non-linearities, high time-delays, signal limitations, the necessity to ensure control robustness, as well as to the fact that processes are multi-dimensional. Many monographs are available that deal with modern automatic control techniques (Åström and Wittemark, 1997; Grimble, 2001; Goodwin et al., 2001; Ackermann, 2002; Duda, 2003; Grega, 2004; Albertos and Sala, 2002; Christofides and El-Farra, 2005; Rosenwasser and Lampe, 2006; Bańka, 2007; Astolfi et al., 2008; Chernous'ko et al., 2008; Zecevic and Siljak, 2010).

At the direct automatic control level, fuzzy control algorithms are more and more widely applied (Kacprzyk, 1996; Palm and Driankov, 1997; Ying, 2000; Piegat, 2001; Tatjewski, 2007), and they are especially useful for non-linear installations. To loops having difficult dynamics, predictive controllers are also applied.

Model Predictive Control (MPC) is an effective way to control multi-dimensional processes with difficult dynamics (Maciejowski, 2002; Camacho and Bordons, 2004; Tatjewski, 2007; Huang and Kadali, 2008). Because of that, it is applied more and more often at the supervisory control level, where slowly changing process variables are to be controlled that have a key significance for obtaining the required product quality. The advantage of this technique is the possibility of taking into account the limitations of control and output signals.

At the optimization level, algorithms for the optimization of operational points are applied. They should define set-points for the control loops implemented at lower levels (Brdyś and Tatjewski, 2005; Tatjewski, 2007).

Figure 1.3 presents classic single loop control systems and multi-dimensional predictive control. Figure 1.4 shows the possibility of reaching the optimum point of operation with the use of advanced control algorithms.

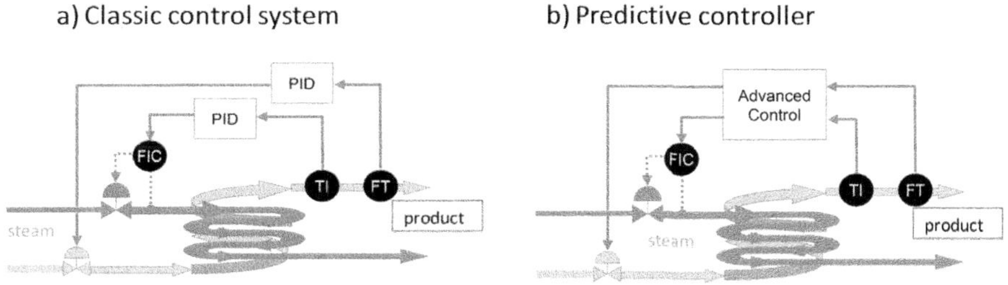

Fig. 1.3 Classic and advanced control (based on the Foxboro company's materials)

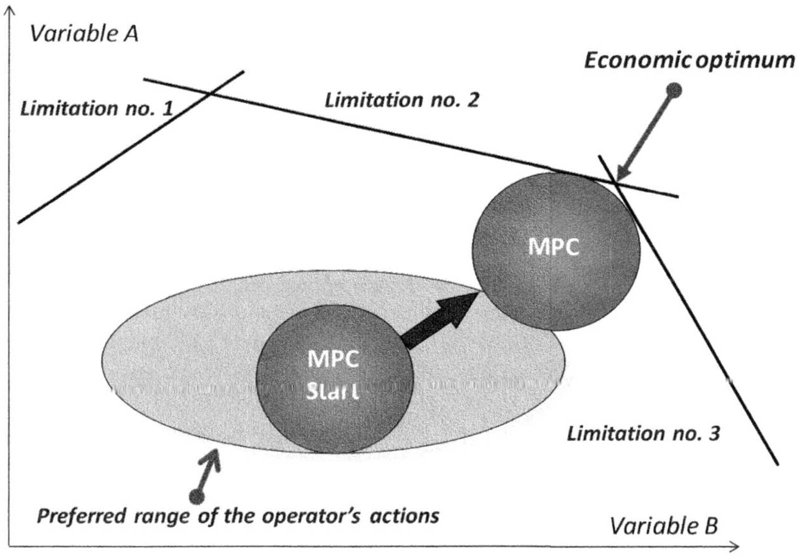

Fig. 1.4 Reaching the optimum point of operation (based on the ABB company's materials)

Fault-Tolerant Control (FTC) algorithms (mainly of measuring loops and actuators) define a different dimension in research and development. The ability of fault tolerance (Mahmoud et al., 2003; Hajiyev and Caliskan, 2003; Blanke et al., 2003; Héctor and Fabián, 2005; Isermann, 2006) is obtained first of all by the implementation of current fault diagnostics and the re-configuration of the hardware or software structure of the automatic control system with states with faults (Fig. 1.5).

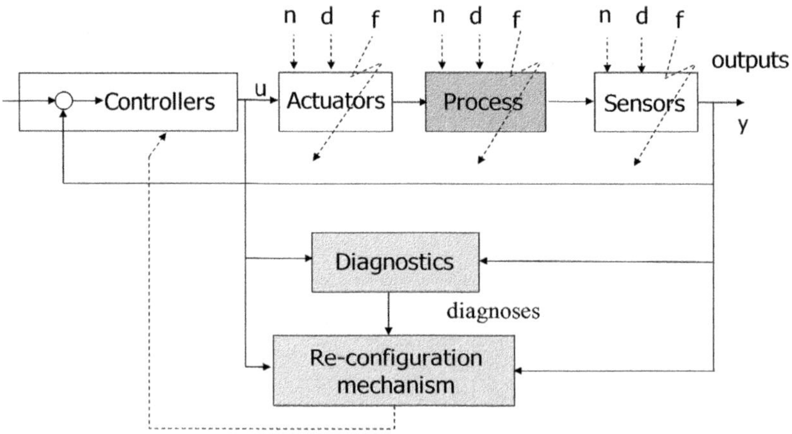

Fig. 1.5 Structure of the fault-tolerant system (n: measurement noise, d: disturbance, f: faults)

Modern computer systems allow us to apply complex (in the sense of calculations) algorithms developed on the grounds of computer science, automatic control, diagnostics and knowledge engineering. They use artificial intelligence techniques such as artificial neural networks (Tadeusiewicz, 1993; Hrycej, 1997; Osowski, 2000; Gupta et al., 2003; Dreyfus, 2005; Rutkowski, 2004a; Talebi et al., 2010), fuzzy logic (Kacprzyk, 1996; Palm and Driankov, 1997; Fuller, 2000; Piegat, 2001; Rutkowski, 2004a), rough sets (Pawlak, 1991), evolution algorithms (Tenne, 2010; Rutkowska et al., 1997), as well as expert systems (Mulawka, 1996; Ligęza, 2005), methods for knowledge detection using databases (Moczulski, 2002a). Adequate software has the form of special software modules being an element of the automatic control system, or expert systems integrated with the automatic control system. During the last dozen of years we have observed significant development of current diagnostics of industrial processes. There have been published monographs that present methods for fault detection and isolation in measurement instruments, actuators, and components of the process (Basseville and Nikiforov, 1993; Chen and Patton, 1999; Gertler, 1998; Patton and Frank, 2000; Kościelny, 2001; Chiang et al., 2001; Simani et al., 2003; Korbicz et al., 2004; Isermann, 2006; Witczak, 2007; Sobhani-Tehrani and Khorasani, 2009), along with many survey papers. The developed methods make the basis of modern diagnostic systems. The need for early and accurate fault recognition arises from the aspiration for increasing process safety as well as the limitation of losses in abnormal and faulty states. In technical installations of the power, chemical, iron and steel, food and many other industries, faults of components of the technical installation, measuring devices and actuators still appear despite the application of high reliability elements. In systems of automatic control of industrial processes, the module for alarm signaling is applied to the recognition of abnormal and faulty states. It is a simple version of the diagnostic system. The main disadvantage of alarm systems is the high number of signaled alarms. According to the data collected by the Engineering Equipment & Materials Users' Association (EEMUA), the average daily number of faults should not exceed 144 but, in reality, it is much higher, e.g., about 1500 in the oil industry and nearly 2000 in the power industry. Alarms appear especially often in states with faults. The interpretation of such a high number of alarms is a very serious problem for the operator. The information overload phenomenon comes into being here, as well as the stress as its result. This can lead to the existence of additional operators errors that, cumulating with earlier faults, cause serious breakdowns. The mechanism of such unfortunate (positive) feedback is presented in Fig. 1.6. It was the cause of many serious faults in nuclear power and conventional plants as well as in chemical industry installations.

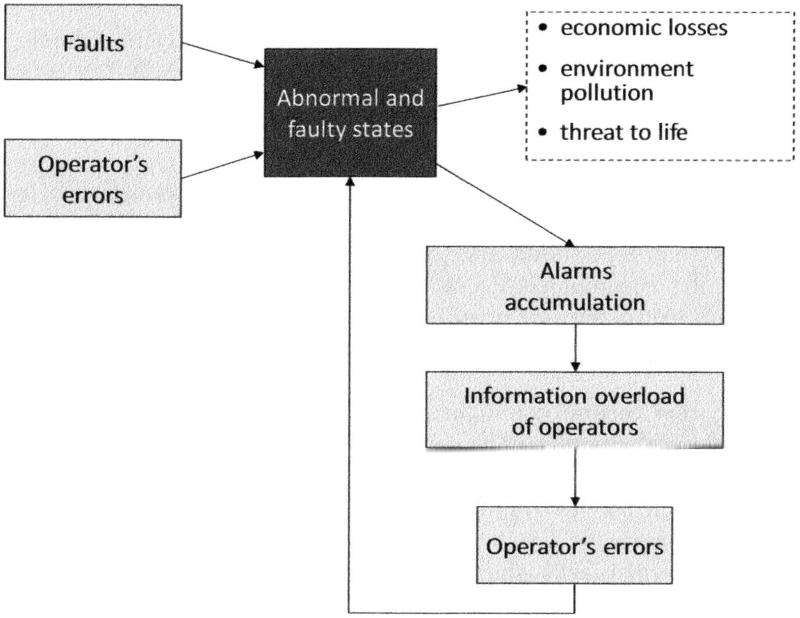

Fig. 1.6 Causes and results of abnormal and faulty states

However, classic methods applied to alarm detection and signaling have many disadvantages, including the following:

- a lack of the possibility of detecting some faults whose symptoms are masked by control loops (see Fig. 1.7);
- high detection delays;
- a lack of inference mechanisms that allow us to formulate diagnoses on faults;
- the alarm presentation method has its disadvantages, too: a fault manifests itself usually as the appearance of many alarms at different levels of the process, but alarms being the result of different faults are signaled together in the same picture.

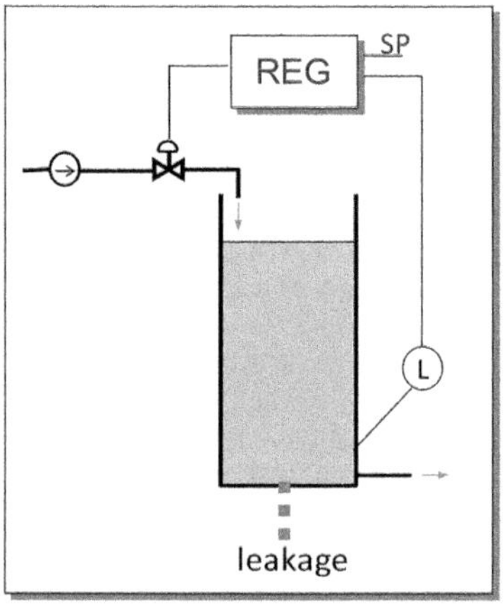

Fig. 1.7 Toxic substance leak masked by the level control system

The above-mentioned disadvantages make diagnosis formulation difficult for operators, i.e., the recognition of the cause of alarm set existence, which is necessary in many cases to undertake the proper protection action. Therefore, proper diagnosis formulation depends only on the operator's knowledge and his/her psycho-physical state.

Alarm system imperfection is the cause of the development of diagnostic systems for industrial processes. The diagnostic system's task is to detect and recognize faults understood as various events that destruct the process run. Diagnostic systems can implement the following functions: the detection, isolation, identification of faults, logging the data on faults into archives, diagnostic reports generation, visual presentation of faults, diagnoses justification, helping operators to arrive at a decision in states with faults.

Diagnoses that were accurately and quickly obtained make it possible to undertake necessary protection actions. The diagnostic system, together with adequate protection actions, makes therefore the second, higher level of the process protection system, while *Safety Instrumented Systems* (SISs) make the first, lower level of process protection system. The higher level of the process protection—thanks to accurate and quick diagnosis—gives us the possibility to reduce or even eliminate fault consequences (process run presented as a dashed line in Fig. 1.8). It also allows us to avoid the action of the lower-level protections, which in many cases causes a process shutdown or lowers its efficiency.

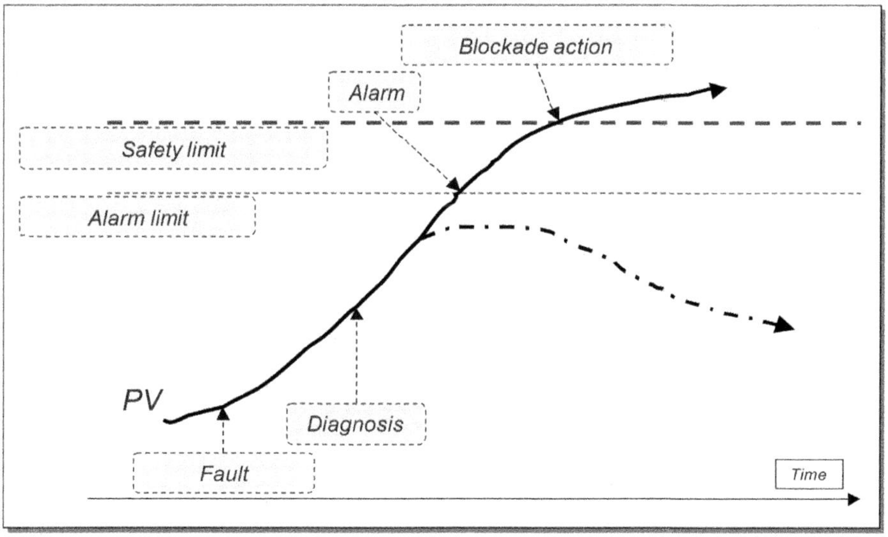

Fig. 1.8 Process changes in a system without and with diagnostics

The above-characterized new modeling, advanced control and diagnostic functions in processes were implemented in the *DiaSter* system. This monograph contains a description of the system and the characteristics of modeling knowledge invention, simulator construction, diagnostics and supervisory control methods applied in the system, as well as methods of their implementation. Examples of simple applications of the system are also given.

Chapter 2
Introduction to the DiaSter System

Jan Maciej Kościelny, Michał Syfert, and Paweł Wnuk

2.1 Introduction

The *DiaSter* software platform is a set of programs working together to realize functionality related to process modeling, variables processing, fault detection and diagnosis, advanced control, and decision support. This chapter begins with short descriptions of main applications of the *DiaSter* system as a convenient tool to realize multiple functions. Then a more detailed description of the *DiaSter* platform is given, starting with a presentation of the information model handled in the system. The possibility of data exchange between modules, different work modes and system extension mechanisms are presented. Finally, all main *DiaSter* modules are described, including off-line and on-line variables processing and visualization.

2.2 System Structure and Tasks

2.2.1 Main Uses of the System

The main application areas of the *DiaSter* system are advanced process modeling, supervisory control, process optimization and diagnostics (Fig. 2.1). The system gives a possibility to implement advanced supervisory and control algorithms and set-points optimization. In additionally, tools for control loop adjustment and on-line tuning are included, with a possibility to work with control loops embedded in a DCS and/or PLC controllers.

Jan Maciej Kościelny · Michał Syfert · Paweł Wnuk
Institute of Automatic Control and Robotics,
Warsaw University of Technology,
ul. Św. Andrzeja Boboli 8,
02–525 Warsaw,
Poland
e-mail: {jmk,msyfert,p.wnuk}@mchtr.pw.edu.pl

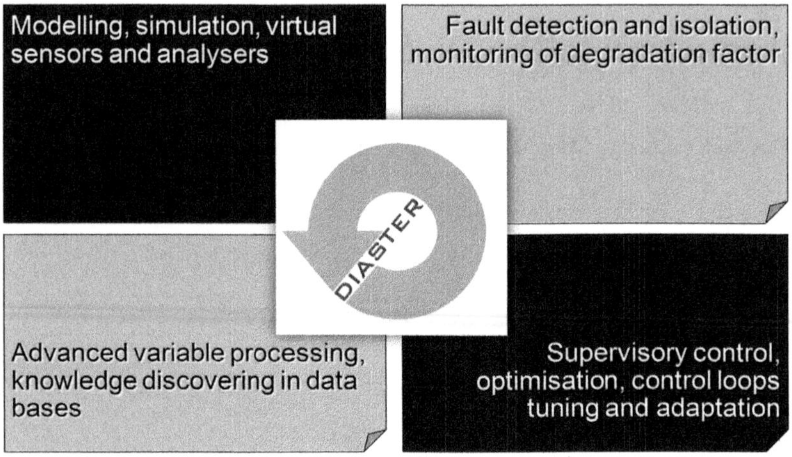

Fig. 2.1 Main uses of the *DiaSter* system

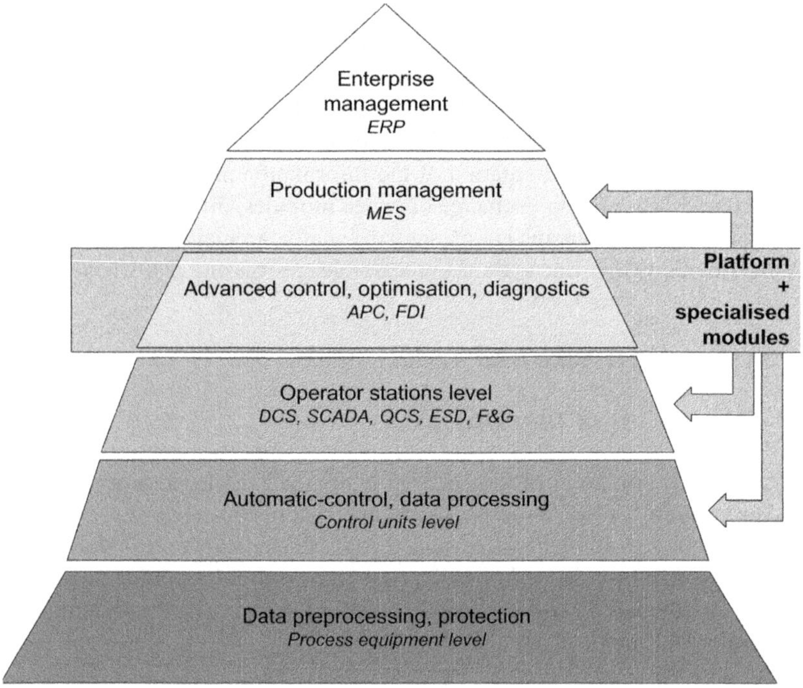

Fig. 2.2 *DiaSter* in industrial processes management tasks

DiaSter provides tools for early and precise detection and recognition of improper states of industrial processes as well as technological units, actuators and measurement faults. In abnormal and faulty states, *DiaSter* supports process operators through diagnosis presentation with optional advisory messages and operating instructions. The diagnosis given by the system is much more precise than the standard alarm sequence.

The system is intended for use in the power generation, chemical, pharmaceutical, steel or food production and many other industries. It is fully prepared to work in cooperation with other systems used by the industry (Fig. 2.2). The development of the system was supported by a Polish government grant entitled *Intelligent diagnostic and control support system for industrial processes DiaSter*. The work was conducted by a research team composed of specialists from the Warsaw University of Technology, the Silesian University of Technology, the Rzeszów University of Technology and the University of Zielona Góra. *DiaSter* is a new, extended version of the *AMandD* system, developed at the Institute of Automatic Control and Robotics of the Warsaw University of Technology.

The system is a world-scale unique solution. It includes the implementation of a wide range of the latest algorithms in the field of intelligent computation, used in system modeling, supervisory control, optimization, fault detection and isolation. Thanks to its open architecture, connections to virtually any automation system are possible and easy to implement.

2.2.2 *System Functionality*

2.2.2.1 Process Variables Processing

The system gives a possibility to freely design processing paths for each variable. This objective is realized by the module *PExSim*, similarly to the solution used in the *Matlab Simulink* package (Fig. 2.3). The user has access to several blocks

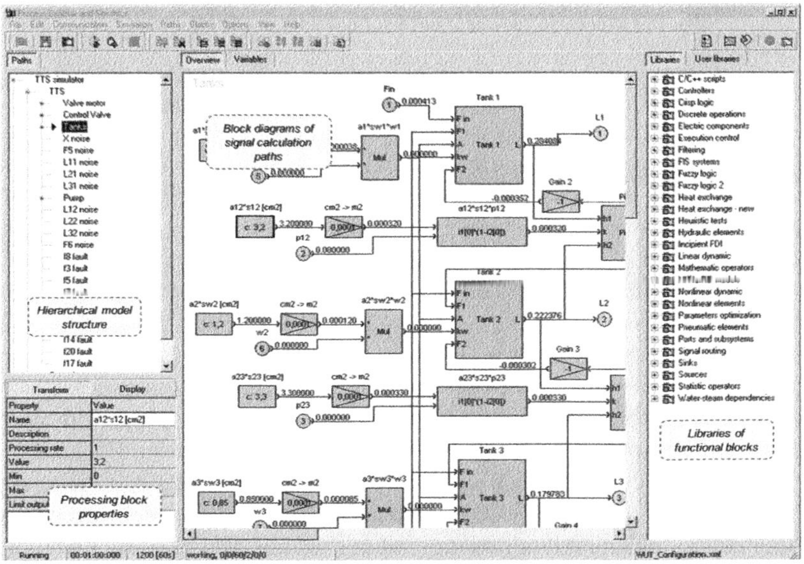

Fig. 2.3 Signal processing defined as paths: *PExSim* module

with predefined core operations: mathematical and logical operations, IO blocks, flow control, operators, integration and differentiation, filtration and many others. Moreover, thanks to the system's open architecture, easy expansion and creation of new specialized units compiled as plug-ins are possible.

2.2.2.2 Simulation and Modeling

Applying new management strategies, control, optimization and diagnostic processes requires the whole process model or its part. To create a global model of an industrial installation, the system uses a module called *PExSim*. Partial models of the process or its fragments are usually obtained through identification, carried out by using the existing archival measurements database. To identify such models in *DiaSter*, the *MITforRD* (Model Identification Tool for Reconstruction and Diagnosis) module is provided.

Identification is conducted in the off-line mode, with the use of archival datasets imported from a DCS or a SCADA system. *MITforRD* provides a possibility to identify static as well as dynamic models. Currently implemented modeling techniques include the linear transfer function, neural networks and fuzzy models.

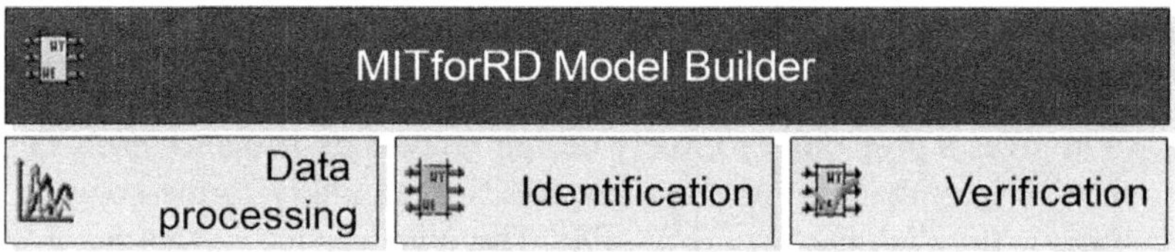

Fig. 2.4 *MITforRD Model Builder* functions in the off-line mode

The software facilitates comfortable work with measurement data. In addition to identification, *MITforRD* allows us to preprocess and analyze process variables as well as verify the obtained models (Fig. 2.4). Data analysis is possible with the use of a series of plug-ins, allowing, among others, the calculation and display of correlation, the power spectrum, histograms, trends, freely configurable filtration and the transformation of signals according to the user-defined mathematical expression (Fig. 2.5). Additionally, the set of data import/export filters is also provided as separate plug-ins.

The identification process is based on well known wizards helping the user with parameter selection (Fig. 2.6). Usually, the user only needs to answer some simple questions and give basic parameters (or accept the default ones), e.g., indicate input signals or select the model type (linear, neural, fuzzy, etc.). In the *MITforRD* module, sophisticated computational intelligence algorithms (genetic, particle swarm, machine learning) are used for the automatic search of the model structure. They allow achieving high quality modeling even in the case of insufficient knowledge of the modeled process.

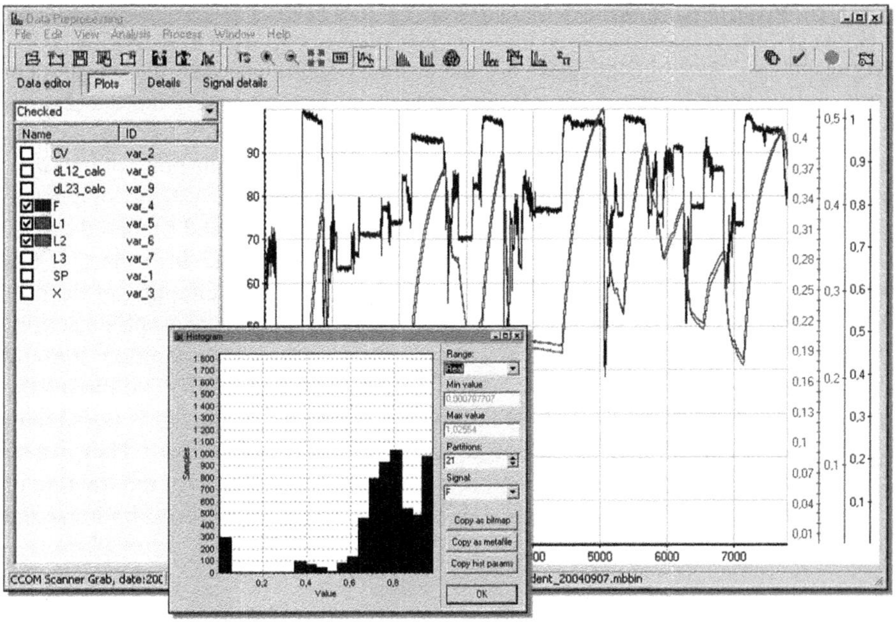

Fig. 2.5 Data analysis and preprocessing in *MITforRD Model Builder*

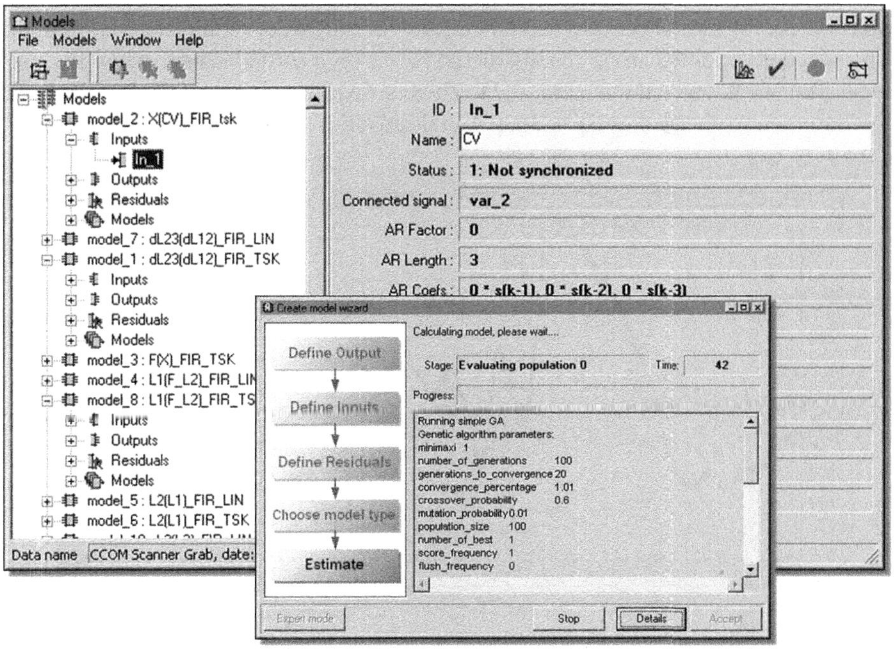

Fig. 2.6 Identification and model management in *MITforRD Model Builder*

2.2.2.3 Process Simulators

At present, industrial installations are increasingly saturated with modern process control devices and technology solutions, which are also becoming more and more complex. At the same time, requirements for installation reliability, operation continuity as well as maintaining the process working point close to the point of maximal efficiency increase. Such demands require reliable automation systems, diagnostic procedures and process optimization along with well trained and prepared operators. The requirement for operators training is especially difficult to meet. It has been found that, with the increasing degree of automation and reliability of the process installation, the competences of the operators to undertake appropriate actions in the case of unusual and emergency situations are reduced. In addition, training the operators with the use of a real, running installation is ineffective due to the lack of the possibility to simulate abnormal and emergency states. It is possible to avoid the above-mentioned disadvantages by using process simulators. They can be realized in the *DiaSter* system. The process simulator developed for the training of the operators can be later used to test new strategies of control or to optimize the process.

2.2.2.4 Virtual Sensors and Analyzers

Virtual sensors based on analytical, neural network or fuzzy models can be treated as information redundancy for real measurements. In many situations, the preparation and use of partial process models instead of hardware redundancy is cheaper. The process variable can be reconstructed based on a virtual sensor when the real measurement is unavailable (Fig. 2.7). Thus, software redundancy exists for each measurement with the corresponding virtual sensor.

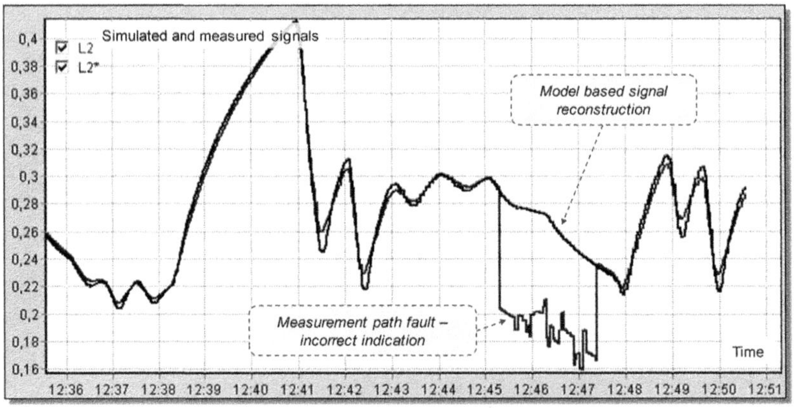

Fig. 2.7 Example of measurement validation with soft sensoring

2.2.2.5 Fault Detection

In the *DiaSter* system, methods based on analytical, neural network or fuzzy models as well as heuristic tests utilizing different kinds of relations between process variables are used for fault detection (Fig. 2.8). Such detection methods give a possibility of early detection of a much higher number of faults than is the case with a classic alarm system.

During on-line work, the system calculates residual values, i.e., the differences between measured and modeled process variable values. The values of the residuals significantly different from zero are fault symptoms. The main advantage of the employed approach is the capability of early detection of faults, including those of a small size. On the other hand, heuristic tests are based on simple as well as more complex relations between process variables. The knowledge of these relationships is usually held by process engineers as well as automation and process operators, even if the process model is unavailable. Its skilful use enables robust fault detection.

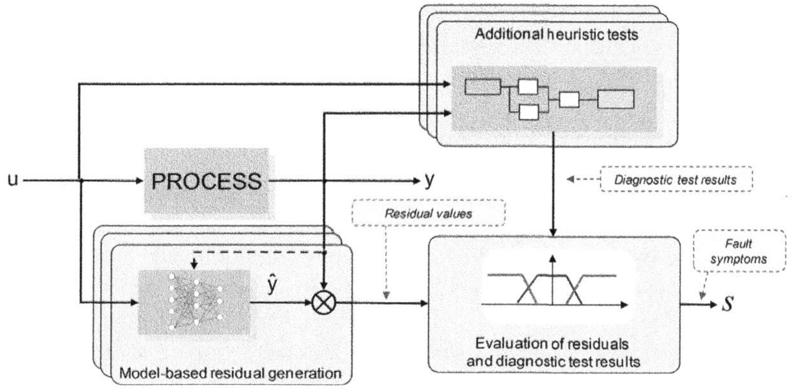

Fig. 2.8 Fault detection schema with informational redundancy and heuristic tests

During the configuration of a diagnostic system for industrial processes it is hard to define proper threshold residuals, whose exceeding values testify that a fault occurred. One of the known and effective techniques used to handle imprecise and/or uncertain information is fuzzy logic. In the *DiaSter* system, fuzzy logic is used to evaluate residual values (Fig. 2.9). The parameters of fuzzy sets can be obtained automatically by the analysis of statistical description of the residual signal in a fault-free process state, or they can be assigned manually by a system engineer.

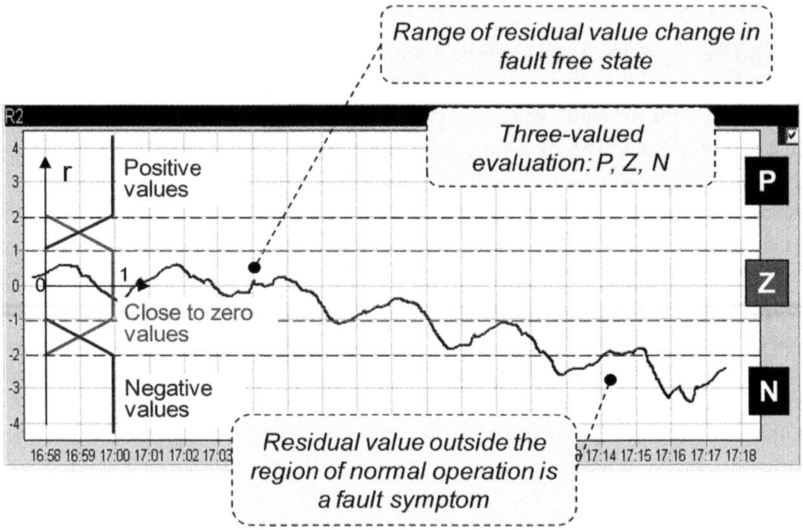

Fig. 2.9 Three-valued fuzzy residual evaluation

2.2.2.6 Fault Diagnosis

In the *DiaSter* system, two independent diagnostic inference mechanisms are implemented. The first one allows fault isolation based on the analysis of the set of actual fuzzy diagnostic signal values and the relation between faults and symptoms stored in the base of knowledge. This relation takes the form of rules: if a symptom s_j occurred, then the faults $f_k, f_m, \ldots, f_s$ are its possible cause. The inference is conducted according to the industrial Dynamic Table of States (iDTS) method developed at the Institute of Automatic Control and Robotics of the Warsaw University of Technology. The elaborated diagnoses point out faults together with certainty factors of their existence.

A diagnosis given by the system can be either presented directly in *DiaSter* or sent to a DCS or a SCADA system. *DiaSter* provides a native visualization module called *InView*. The main task of this module is to show faults on process mimics and specialized diagnostic windows (Fig. 2.10).

The second diagnostic mechanism uses belief networks and multi-faced models. In those networks the nodes represent statements while direct graph arcs between nodes are interpreted as conditional probability counterparts. The solution of inference is achieved when the network achieves an equilibrium state (nodes values are equal to the corresponding statements values). A value change in any of the nodes (statement), caused, e.g., by system observation, results in corresponding changes of other nodes values and network transition to a new equilibrium state.

The methodology of diagnostic system construction based on multi-faced models is currently under development and it is expected that it will be of particular interest in near future. It is anticipated that systems using such models will be a convenient

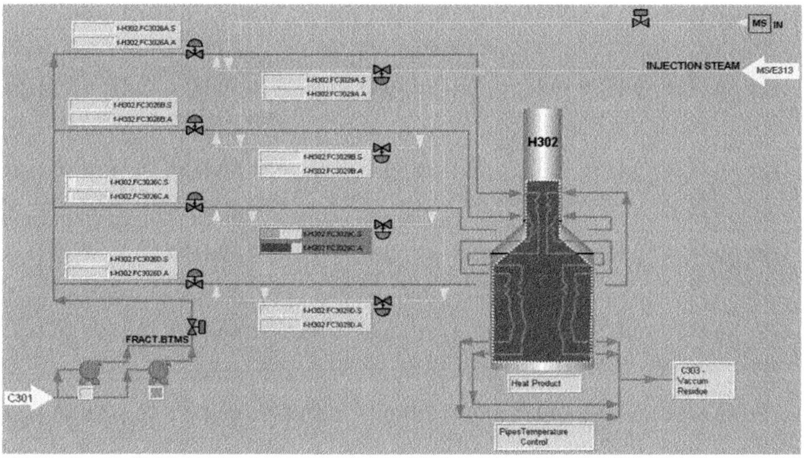

Fig. 2.10 Faults visualization in *DiaSter*: example of control valves and measurements in the H302 furnace. Each fault considered has its own indicator showing the certainty factor regarding its existence. If the factor value is high, then the corresponding display is red. When the value is lower, the color is changing accordingly to purple, yellow, up to white for low values.

tool for both the representation of knowledge from various fields and its application to reasoning about the state of the system and the degree of risks associated with its operation. The results of a comparison of a multi-faced model with the system are written down in the form of statements. This allows building a hybrid system, in which statements describing the results of the comparison are introduced to the statements network. Then, the network based on such inputs designates conclusions about the state of the process together with their explanations. The main advantage of such systems is the fact that there is a possibility to use knowledge obtained in different forms. This knowledge can be made available in the form of different kinds of models as well as in the form of relations directly defined in the sets of statements with the use of belief network.

2.2.2.7 Monitoring of the Degradation Degree of Technological Equipment

In technological devices, slowly developing destructive changes often take place. The reasons for these changes are the processes of materials wearing, sedimentation of various substances on equipment elements, etc. If there exist measurable symptoms of such changes, the system allows their early detection and tracking the degradation degree (Fig. 2.11). Such a procedure permits replacing periodical inspections and renewals with the strategy of carrying out the maintenance based on on-going evaluation of the technical state of the process or device.

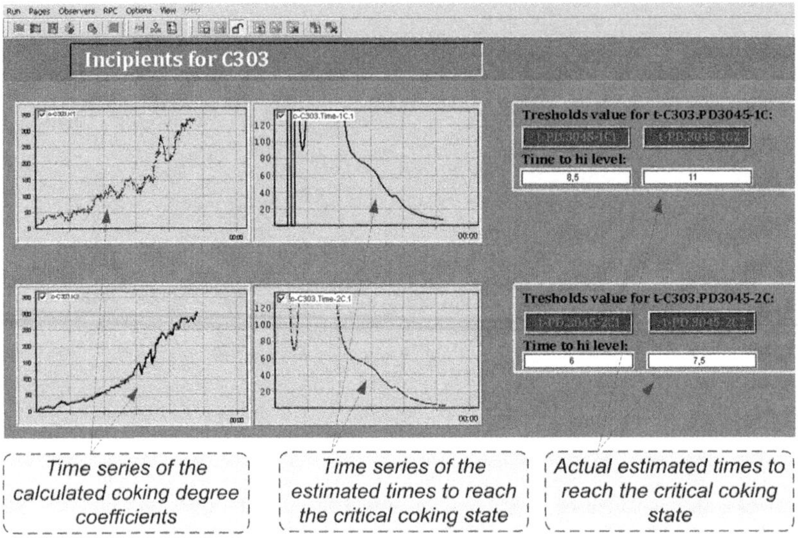

Fig. 2.11 *Operator Display*: coking rates visualization and the estimated time for repair

2.2.2.8 Support of Process Operators' Decisions

On the basis of the elaborated diagnosis, the *DiaSter* system can additionally support process operators' decisions in abnormal and emergency states. The management strategies for some or all system faults can be developed during the design phase. After fault isolation, these strategies, together with the diagnosis, are, in realtime, displayed for the operators. They take the form of any document prepared in advance by the designers of the system and may contain the procedure to be followed, the diagrams of faulty equipment or reconfiguration procedures.

2.2.2.9 Knowledge Discovery in Databases

The databases of DCSs and SCADA systems contain a lot of data carrying information about the process, its attributes (process variables, parameters) and the relationships between them. The system is equipped with mechanisms for knowledge discovery that may be used in diagnostic reasoning. The subject of knowledge discovery useful for constructing dynamic models can be widely understood regularities in the data. Some pattern and the scope of its validity determined by a subset of the set of attributes and/or records are understood as regularity. Such understood regularities can be represented as trees, rules, relationships, associations, and fuzzy and neural networks representations. The employed algorithms are based on the selection of attributes, which applies Support Vector Machines (SVMs) and genetic algorithms. The new knowledge represented by both dynamic qualitative and quantitative models is a result of dedicated module operation. Such models can be used in case-based and model-based diagnostic reasoning.

2.2.2.10 Advanced Control and Optimization

Predictive control algorithms are currently the dominant advanced automatic control technique used in modern industrial control systems. A algorithms of superior predictive control (DMC—Dynamic Matrix Control, GPC—Generalized Predictive Control) using linear models, as well as algorithms based on non-linear models, in particular on fuzzy and neural networks, are implemented in the *DiaSter* system modules. Moreover, the procedures of current optimization of the set-points for control systems are also implemented. Such design of predictive control algorithms and procedures of set-point optimization enables setting up modern optimizing control structures for a practically important situation when the variability of disturbances is comparable with the system dynamics.

2.2.2.11 Superior Tuning and Adaptation of Control Loops

Automatic tuning of control loops consists of two stages:

- preliminary tuning (pre-tune) in an open or a closed loop,
- precise tuning (fine-tune) in a closed loop.

Both pre-tuning and fine-tuning in an open system can be carried out with the use of the step response or frequency methods. The adaptation is performed by the step response method. The user interface allows selecting pre- or fine-tuning versions or the adaptation together with the required parameters (pitch, amplitude, frequency range, realignment, recovery time, etc.). The calculation results are presented to the user in the form of the table of values and the graphs of dynamic and frequency characteristics. After acceptance, the new settings are sent to the automation system.

2.2.3 *System Structure*

The *DiaSter* system is a *freely configurable* and *extensible* computing environment that provides, in addition, the ability to store and visualize processed information.

The "configurability" is realized by

- the availability of a number of tools allowing the design of advanced processing of process variables in a flexible way;
- scalability, i.e., the possibility of implementing both simple (e.g, dedicated to individual devices) and complex applications (e.g., for complex industrial processes);
- the ability to add new interfaces to external systems in an easy way;
- the ability to operate in a distributed environment (important for the implementation of computationally complex algorithms) using separate computers and Ethernet-based communication.

The "extensibility" is realized by

- the possibility to easily expand the functionality of the system by creating specialized plug-ins and modules for archival and on-line data processing, modeling, visualization;
- the possibility to create user-defined data types and dedicated tools for their processing and visualization;
- the ability to create user-defined (dedicated to a particular application) description of the configuration elements in order to facilitate the management of the information processed in the system.

The system consists of a *software platform* and a number of *specialized packages* cooperating within the platform (Fig. 2.12).

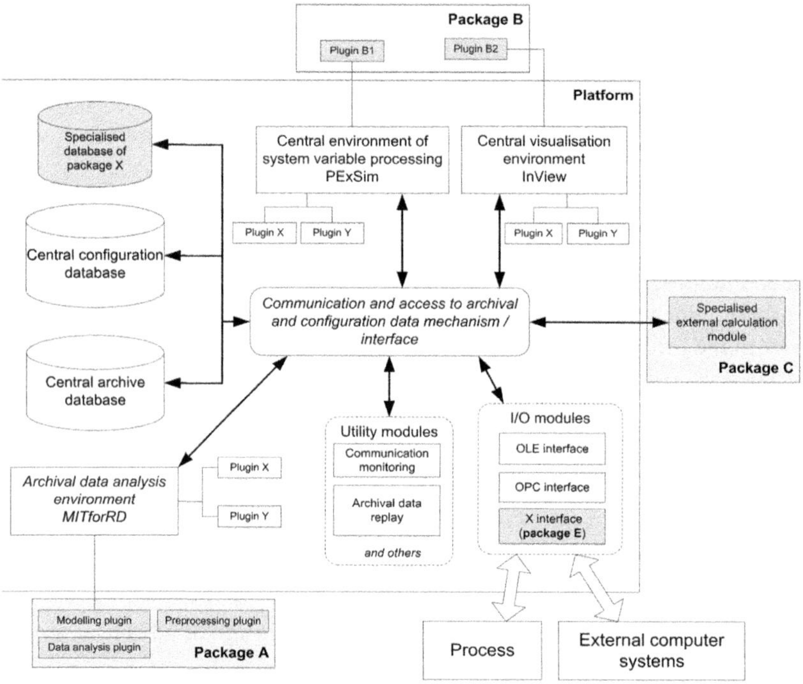

Fig. 2.12 *DiaSter* system structure. Dark blocks symbolize various software components that interact with the system platform

The software platform is a key element of the system. It includes tools and environments for the implementation of the system applications and for running specialized packages. The main platform components include a *module of archival data processing and model identification based on these data* (*MITforRD*), an *on-line system variables processing module* (*PExSim*), a *visualization module* (*InView*), *central archival/configuration databases* and a *communication server*.

The basic system modules provide the appropriate environment for the implementation of advanced tasks related to the modeling, advanced control, and diagnostics of processes and systems. These modules consist of a set of standard elements allowing only the basic processing. Advanced functionality is provided by *specialized packages*. These packages are implemented as independent software modules cooperating with the rest of the platform and other modules through the communication mechanisms provided by the platform, or in the form of dynamically loaded plug-ins of modeling, variables processing and visualization modules.

2.3 Software Platform

The software platform of *DiaSter* is an information system providing mechanisms for processing information of a specified structure and a set of tools for easy implementation of additional functions performed by specialized modules (called "user modules"). The structure of information processed by the platform is determined by its *information model*. This model provides common definitions of the system components and their mutual relations. This is a necessary element to ensure data exchange between the system elements and their consistent interpretation. The system semantics and the data definition are common for all elements of the system. This feature allows the implementation of the standard platform mechanisms responsible for basic operations such as archiving, process structure analysis, visualization and communication. The basic modules and mechanisms of the system are shown in Fig. 2.13.

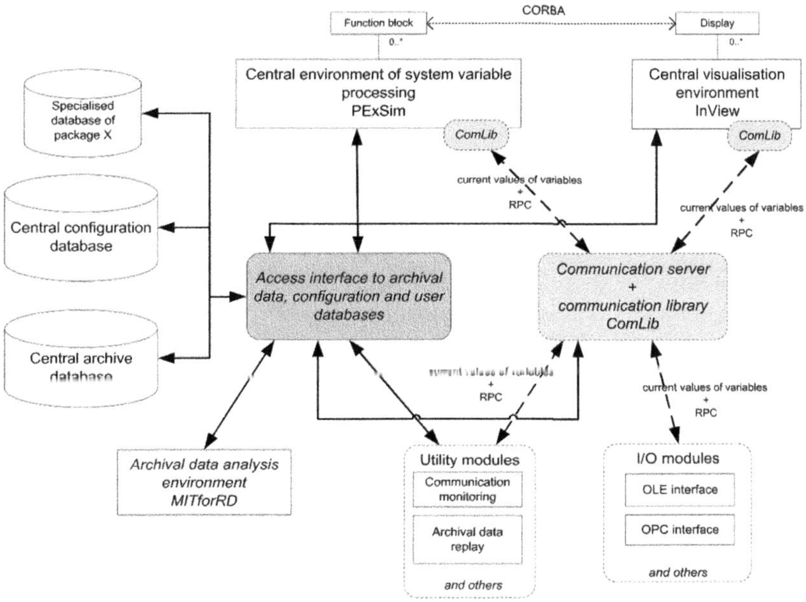

Fig. 2.13 Method of data exchange and access for the components of the *DiaSter* system software platform

Central configuration and archival database. The configuration database is used as a single repository of configuration data used by the system modules, whereas the archival database is a center for storing and sharing both measured and calculated variables in the *DiaSter* system. The structure of the data stored in these databases is consistent with the platform information model. Each system component (including specialized packages in the form of independent software modules or plug-ins of the basic modules) can access the configuration and archival data through specialized interfaces.

Server and communication libraries. The platform provides a communication server together with a communication library to enable data exchange between all system modules. The communication mechanism is developed to be used on-line, mostly for periodic exchange of processed variables values. Additionally, it enables Remote Procedure Calls (RPCs) of particular modules, including the control of their behavior (suspending and resuming calculations, reconfiguration, etc.). Communication works over any TCP/IP network but can be used also locally on a stand-alone PC.

Module for archival data analysis, processing and model identification. This module works mostly off-line and is designed to fulfill the tasks connected with the visualization, selection and processing of archival process data. Another important task is to provide an interface and environment to build parametric models on the basis of archival measurements, i.e., model identification. The module can be easily extended with the use of plug-ins that can provide almost any kind of static as well as dynamic parametric model. It is possible to use the *DiaSter* databases to read archival data as well as to store identified models. The configuration information (e.g., relations between variables) can also be used during data analysis.

On-line variables processing module. The on-line calculation module executes system variables processing according to algorithms written in the form of block diagrams. Processing paths (schemes) are created in a specially designed graphical environment with the use of standard or dedicated (supplied by specialized packages) function blocks. The system can run several processing modules in parallel. They communicate via the communication server and/or the database. In this way, it is possible to implement distributed calculations on multiple computers.

Visualization. This module is designed to implement a graphical user interface, which presents information processed by the system. An important feature of the visualization module is its consistency with the platform information model. This allows creating dedicated displays for user data types and automatical linking of the displayed information to the system configuration.

Communication bridges. This is a group of modules responsible for data exchange with external systems (single devices as well as complex systems such as DCSs or SCADA systems). New modules can be easily added to the system. These modules communicate with the rest of the system through the communication server.

In addition, a series of utility modules, used during application development, are provided. These modules include an implementation of additional features, such as communication monitoring, real time archives playback from the archive database (including the simulation acceleration factor), archiving system variables in the archival database and others.

2.3.1 Information Model and the System Configuration

The *DiaSter* system software platform consists of several modules which can work in a distributed environment, carrying out various tasks in the field of advanced control, modeling, simulation and diagnostics. At the same time, the platform is a computational runtime environment for different types of specialized software packages. To enable the cooperation of all system components, it was necessary to develop

- a system information model,
- a central configuration database.

Information model. The information model defines the common part of the information processed by the platform modules and used to describe supervised processes and realized tasks. This model is a foundation for information exchange inside the system, including the configuration as well as processed signals. Its implementation by particular system modules ensures common understanding of processed data and allows, among others,

- data exchange via the *DiaSter* communication channels;
- using a single description of signals transmitted/processed by various elements of the platform/system;
- providing a uniform mechanism for archiving system variables;
- creating a central system configuration, which enables the analysis of the data stored in it by various system modules.

The information model primarily defines the elements associated with a description of the process/system (division into component parts, the set of process variables, etc.), logical objects related to the processing algorithms implemented by the platform (models, residuals, modeled and calculated variables, faults, etc.) and their mutual relations. In addition, it contains a description of the objects necessary for the operation of the system but not directly connected with implemented algorithms (definition of the subjects of distributed messages, projects definitions, system modules, etc.).

The information model also includes a definition of the types of system variables supported by the platform. There are the following built-in variable types: floating

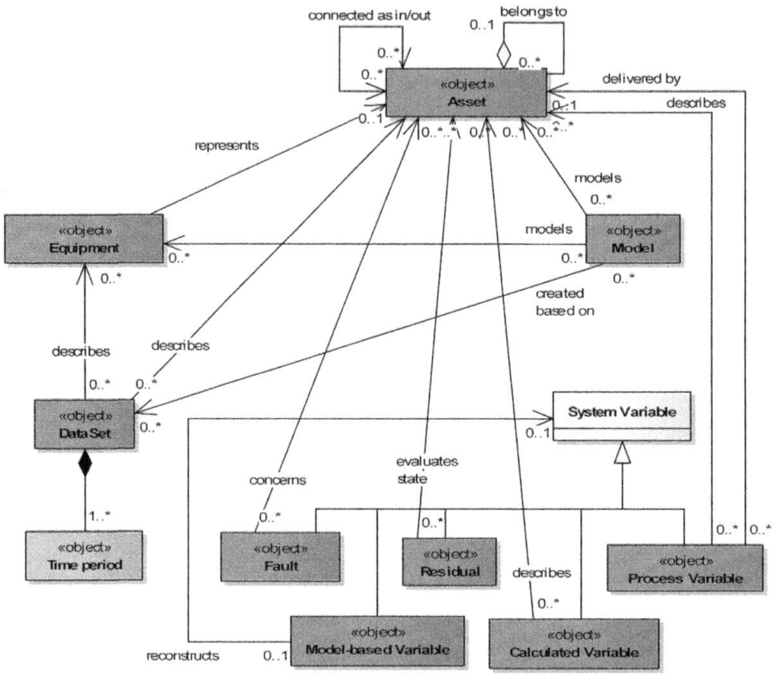

Fig. 2.14 Basic information model objects and their links related to the description of the process and implemented algorithms. Light-grey boxes represent superior classes, while dark ones represent objects and concepts.

point and integer numbers, binary signals, text strings, binary data fields and vectors of any types of constituent elements (each vector element can be of the mentioned types, including the vector itself). An overview of information model objects is presented in Fig. 2.14.

The information model is the backbone of process and processed variables description. It was designed to be extendible according to the needs of particular packages (Fig. 2.15). In relation to this, the platform allows

- creating *user-defined data types*. A user-defined data type is virtually equivalent, in terms of its use by the platform, to built-in data types. This means that the variables of such a type can be fully supported by the platform, i.e., stored in the archives and in the configuration, transmitted through the communication server and processed by the function blocks of the calculation module;
- creating *user-defined types of system variables*. Such variables can be stored in the central configuration; however, their full utilization is possible only by system elements familiar with their definitions;
- defining *user-defined groups of variables* and *types of process components* for the purpose of better process/system structure management;
- defining *user-defined relationships* binding the elements of process description with the objects of processing algorithms implemented by the platform modules.

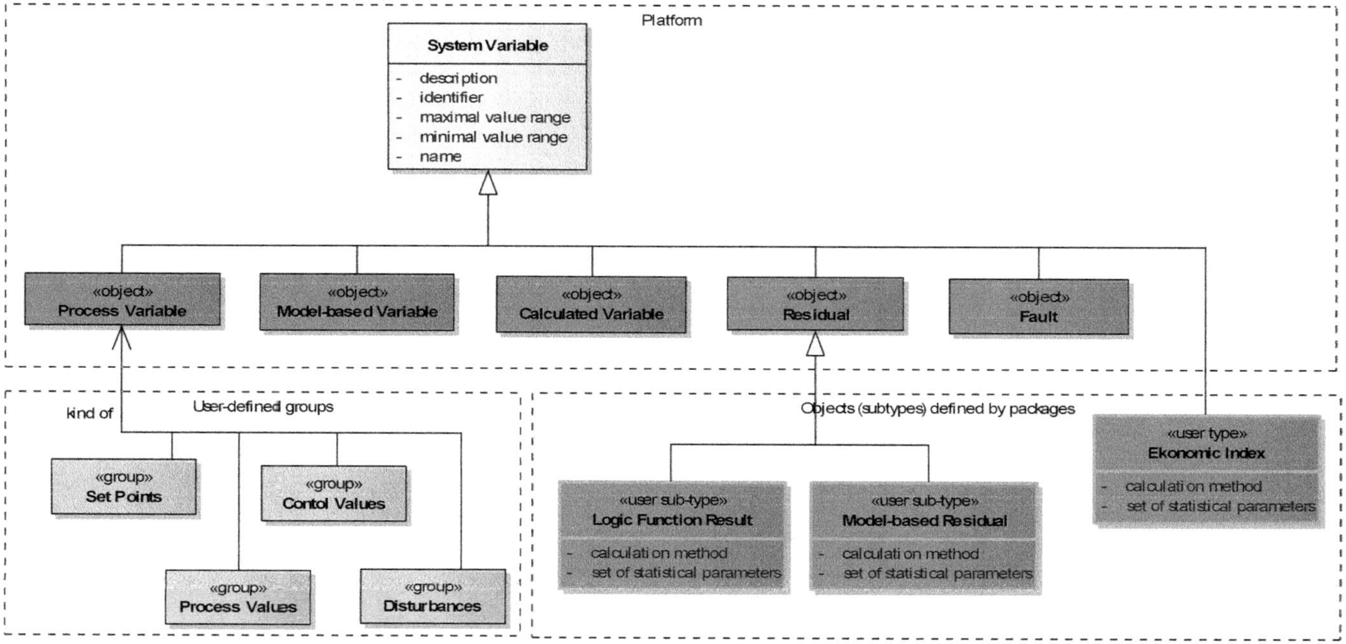

Fig. 2.15 Example of flexible configuration management and the development of the information model by creating user groups and types of system variables. Light-grey boxes represent superior classes, dark ones represent objects and concepts, while middle-grey represent groups

Central configuration environment. This is a repository of configuration data for all modules and computing elements of the platform used in a particular application. Such a repository is necessary to ensure configuration data consistency, enable their automatic analysis (at a general level) and process data storage in the central archives.

The central configuration environment consists of

- the *central configuration database*: it is a relational database that can be accessed by individual modules through specialized interfaces (Fig. 2.16). The data model of this database is consistent with the platform information model;
- the *set of configuration interfaces* through which the platform modules (and their components, including plug-ins) have access to the system configuration. This layer provides separation of user data from the physical layer. It also validates the stored configuration. The interfaces provide a set of necessary and convenient functions for manipulating objects in the configuration;
- the *set of interfaces to create and access user databases*: such databases can be used to store any data specific to a particular package (or a group of packages). Such data do not need to be consistent with the information model. The platform does not analyze those data;
- the *central configuration module* called *SysConfig*: this module allows managing the system configuration at a general level consistent with the information model. Individual packages are responsible for managing additional information introduced by them.

2.3.2 Central Archival Database and User Databases

Historical values of the system variables (both process variables and other variables processed in the system: model outputs, residuals, faults, and user variables) are stored in the system with respect to

- the need to create archives of process variables for the subsequent search and analysis. Mainly, the sets of data used for model identification are created based on available archives;
- the need to back up the information produced by the components running on the platform. Such data are further used for the purpose of the analysis of historical records and the operation of the system in order to improve its future work.

The software platform provides two mechanisms of creating archives of processed system variables:

- in a central archival database,
- in the databases of user-defined packages.

Central archival database. Any system variable defined in the system configuration can be stored in the central archival database. Data of this type are transmitted (usually periodically) during normal system operation among computing modules via the communication server. There is a possibility to store variables of built-in

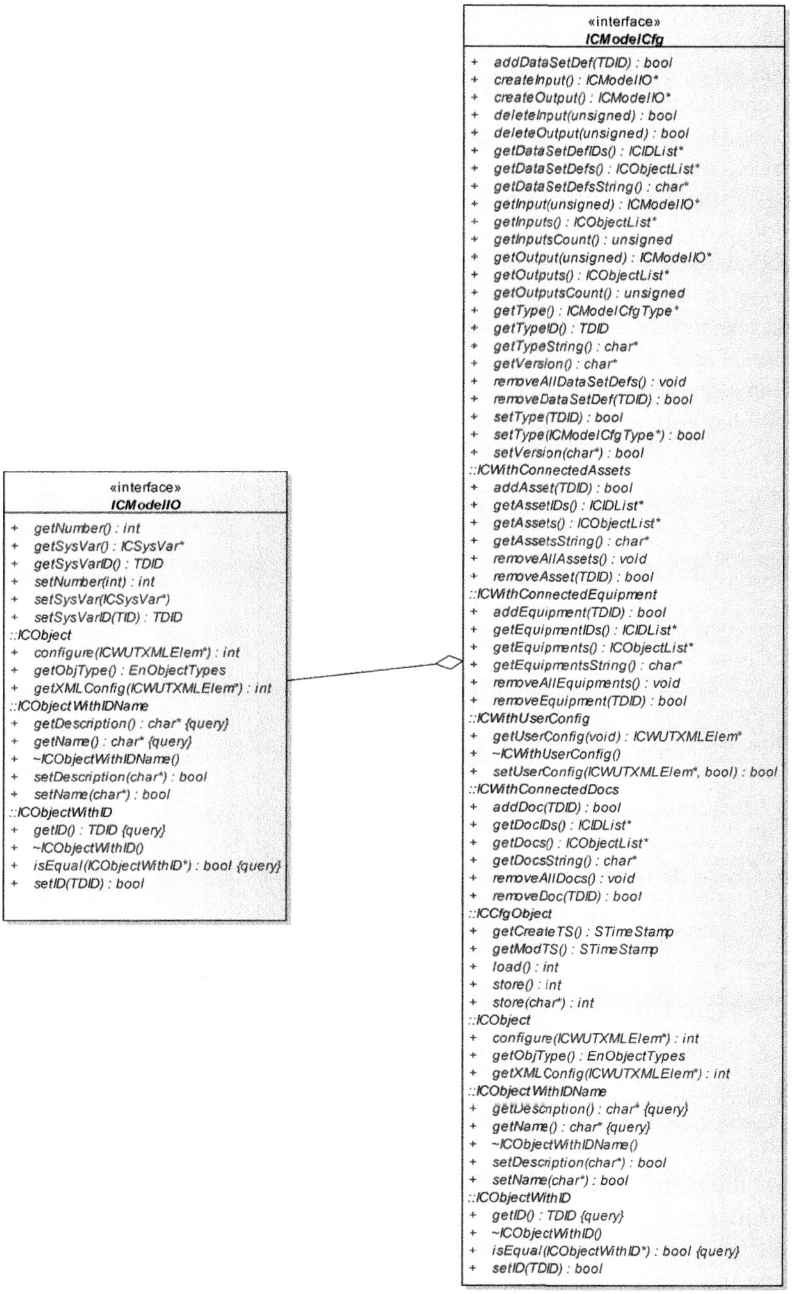

Fig. 2.16 Sample interface to process variables stored in the common configuration

types (process variables, residues, faults, etc.) as well as user-defined ones. Also, the type of values is unrestricted. Both built-in (floating point, integer, Boolean, etc.) and user types are handled. Values in the archive database are stored with a time stamp and the value status. The configuration parameters of the system variables (name, range of values, physical units, etc.) are stored in the system configuration.

Each system component working on the platform (plug-ins for modeling, processing, visualization as well as user-defined modules) has access to the uniform, common interface to archival data (Fig. 2.17: *ICArchivesCont*). The user has access to functions such as downloading the list of variables available in the archives, identifying the time period from which each variable is available, etc. The real exchange of data takes place through a specially prepared, at the user's request, temporary recordset (Fig. 2.17: *ICRecordSet*). After determining the parameters of a recordset (defining a set of system variables, the time period, the sampling rate, etc.), the user can download/upload the entire data set or read sample by sample.

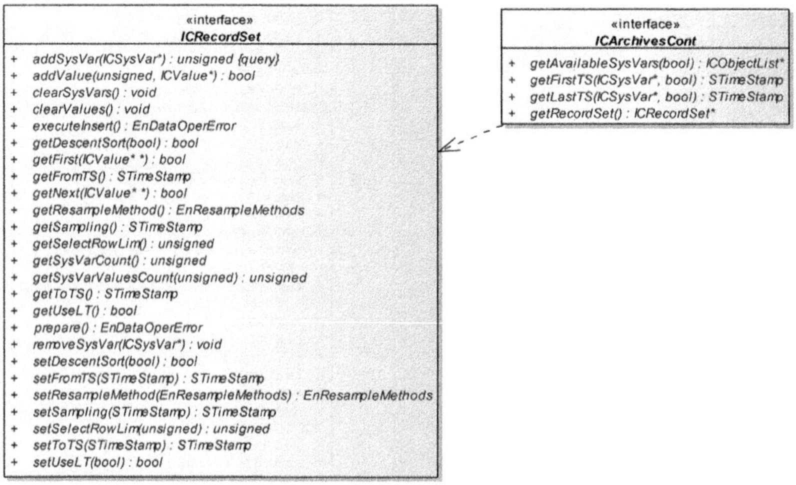

Fig. 2.17 Interfaces and objects for accessing (read/write) archival data

The central database supports both short- and long-term archives (Fig. 2.18). The system administrator determines which variables have to be archived, and which are to be placed in long-term archives. From the user's point of view it does not matter whether the data come from short- or long-term archives. If the case of the need to download the "old" data, the database interface automatically searches for the appropriate long-term archive and retrieves the data if they are available. The user can limit the search to short-term archives by setting the appropriate option in the system configuration.

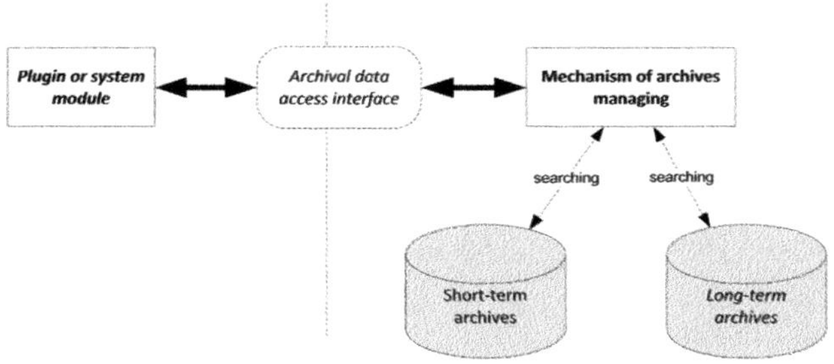

Fig. 2.18 Support for short- and long-term archives

User-defined databases. Data specific to a particular package (or a group of packages) can be stored in user-defined databases. These data do not need to be consistent with the platform information model as the platform does not analyze them. It only provides a mechanism of automatic creation and access to such databases for each package (Fig. 2.19: *ICUserDBFactory* interface). The package must be registered in the system to be able to use this mechanism. The provided interface allows packages to execute any queries on their dedicated databases (Fig. 2.19: *ICUserDB* interfaces and *ICQueryResult*).

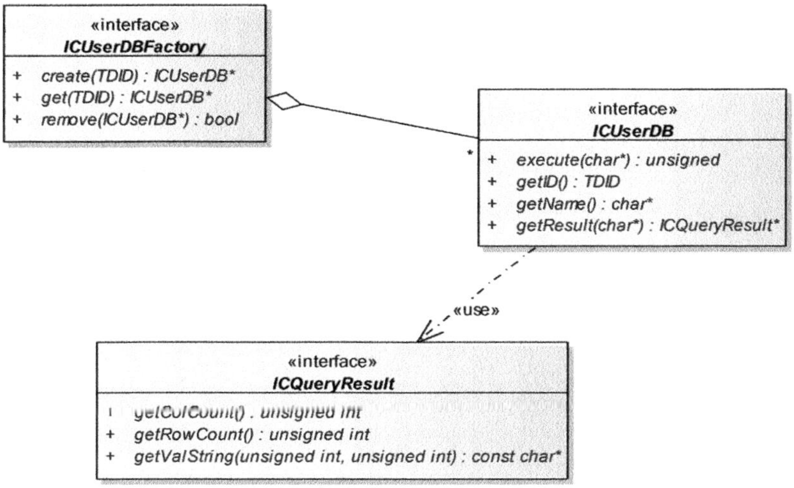

Fig. 2.19 Interfaces and objects for accessing user databases

An example of such a database is the database of diagnostic messages generated by the specialized diagnostic module *iFuzzyFDI*. The generated diagnostic messages stored in the database are further used by dedicated components of the visualization module (Fig. 2.20).

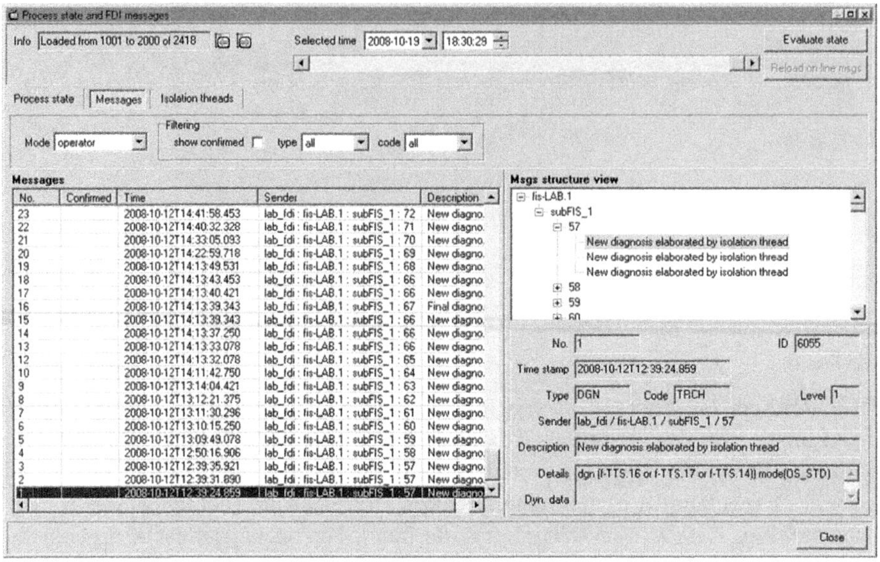

Fig. 2.20 Example of a dedicated interface in the visualization module intended for viewing and managing diagnostic messages generated by the specialized diagnostic module *iFuzzyFDI*

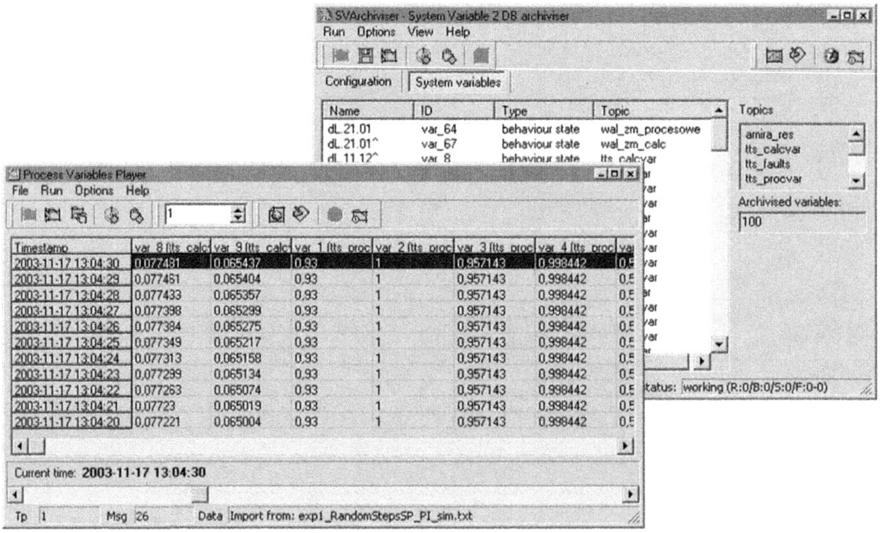

Fig. 2.21 *DBArchiviser* and *PVPlayer* modules used for archival data grab/play

The platform also provides additional useful modules designed to work with the central archives:

- the *PVPlayer* module is used to retrieve data from the database and send them through the server in real (or accelerated) time (Fig. 2.21). This module reproduces the data from a particular time period. It is used to simulate the data received from external systems in order to replay system operation from the past;
- the *SVArchiviser* module is responsible for creating archives (Fig. 2.21). Current values of chosen system variables are automatically grabbed from communication and placed in system archives.

2.3.3 Data Exchange

Communication modes in the *DiaSter* system can be divided into two main groups:

- internal communication between system modules,
- external communication with the outside environment.

Communication between system modules. *DiaSter* is a fully distributed system, with the possibility of working on multiple computers connected via the standard TCP/IP protocol to a local network of any kind. The communication subsystem used within the software platform is a native solution, using a specially designed communication protocol, without the need to employ any external software.

The primary method of communication in the *DiaSter* system is based on the transmission of messages between the modules. This can be done in two modes: a direct connection between the two modules and broadcasting. The messages are always exchanged via the *MRIaS* server, a part of the *DiaSter* software platform. *MRIaS* is only used to transmit information between modules and to provide transmission errors maintenance. It is not responsible for message analysis or storage. From the user point of view, this type of communication can be described as stateless. It is done via the TCP/IP socket, regardless of whether the modules work on a stand-alone computer or on multiple networked PCs. When any program that uses the *MRIaS* server starts communication, a bilateral connection between the program and the server is established and maintained throughout all the working time. The communication protocol (structure and sequence of messages sent between the server and each of the modules) is designed especially for the system. A typical communication cycle is presented in Fig. 2.22.

Publish/subscribe mode. In the publish-subscribe mode, each message is sent under a specific topic. Any number of other modules can be subscribed to topics. Each subscriber will receive all messages published under the subscribed topics. Current values of measurements collected from the automation system are distributed in this mode.

Direct mode. In the case of sending a message to a specific recipient, the sender needs to give the recipient's name (system-wide unique). The message is sent via the *MRIaS* server, which delivers it to the recipient, or, in case of failure, informs

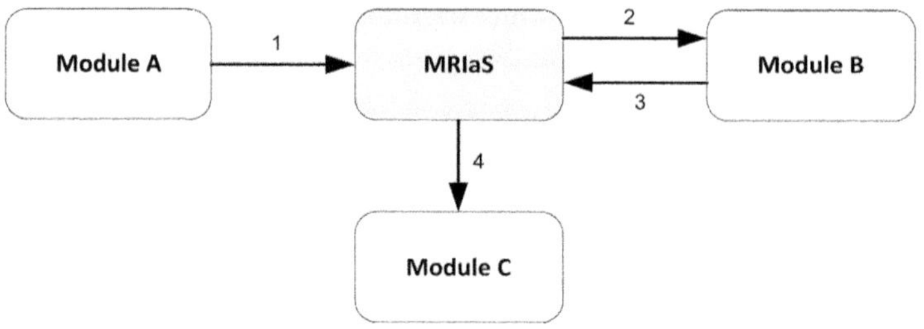

Fig. 2.22 Diagram of communication between the *DiaSter* modules

the sender. The recipient's name comes from the system configuration and is independent of its IP address or DNS name.

The mechanisms discussed above are mainly used to transmit rapidly and frequently changing data, e.g., the values of measurements, the result of current calculations, etc. Any type of value acceptable by the software platform (floating point numbers, binary signals, fuzzy values, arrays, etc.) can be transmitted this way. Additionally, it is possible to call remote procedures—both via direct communication (a module A calls a procedure executed by B and receives the results) and broadcasting (e.g., the configurator module sends a request to "reconfigure" to all modules within a single RPC).

More sophisticated communication mechanisms are provided by some modules. Additional communication interfaces built on the basis of the CORBA standard are used to access the properties and methods of each function block embedded in the *PExSim* module, and to remotely control the process of simulation/calculations realized by this module. A similar mechanism is used in visualization. It allows transmitting complex as well as infrequently changing information and direct access to system objects.

Interfacing external systems. *DiaSter* is able to cooperate with various decentralized automation systems (DCSs) as well as SCADA systems. The measurements are collected with the use of digital transmission between *DiaSter* and the automation system using the standard OPC link or other communication solutions presented in Fig. 2.23.

In the core *DiaSter* platform, the following modules, among others, are included:

- *OPCLink*: communication with OPC servers (most automation systems),
- *OleDBLink*: communication via OLE DB (most databases),
- *PILink*: a dedicated link to PI OSI-Soft,
- *DTMLink*: a dedicated link to the DSS-CHEM system.

New communication modules can be quickly developed if there is a need to connect to other software platforms.

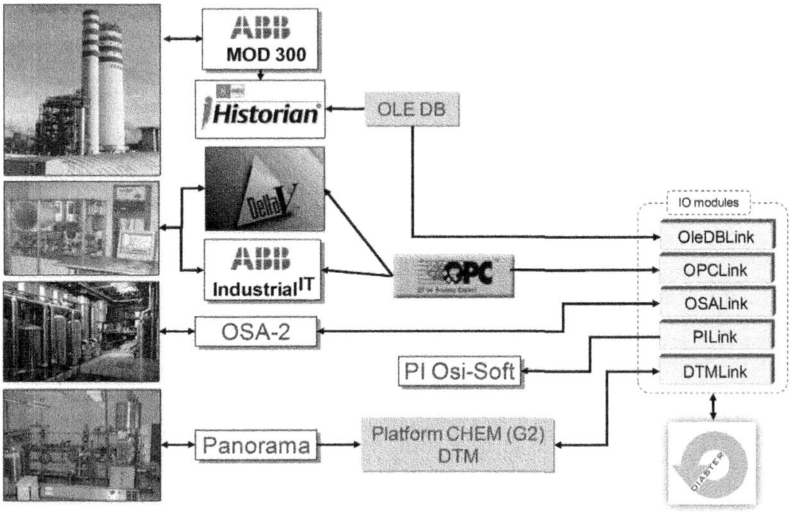

Fig. 2.23 Examples of communication between *DiaSter* and external automation systems

2.3.4 Modeling Module

For identification purposes, the *MITforRD* module is designed in the *DiaSter* system. It allows creating models without the knowledge of the analytical form of the relationship between the modeled variables. *MITforRD* allows identifying both static and dynamic models of different types, starting from well-known linear transmittances to neural networks or fuzzy-logic-based models. Identification is carried out off-line using measurement data collected from automation systems (DCSs or SCADA systems).

Module structure. The *MITforRD* module facilitates quick and easy development of the number of models based on the values of archival measurements. The models implemented in *MITforRD* belong to the group of partial parametric models of process variables (time series). The models are obtained in a semi-automatic identification process. A deep knowledge of processes and their physical characteristics is not necessary. The module effectively supports users unfamiliar with identification techniques during the whole identification process.

The main features of the software include the following:

- a common interface to many kinds of models;
- the user does not necessarily need to be an expert in system identification;
- a flexible, self-configurable distributed calculation environment allows the use of free computing power of office PCs;
- a plug-in-based architecture allows easy extension of module functionality by independent software vendors.

The *MITforRD* module consists of the following programs:

- *Model Builder*: a stand-alone tool for model identification,
- *Calc Server*: a calculations server running on multiple networked PCs.

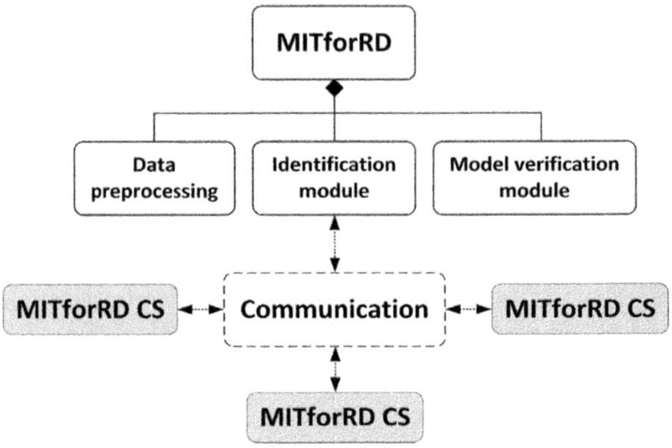

Fig. 2.24 General package structure

Model Builder is a core of the *MITforRD* module. It is split into parts supporting users in each step of the identification process. A general schema of the module structure is shown in Fig. 2.24.

There are four main parts of the *MITforRD* module:

- *data preprocessing*, used to import and preprocess measured values;
- *model identification*, responsible for the identification process;
- *model verification*, which allows validating identified models. The *MITforRD* module allows simulating models in two modes: standard (based on measured output values in the past) and the Model Reference Output (MRO), based on internal autoregression. The quality indexes are calculated during the validation. Also parallel simulation and displaying output signals are possible;
- *MITforRD CS*, which represents cooperating calculation servers.

The whole *MITforRD* module is designed for users who are not experts in the field of identification.

Data preprocessing. *MITforRD* allows editing process archives inside the embedded editor without restrictions on file/data size. Built-in data visualization is capable of displaying one or more time series with zooming and moving charts, change scales, and comparing signals with common or independent scales. Additional features are available via attached plug-ins: data import and export (currently from text or CSV files and the *DiaSter* archival database), data analysis (histograms, statistical parameters, frequency-domain analysis, correlations), data preprocessing (transformation according to a given mathematical expression, filtering, numerical differentiation and integration). Extra functionality is provided to work with process archives (filling-in missing samples, signal validation). New features can be added by the user by creating a new plug-in.

Model identification. One of the main assumptions behind the *MITforRD* module was to make it flexible and expandable. Therefore, the main program, *MITforRD Model Builder*, does not implement a possibility of estimating models of any type. All model types are implemented as plug-ins, dynamically loaded into the program. In the *DiaSter* platform, two plug-ins of linear models are included:

- a simple version, which allows estimating linear difference equation parameters with the LS algorithm and Singular Value Decomposition (SVD). All parameters of the equation structure (model order, delays) need to be given by the user;
- a more sophisticated version, which allows an automatic search for difference equation structures through optimization with a user-defined quality index.

To make the implementation of different model types (static as well as dynamic) possible, an additional wrapper for the Multi-Input Multi-Output (MIMO) model is placed inside the *MITforRD* module. A block diagram of the wrapper is presented in Fig. 2.25.

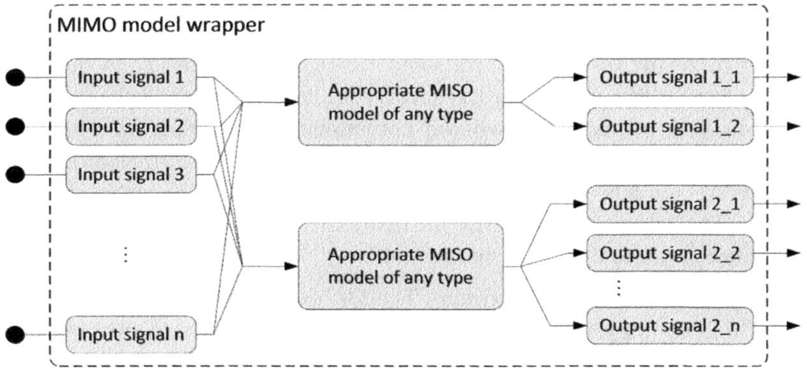

Fig. 2.25 MIMO model structure

Each MIMO wrapper can consist of one or more MIMO models of different types. Individual plug-ins include the implementation of the MIMO models. The task of the model wrapper is to make a common, uniform envelope for all MIMO models. Thus, the model exchanges data with the surroundings (collects data and sends the results) only by its input and output connectors. Input connectors gather data from relevant sources, either from the archival data set during the identification phase or from the actual process in the on-line mode. They also validate and scale the values if necessary. Short-term buffers containing the last measurements for dynamic models are kept inside the connectors. The output connectors are responsible for values re-scaling (optionally) and transmitting the results to the automation system or display. In addition, there are special connectors for automatically calculated residuals associated with the output connectors. They can be used, among others, to generate diagnostic signals.

The following tasks of the identification part of *MITforRD Model Builder* can be specified:

- providing the input data to the model including compression without loss of information about the system dynamics;
- a common definition of the MIMO model structure (selecting input and output signals with the possible value range, selecting the internal models type);
- providing routines for saving and loading the model, process synchronization, data buffering and managing the list of models;
- providing access to the distributed calculation environment for identification algorithms.

Each plug-in needs to realize pure computational tasks only. Therefore, the programmer responsible for plug-in creation can focus on the main task—the development of an identification algorithm.

Plug-ins mechanism. The plug-in mechanism implemented in *MITforRD* as well as in other *DiaSter* modules is based on Dynamic Link Libraries (DLLs). In the basic specification of DLLs, only the export of functions is allowed. Thus, in the *DiaSter* system an additional layer is added that allows working with objects.

Because *Windows* always creates the same virtual functions table (regardless of the programming language used), it is possible to load the object from the DLL only if it inherits the abstract base class and is created and destroyed by the functions exported from the library. The abstract class is a class having all methods virtual and abstract. In order to ensure the two-way communication between the main program and the plug-in abstract, classes are used as interfaces for core *MITforRD* services.

The MIMO models implemented for the *MITforRD* module can be divided into two types. The first one does not use the distributed calculation environment during the identification, while the second one does.

Distributed calculation environment. The *MITforRD* module may employ a wide range of evolutionary algorithms in order to explore the structure of linear or fuzzy models. Such algorithms, besides many advantages, also have one serious disadvantage—they are usually very time consuming. Genetic optimization requires repeated calculation of the objective function. In the case of the model structure search, each objective function calculation requires the inverse of large matrices. This means that time for calculating the objective function clearly dominates during optimization—over 95% of the total optimization time. To reduce the computation time, most important is the reduction of the time of single objective function calculation.

There is provided a distributed computing environment with the *MITforRD* module. It can be used to utilize free computation power of classic office PCs running *Windows* and connected to a local area network. The environment does not require any change in the local PC's configuration and does not disturb its normal operation. The idea is to provide the environment with a configuration as simple as possible, and with minimal requirements for additional software to be installed on the computers comprising the cluster. The structure of the provided distributed environment is presented in Fig. 2.26.

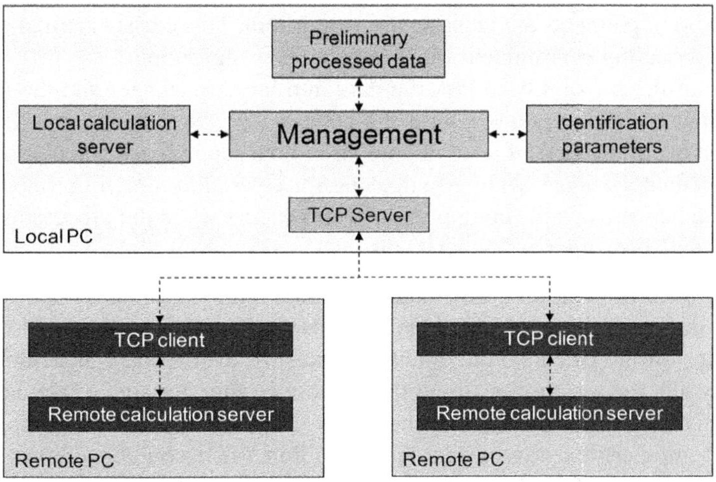

Fig. 2.26 *MITforRD* distributed calculation environment

The main PC running *MITforRD Model Builder* is a control center for the calculations. There is a built-in calculation manager inside *Model Builder*, capable of distributing the calculations among the local host as well as remote PCs available at the moment. On the remote PCs, a simple program called *MITforRD CS* (Calculation Server) is running. When *MITforRD CS* is started, it publishes over the network its willingness to perform calculations. After receiving such a datagram, the calculation manager tries to connect to the calculation server. If the connection is successful, the new client is added to the list of available servers. Then, before the real calculation starts, the configuration stage has to be performed. The first step of the configuration stage is checking if the calculation server is ready and will accept new request. Then, the identification data are sent together with the general identification parameters. During the real identification process, the manager sequentially distributes calculations generated by the identification algorithm between the configured servers. *MITforRD CS* performs calculations and sends the results back to the manager, then starts to listen for the next commission. The communication between the servers and the manager is done via TCP/IP with the native protocol. The computation environment built-in into the *MITforRD* module has the following features:

- no configuration phase is required. During the start-up of the server program (*MITforRD CS*) on the user computer, it is automatically attached to the list of available servers. When the server is stopped, its work is automatically transferred to other servers;
- the communication mechanism used allows avoiding any possible overhead caused by technologies like COM, CORBA, etc. This feature allows achieving high speed of data transfer and low delay (less than 10 ms for 100 MB Ethernet);

- data transfer and the management of the calculation servers are invisible to the user who implements the identification algorithm. This clearly facilitates the use of the modeling environment while creating new algorithms;
- a lack of the possibility to link the data into larger packages and the ability to set synchronization points within the code makes the calculation environment ineffective in the case of a small amount of data sent to each calculation server. Approximately linear scaling (calculation acceleration directly proportional to the number of available computing servers) occurs when the processing time of one calculation order exceeds 100 ms.

The main goal of creating a distributed computation environment was to use it to estimate the coefficients of individual models (one model is calculated on a single server), and to calculate the quality indexes for evolutionary algorithms. With this approach, the assumption about the processing time for single calculations order usually meets the limits mentioned above. Thus, n-times computations speed up on n computers becomes available. In addition, this technology allows utilizing dual, or more complex, processor computers, or multi-core processors, increasingly available on the market.

2.3.5 On-line Calculation Module

The main element of the *DiaSter* platform for on-line use is a calculation module called *PExSim* (Fig. 2.27). It is dedicated to advanced system variables processing. The processing algorithms are written and stored in the form of configurable function block diagrams.

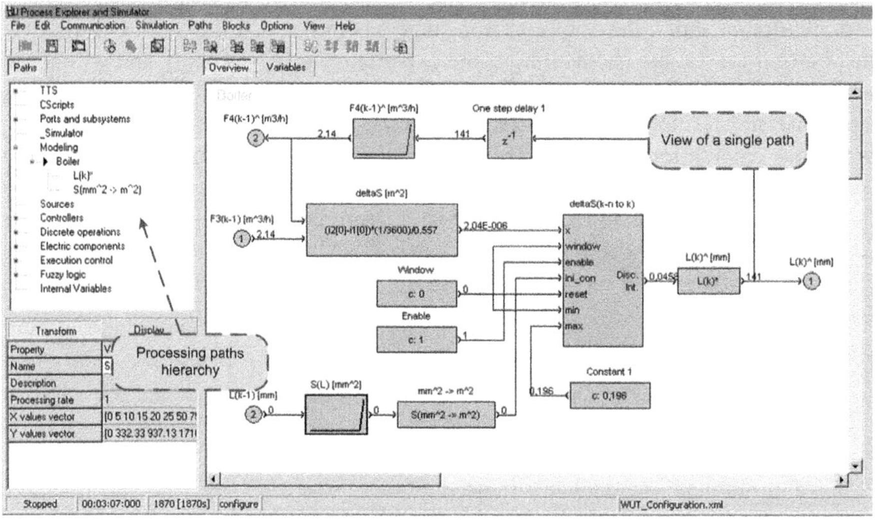

Fig. 2.27 *PExSim* main window

The primary task of the *PExSim* module is to process the information circulating in the system in a way defined by the user. From this point of view, the *PExSim* module can be treated as a specialized programming language. The algorithm of information processing is defined graphically by creating the so-called *processing paths*. Each path consists of a set of interconnected *function blocks*. Each function block carries out various tasks on *signals*. The signal is treated as a "medium for information changed in time". The proper signal flow is carried out by connecting block inputs and outputs.

Function blocks. Algorithms for variables processing are made of blocks responsible for different operations on signals. The following types of blocks can be distinguished:

- **sources:** elements used to collect (from outside the *PExSim* module, e.g., buffers, files, external systems) or generate input data for processing paths. Such blocks are placed at the beginning of the path;
- **processing elements:** core elements responsible for processing input signals into output signals according to a given algorithm;
- **sinks elements:** these are used to send (outside the *PExSim* module, e.g., to files, buffers, external systems) or display (e.g., via numeric displays, different charts, etc.) output data from the processing paths. Usually such blocks are placed at the end of the path;
- **subpaths:** special elements used to create subpaths, i.e., subsystems. They allow building a hierarchical structure of the processing paths.

All the function blocks are provided as *PExSim* plug-ins. None is embedded in the calculation module. Single blocks are grouped into libraries according to fulfilled tasks. The standard installation of the *DiaSter* platform includes the set of the following basic libraries:

- *mathematic operators*: basic mathematic operators,
- *discrete operations*: basic discrete operations (with time dependent and independent parameters),
- *execution control*: the control of paths execution,
- *linear dynamics*: basic blocks implementing linear transfer functions,
- *crisp logic*: binary logic operations,
- *non-linear elements*: non-linear, static elements (e.g., a loop-up table),
- *ports and subsystems*: subsystems and related tools,
- *signal routing*: the control of signal routings,
- *sinks*: signal sinks (path outputs),
- *sources*: signal sources (path inputs),
- *statistic operators*: basic statistical functions,
- *scripts*: additional blocks that allow creating user scripts coded in C++.

The above list of standard libraries can be easily extended with the use of the plug-ins mechanism. There is also a possibility to create user libraries, without access to the whole code of the *PExSim* module. User blocks and libraries are treated in the same way as the original standard *PExSim* libraries. Each block can have a set of configuration parameters of different types. The parameters are split into two groups:

- *transform*: parameters characterizing the processing algorithm, e.g., the number of inputs/outputs, gain, time constant, etc.;
- *display*: parameters that control the way of block displaying on the screen in the *PExSim* graphical user interface, e.g., the width or height of the block, its color, position, etc. Parameters from this group do not influence signal processing conducted by the block.

Each function block has a set of general parameters common for all the blocks as well as a set of its own parameters. The latter is determined by the plug-in creator.

Individual function blocks are combined into a signal processing structure by connecting appropriate outputs to the inputs of subsequent blocks. Each input can be connected to a single output only, but one output can be connected to multiple inputs. Various inputs and outputs can transmit data of various types (e.g., floating numbers, fuzzy values, vectors of floating numbers, etc.). The types of inputs and outputs depend on the nature of the block. Only inputs and outputs of compatible types can be connected.

Processing paths. The group of connected blocks working together in order to realize the defined task create the so-called "processing path" (Fig. 2.28). The following path types can be distinguished:

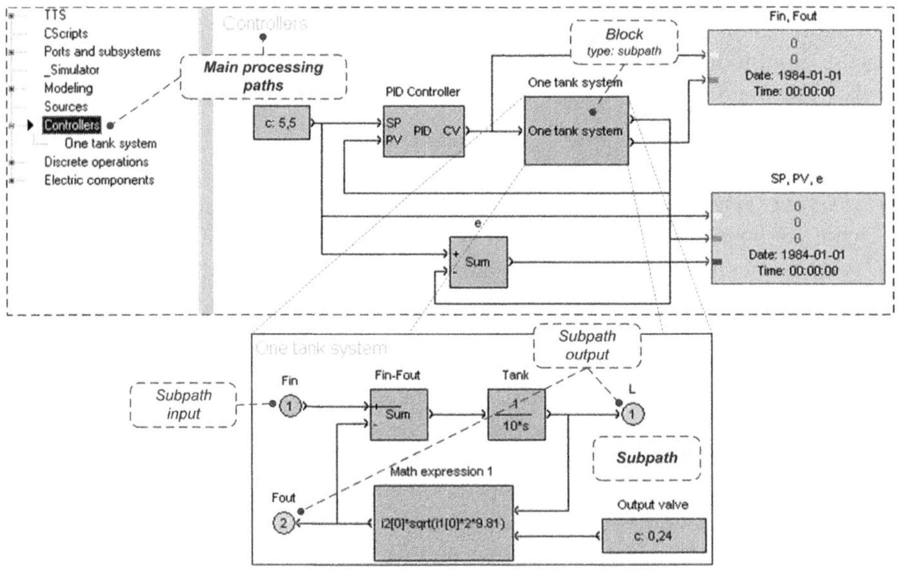

Fig. 2.28 Tree-like structure of processing paths with function blocks

- *main processing paths*: they are placed on the top of the hierarchy. The user attributes the path to one of the existing groups. The type of calculations (synchronous or asynchronous) and the sampling rate (multiple of the main module sampling rate) are attributed to each group. Paths belonging to the same group are processed cyclically—from the first one to the last;
- *subpaths*: they are used to divide the whole system into subsystems. Subpaths can be embedded in main paths or other subpaths. In this way the hierarchical structure of the processing is built. Each subpath has exactly one parent path. Each path can have multiple subpaths embedded at any depth.

The main paths play the main role from the perspective of calculations. Subpaths are only a convenient tool to arrange the algorithm structure inside the *PExSim* application. The main paths can be calculated in different modes, depending on the source of processed data (generated or acquired from external sources) and the time limits for processing (defined or not).

Internal variables. Except the standard inputs and outputs, the internal buffers can be used to exchange data between the paths. These buffers are called "internal variables". This mechanism is provided by two functional blocks: *Internal Variable* (*Sinks* library) and *Internal Variable Input* (*Sources* library). After placing the *Internal Variable* block on a path, the internal variable is automatically created. The list of internal variables is visible on the *Variables* tab. The user can place *Internal Variable Input* in any place on other paths and connect it with the appropriate internal variable. The length of each buffer can be set individually by the user.

Simulation and data synchronization modes. The *PExSim* module can run as a stand-alone tool (*simulator mode*), or as a module working in a distributed system (*multi-module mode*). The multi-module mode allows exchanging data with other *DiaSter* modules. The possibility to exchange data rises the question about synchronizing external data with the internal simulation clock. Thus two additional options for simulation triggering are available in the multi-model mode (Fig. 2.29): *external* and *mixed* triggering.

The process of executing calculations in the paths is controlled by the simulation kernel. The kernel is responsible for executing successive simulation steps, synchronization with external systems and proper simulation timing.

The order in which the blocks are executed inside a path is controlled by signal propagation. First, the source blocks in the path are triggered. After the block finishes its calculation step, it sends a triggering signal to each consecutive block (connected to its outputs). Such a mechanism protects against infinite, algebraic loops.

The *PExSim* module can speed up/slow down the simulation time when acting as a simulator. When it is connected to an automation system, it can be triggered by this system or use its own internal clock to initialize the consecutive steps of the calculations.

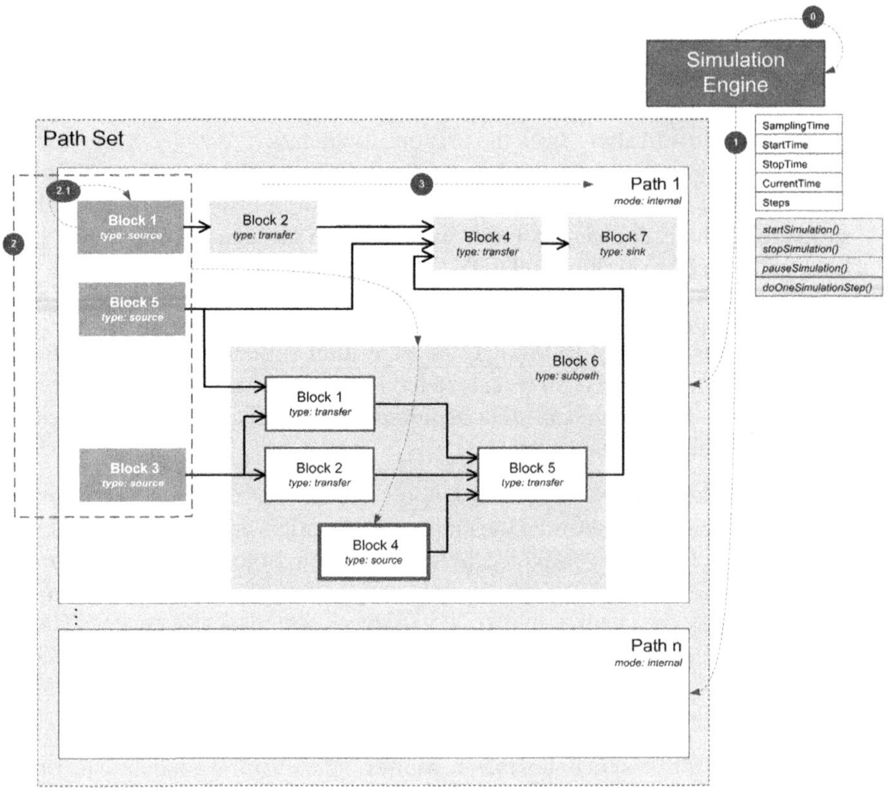

Fig. 2.29 Simulation triggering in *PExSim*

Each main path of the calculation module can be assigned to one of the following groups:

- *synchronous paths*: in this case the simulation kernel triggers a given path with a constant, previously defined time period, or its multiplicity. Such a time period is called the "sampling time";
- *asynchronous paths*: these paths are triggered by events generated internally (via a specialized block) or coming from the outside world (message, RPC, etc.).

2.3.6 Visualization Module

Information generated by the *DiaSter* system is usually used in two ways:

- it is send back to external systems (usually to the control or monitoring layer), e.g., new settings for controllers;

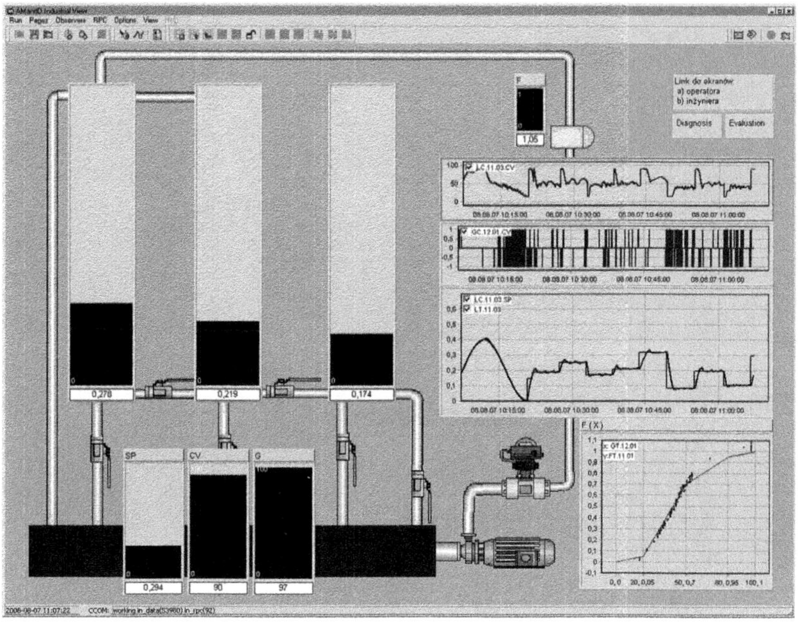

Fig. 2.30 Example process mimics of the *InView* module with standard display components

- it is presented to the operators, usually in a graphical mode, in order to inform them and take appropriate actions related to the supervised process control or reconfiguration (Fig. 2.30).

  The graphical presentation can be implemented in two ways:

- through the graphical interfaces of external systems (e.g., process mimics in SCADA systems or DCSs),
- with the use of the built-in *PExSim* visualization module.

Which of the mentioned presentation methods is used depends on the specific application and engineers' requirements. In some cases only external information systems are used while in others the major operator interface is realized in the native *DiaSter* environment. However, the second approach allows displaying more advanced information, e.g., fuzzy signals or values of process variables related to the process information model.

The *InView* module is designed to create graphical user interfaces, i.e., a different kind of process mimics. The module has a structure similar to graphical tools used in DCSs or SCADA systems. A set of synoptic screens is prepared during system configuration for a particular application. Several displays can be placed on such screens. The displays present values of variables or a set of variables in a predefined way depending on the type of display. Additionally, static elements (backgrounds, graphics, text areas, labels, etc.) as well as navigation buttons can be added.

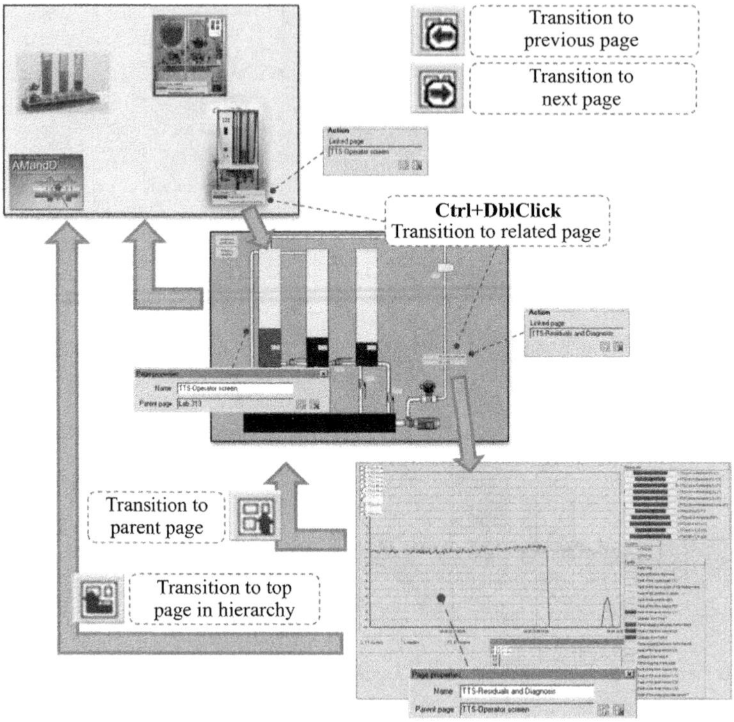

Fig. 2.31 Hierarchical structure of synoptic screens with base navigation methods

For easy navigation, the mimics (process and dedicated screens) are combined in a hierarchical structure (Fig. 2.31). There is available a built-in, automatic mechanism of navigation is such a hierarchical structure, e.g., buttons for transition to a higher or the highest level screen.

The *InView* visualization module exchanges data with other system modules with the use of several communication channels available in the *DiaSter* platform (Fig. 2.32):

- the current values of the system variables are received and send via a native communication server;
- historical plots are created based on the data retrieved directly from central system archives;
- it is also possible to use a direct connection between displays placed on the mimics and the corresponding calculation blocks in the *PExSim* module. The connection is made according to the CORBA standard. It is available to the block–display pairs that implement this standard. Such a mechanism is used mainly to realize atypical tasks, e.g., the connection between the operator station for a PID controller and a block that implements the control algorithm itself.

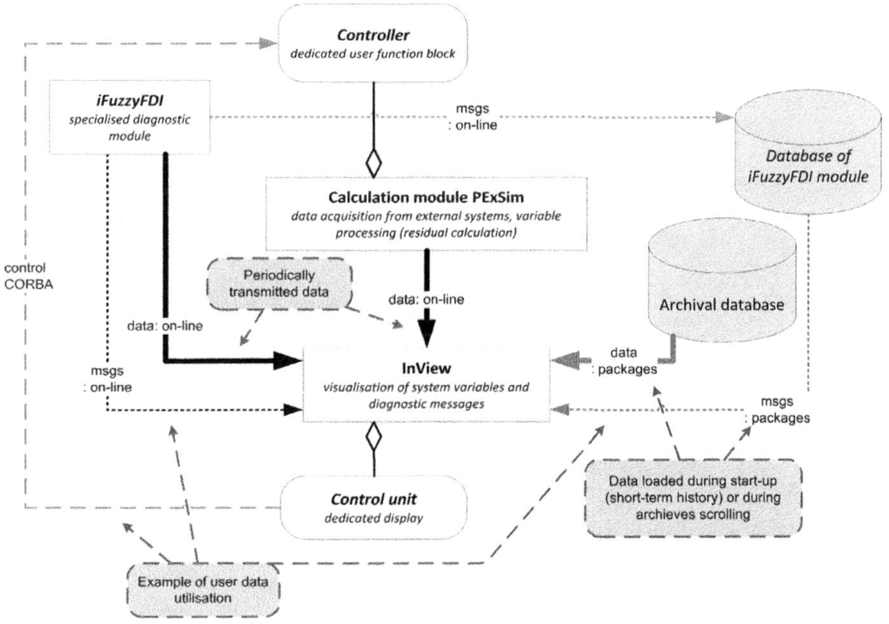

Fig. 2.32 Data exchange mechanisms implemented in the *InView* module

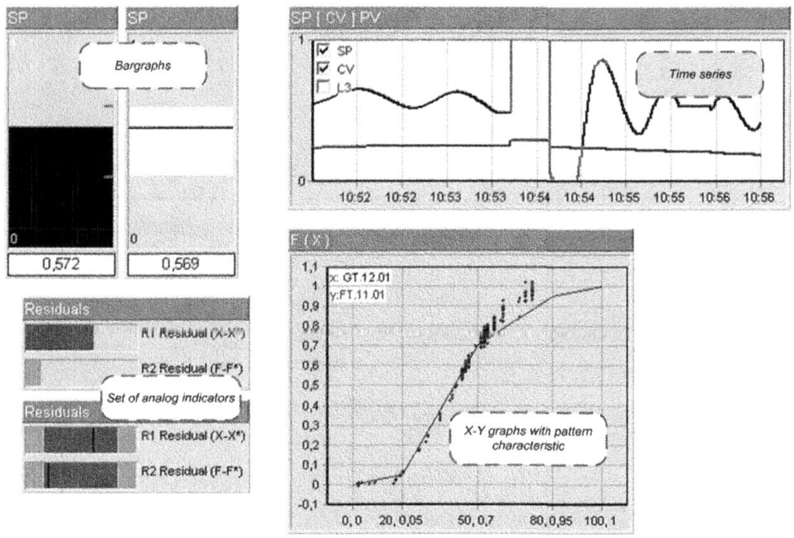

Fig. 2.33 Example displays in *InView*

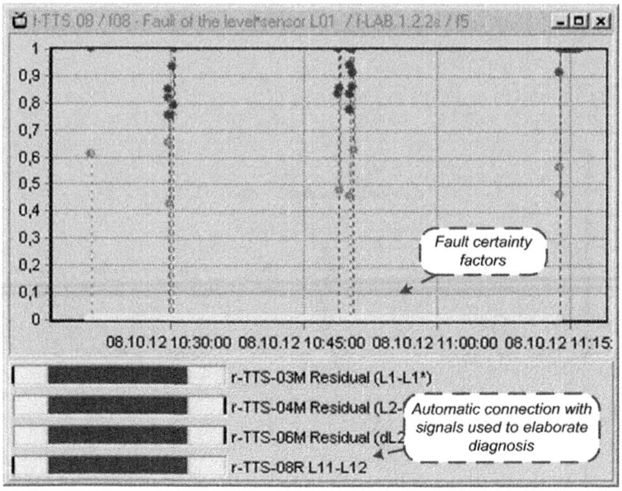

Fig. 2.34 Dedicated display for faults presenting the certainty factor value as well as diagnostic signals values

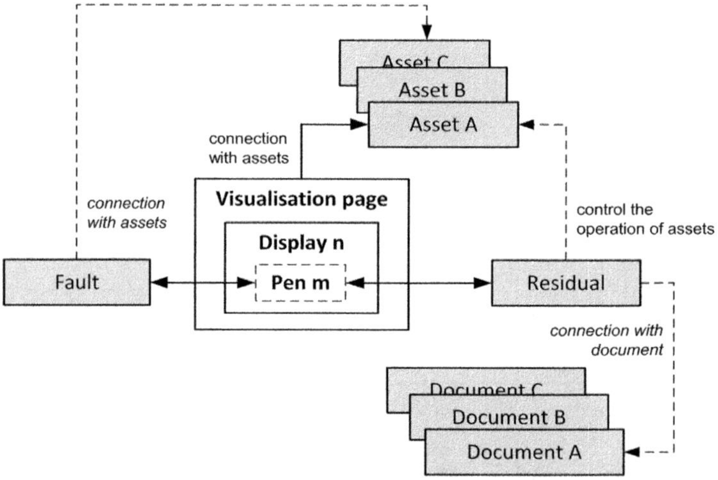

Fig. 2.35 Example of display markers linkage to different elements of the system configuration: faults, residuals, objects (system components), and documentation

Each graphical display is implemented as a plug-in. The set of standard displays is provided (Fig. 2.33). The plug-ins technology makes it possible to develop and use specialized displays, e.g., to display fuzzy signals.

The *InView* visualization module is compliant with the information model of the *DiaSter* system. It makes it possible to use many unique features of the *DiaSter* system, not available in the case of displaying the results through external systems. The system user has the possibility

- to design specialized visualization displays for user-defined data types (Fig. 2.34);
- to create connections between the components of the mimics (displays), the supervised process structure and system elements (Fig. 2.35). The connections make the navigation easy and help us understand (explain) the presented information, e.g, dedicated displays of faults with automatically created explanation panels linked;
- to make direct connections to processing blocks from visualization. This feature allows controlling the execution of a block and displaying its parameters;
- to make a direct connection to system archives and/or user databases. This way each display can have access to virtually an unlimited number of data.

Another way of displaying the information provided by the system is to use specialized modules developed by the user. Such modules are software units cooperating with the *DiaSter* platform though one or all of the communication channels.

Chapter 3
Process Modeling

Krzysztof Janiszowski, Józef Korbicz, Krzysztof Patan,
and Marcin Witczak

3.1 Introduction

Process models in the problems of advanced automatic control, diagnostics or design and investigation of alternative solutions are the grounds for the implementation of intended aims. This chapter is devoted in general to the creation of the investigated process model in a form that would be useful for the conducted investigations or designs. It could be implemented by *modeling* or *identification*. Taking into account practical requirements of the diagnostic system designed for more or less complex processes, analytical, fuzzy, neural and nero-fuzzy models are considered.

For automatic control purposes it is required to know the reaction of the process which is to be controlled, so accurate that one could evaluate the reserves of operation of units (programmer/controller, actuator, process) in order to avoid changes after the design is completed and the system is ordered or manufactured (Åström and Hägglund, 1995). Such changes may be very costly. In the problems of diagnostics, the adequate model determines the success of the intended aim; the proper reaction of the model to the activation is a requisite of the determination of the error residual, without which the diagnosing cannot be implemented in a proper way. During the design and investigation of alternative solutions, the *fast prototyping*

Krzysztof Janiszowski
Institute of Automatic Control and Robotics,
Warsaw University of Technology,
ul. Św. Andrzeja Boboli 8, 02–525 Warsaw,
Poland
e-mail: kjanisz@mchtr.pw.edu.pl

Józef Korbicz · Krzysztof Patan · Marcin Witczak
Institute of Control and Computation Engineering,
University of Zielona Góra,
ul. Podgórna 50, 65–246 Zielona Góra,
Poland
e-mail: {j.korbicz,k.patan,m.witczak}@issi.uz.zgora.pl

technique is the base for the selection of two/three alternative solutions that are, in turn, examined with great care with the use of system elements. In this case, only the modeling-based approach leads to the acquisition of proper decisions and allows us to radically shorten design preparation as well as examination conducting times. A vital advantage of this approach is its repeatability. Examinations with the use of a real process are rarely conducted always in the same conditions. *Modeling* has the advantage that it allows examining step-by-step the successive phases of the investigated process in the same conditions, as well as correcting the procedure (automatic control, diagnosis, design, control) in a way that would be profitable for the intended aim. Model acquisition methodology (modeling or identification) is not the basic problem. The most important is the aim—the functionality of the model in the intended task. Therefore, different kinds of approaches are applied in the case of the above-mentioned tasks. Each of the approaches has its properties that may create some additional possibilities for the user. Thus, it is good to have general information on the employed methods before the decision is made on which method is selected and given most of our efforts and time.

Modeling leads to the creation of analytical models based on the known description of the phenomena that exist in the process. The models are very attractive: they can be applied for hierarchical control aims, start-up process analysis, changes of the point of operation or the implementation of static interaction analyses. It can also be applied to the design of automatic control algorithms in selected points of operation after adequate simplifications (dynamic linearization) to the transfer function form have been conducted. Such models are usually complex in the calculation sense but they allow us to investigate the problems of open loop control, static optimization and automatic control structure changes, or diagnostics and advanced automatic control tasks in the case of the lack of available, adequately implemented experiments. One should remember that many critical states of the process such as overloads, extreme loads, failures, etc. are not usually available in the form of measurement data. In such cases, the only way to recreate the potential reaction is the analytical model (Spriet and Vansteenkiste, 1983; Takahashi *et al.*, 1972). Beside the above-mentioned advantages, models have also many disadvantages: their construction requires a very accurate knowledge of the investigated phenomena, as well as many coefficients that are either unknown or require us to implement arduous verification examinations. Widespread different calculation type packages such as *ADAMS, SimulationX* or *Matlab* tools may be applied only when the investigated process has exclusively units belonging to a particular package, the appropriate coefficients are known, and we are able to pay for the purchase of an adequate license. One should also remember that the description of this type is most often based on simplifications that can limit its usefulness, e.g., for purposes of diagnostics, especially when the modeled failure situations are not typical and do not have equivalents in the set of the package applied. The above-mentioned packages, *ADAMS*, *MATLAB* and *SimulationX*, are readily applied in quick prototyping tasks when certain decisions should be quickly arrived at and the lack of information from measurements forces us to choose just such an approach.

Modeling is very often tied to the concept of identification since the aim of both tasks is the creation of a particular model. However, one should clearly differentiate between these two approaches. Modeling methodology relies on the imitation of the operation of a particular process by creating a mathematical description that would recreate known phenomena which occur in the process in a more or less precise way, and measurement data are most often used for the verification of description coefficients that are not adequately known. In the case of simple or well-recognized processes, verification is not necessary, and the model will be good.

Identification is implemented exclusively on the basis of measurement data, and their quality decides on the success of the calculations and the quality of the resulting model. Depending on the number of collected measurements as well as their quality and completeness, that information may be much more precise than analytical description based on approximate assumptions and the initially assumed coefficients since it takes into account real conditions of process operation, existing coefficients, appearing disturbances, etc. The identified parametric model in the form of standard equations or transfer functions estimated from adequately selected or implemented experiments usually yields more analytical models. Therefore, it is more suitable to apply advanced automatic control techniques or diagnostic applications than the analytical description. Parametric models having the form of Takagi–Sugeno–Kang (TSK) fuzzy models are the most precise form of process dynamics description, and—what is very important—they have extrapolation properties. Such models allow us to imitate the process dynamics in areas close to these areas of operation that appeared in the data used for the estimation of the coefficients of the model. If one wants to evaluate parametric models well, one should have the initial information on the process properties, approximate evaluation of the model order, appearing time delays, as well as instructions leading us to an adequate way of dividing variables that cluster areas in which the investigated process has similar properties.

In the last part of the chapter, models in the form of artificial neural networks are presented. They have an important property according to which any continuous non-linear relation can be approximated with arbitrary accuracy. However, the application of neural networks to fault diagnosis of control systems requires taking into account the dynamics of the processes or systems considered. Therefore, state space neural networks, locally recurrent networks with the neuron model containing an additional infinite impulse response filter and the so-called GMDH (Group Method of Data Handling) networks with dynamics neurons are presented. It is important to notice that neural modeling does not require analytical knowledge with respect to the process considered, but it needs a representative set of training data.

3.2 Analytical Models and Modeling

Analytical models are a description in the form of differential ordinary equations (linear or non-linear ones), or partial equations which are derived from balances of quantities that are characteristic for the investigated system.

Balance equations may be directly applied to model creation but empirically obtained laws or dependencies require us to know adequate coefficients that should be obtained with the help of tables, earlier experiments, or verified by examinations.

3.2.1 Basic Relations for the Description of Balance Dependencies for Modeled Physical Processes

This section describes the basic relations existing in processes described by models having lumped constants. The description of the dependencies will be directed to the exposing of similarities existing between different modeling fields in order to prove an oft-used analogy between different dynamic processes. Understanding the analogy allows us to better model complex processes exhibiting phenomena from several fields, as well as to substitute, e.g., the description of dynamic effects in mechanics with the use of a description typical for electric circuits. The basic balance relations will be supplemented with time dependent relations applied to elements of different types of circuits together with the description in the time domain that allows us to obtain Laplace transfer functions.

Considering different kinds of circuits one by one, it is possible to use the following dependencies having a balance character.

Two Kirchhoff laws in electric circuits:

$$\sum_k e_k(t) - \sum_j u_j(t) = 0, \tag{3.1}$$

$$\sum_k i_k(t) = 0, \tag{3.2}$$

where $e_k(t)$ denotes stimulations—electromotive forces, $u_j(t)$ are reactions of the circuit elements for the stimulations, i.e., voltage drops at elements of a selected loop of the circuit (taking their directions into account), and $i_k(k)$ is reaction intensity described by the current values in a selected node of the circuit. Electric energy is dissipated by circuit resistance R, and inductance L and capacitance C store the energy, which is described by the following dependencies (in the domain of time t and operator s):

$$u(t) = Ri(t) = R\frac{dq(t)}{dt} \Leftrightarrow i(s) = Ri(s), \tag{3.3}$$

$$u_L(t) = L\frac{di(t)}{dt} \Leftrightarrow i(s) = sLi(s) - Li_0u, \tag{3.4}$$

$$u_C(t) = \frac{q(t)}{C} = \frac{1}{C}\int_{t_0}^{t} i_C(\tau)d\tau \Leftrightarrow u(s) = u_{C0} + \frac{i(s)}{sC}, \tag{3.5}$$

where the index “0” denotes the initial condition related to the element.

The Ampere and Gauss laws in magnetic circuits:

$$\sum_k \Theta_k(t) - \sum_j l_j H_j(t) = 0, \tag{3.6}$$

$$\sum_k \Phi_k(t) = 0. \tag{3.7}$$

In the selected loop of a magnetic circuit, the laws connect the stimulations—magnetic flow $\Theta_k(t)$ with the magnetic voltage drop at the magnetic circuit section having the length l_j and the magnetic field intensity $H_j(t)$. They also balance values of magnetic fluxes in the selected node of the circuit. Linear description as in (3.3), (3.4) and (4.12) is not applied since they usually are strongly non-linear circuits.

The d'Alambert law in mechanical systems for the linear or rotary motion:

$$\sum_k F_{ek}(t) - \sum_j F_{rj}(t) = 0, \tag{3.8}$$

$$\sum_k F_{ek}(t) - \sum_j F_{rj}(t) = 0, \tag{3.9}$$

where $F_e(t)$ and $M_e(t)$ denote external forces and moments being the stimulations, respectively, and $F_r(t)$ and $M_r(t)$ are the forces and moments of reactions related to the inertia of the system elements or the friction, respectively. The second relation is the result of the momentum conservation law,

$$\sum_k m_k v_k(t) = 0, \tag{3.10}$$

$$\sum_k J_k \omega_k(t) = 0, \tag{3.11}$$

where $m_k(J_k)$ denote measures of the system element inertia, and $v_k(\omega_k)$ are the velocities.

The system energy is dissipated by the friction F_R represented in the form of a damper D, while the potential energy is stored by the elastic element S and the kinetic inertia m or J, producing forces denoted as F_S and F_B, respectively, for the linear motion:

$$F_R(t) = Dv(t) = D\frac{dx(t)}{dt} \Leftrightarrow F_R(s) = Dv(s) = sDx(s), \tag{3.12}$$

$$F_S(t) = Sx(t) \Leftrightarrow F_S(s) = Sx(s), \tag{3.13}$$

$$F_B(t) = ma(t) = m\frac{dv(t)}{dt} \Leftrightarrow F_B(s) = ma(s) = smv(s), \tag{3.14}$$

and for the rotary motion:

$$F_R(t) = D\omega(t) = D\frac{d\alpha(t)}{dt} \Leftrightarrow F_R(s) = D\omega(s) = sD\alpha(s), \tag{3.15}$$

$$F_S(t) = S\alpha(t) \Leftrightarrow F_S(s) = S\alpha(s), \tag{3.16}$$

$$F_B(t) = J\varepsilon(t) = J\frac{d\omega(t)}{dt} \Leftrightarrow F_B(s) = J\varepsilon(s) = sJ\omega(s). \tag{3.17}$$

In hydraulic and pneumatic systems, rules very similar to the description of a state in electric circuits are in force:

$$\sum_k p_k(t) + \sum_j \Delta p_j(t) = 0, \tag{3.18}$$

$$\sum_k f_k(t) = 0, \tag{3.19}$$

where $p_k(t)$ denote stimulating pressures of pumps, $\Delta p_j(t)$ are pressure drops at homogeneous channels, and $f_k(t)$ are flows in the junction node. Unlike in electric circuits, the ideal liquid (gas) flowing through channels has—beside the kinetic energy—also the potential energy related to the gravity g. Due to that, the state of the liquid is described by Bernoulli's principle:

$$p(t) + \frac{\rho v^2(t)}{2} + \rho g h(t) = const, \tag{3.20}$$

where $p(t)$ denotes the pressure value, $v(t)$ is the velocity, and $h(t)$ stands for the height of the column of the liquid.

Within the range of laminar flows, one can apply dependencies that define the pressure drops Δp at elements that dissipate the energy of the liquid (hydraulic resistances R_h) and elements that store the potential energy (capacitance B) and the kinetic energy (inertia of stream having the mass M):

$$\Delta p_R(t) = Rf(t) = R\frac{dm(t)}{dt} \Leftrightarrow \Delta p_R(s) = Rf(s) = sRm(s), \tag{3.21}$$

$$\Delta p_B(t) = \frac{B}{V}\int_{t_0}^{t} f(\tau)d\tau \Leftrightarrow \Delta p_B(s) = \frac{B}{sV}f(s), \tag{3.22}$$

$$\Delta p_M(t) = M\frac{df(t)}{dt} \Leftrightarrow \Delta p_M(s) = Msf(s). \tag{3.23}$$

In thermodynamic systems, in which the gas thermal energy changes, the dependencies are more complex since the third factor in the form of the gas internal energy should be taken into account beside the kinetic and the potential energy. Due to that, descriptions of appropriate changes differ very much from the ones described earlier, and the analogy mechanism is not applied. From the first law of thermodynamics, one can formulate the general law of the gas energy balance:

$$\Delta U = \Delta W + \Delta Q, \tag{3.24}$$

where ΔU denotes the gas internal energy increase, ΔW is the executed work, and ΔQ is the delivered heat.

The gas momentary state is described by Clappeyron's equation,

$$p(t)V(t)M = Rm(t)T(t), \tag{3.25}$$

where $p(t)$, $V(t)$, $m(t)$ and $T(t)$ denote the gas pressure, volume, mass and temperature, respectively, R is the gas constant, and M denotes the gas molecular mass. The internal energy U of one kilomole of gas depends on its temperature and the number φ of degrees of freedom,

$$U(t) = 0.5\varphi RT(t), \tag{3.26}$$

which have integer values depending on the type of gas: $\varphi = 3$ for one-atom gases, $\varphi = 5$ for two-atom particles, and $\varphi = 6$ for others. The internal energy increase ΔU may also be expressed in a form that depends on the specific heat c_v (with a constant volume):

$$\Delta U = c_v m \Delta T. \tag{3.27}$$

The relation between the specific heat c_v value and the specific heat with a constant pressure is defined by the coefficient κ:

$$\kappa = \frac{c_p}{c_v} = \frac{\varphi + 2}{\varphi}. \tag{3.28}$$

Changes in gases are described by dependencies defined for different conditions of the changes and are called Boyle–Mariott's law for constant temperature and slow compression:

$$T(t) = const \Rightarrow p(t)V(t) = const, \tag{3.29}$$

for rapid changes, e.g., in a pneumatic cylinder:

$$T(t) = const \Rightarrow p(t)V^{\kappa}(t) = const. \tag{3.30}$$

For a constant pressure, Poisson's law in the following form is in force:

$$\Delta Q = 0 \Rightarrow V^{\kappa-1}(t)T(t) = const \cup V^{\kappa}(t)p(t) = const \cup \frac{p^{\kappa-1}(t)}{T^{\kappa}(t)} = const. \tag{3.31}$$

Considering the dependencies (3.1) to (3.23), one can notice several relations having similar forms of description, as well as relations existing between quantities appearing in these dependencies. The relations are defined as analogies between

Table 3.1 Analogies between descriptions and variable quantities

Type of system	Type and reaction of system elements				
	Dissipating the energy	Accumulating the potential energy	Accumulating the kinetic energy	Kind of stimulation	
Electric	Resistor R	Capacitor C	Inductance L	Voltage source	Current source
	$u = Ri$	$u = \frac{1}{C}\int idt$	$u = L\frac{di}{dt}$	E	
	$i = Gu$	$i = C\frac{du}{dt}$	$i = \frac{1}{L}\int udt$		I
	$u = R\dot{q}$	$u = \frac{q}{C}$	$u = L\frac{d^2q}{dt^2}$		$\dot{q}$
Mechanic	Friction D	Spring S	Mass m		
linear	$F = D\frac{dx}{dt}$	$F = Sx$	$F = m\frac{d^2x}{dt^2}$	Force F	
motion	$F = Dv$	$F = S\int vdt$	$F = m\frac{dv}{dt}$		Velocity v
Mechanic	Damper D	Spring S	Inertia J		
rotary	$M = D\frac{d\alpha}{dt}$	$M = S\alpha$	$M = J\frac{d^2\alpha}{dt^2}$	Moment M	Velocity ω
motion	$M = D\omega$	$M = S\int \omega dt$	$M = J\frac{d\omega}{dt}$		
Hydraulic Pneumatic	Resistance R	Capacity B/V	Mass M		
	$\Delta p = Rf$	$\Delta p = \frac{B}{V}\int ddt$	$\Delta p = M\frac{df}{dt}$	Pressure p	Flow f

elements applied to the description, as well as between laws that connect the physical quantities. Understanding these analogies allows us to analyze the existing phenomena, as well as apply adequate forms of description. Table 3.1 shows main connections between the quantities and the essence of the function of particular elements of the system.

The table motivates the often-used technique of different types of dynamic systems modeling with the use of the formula for electric circuits. Surely, the presented dependencies do not give all of the laws and models applied to the recreation, but they allow us to notice distinct similarities existing between different kinds of dynamic processes, as well as distinct differences, e.g., in the description of phenomena that accompany dynamic processes and, for example, mechanic or electric ones. The creation of equations describing more complex processes requires us to conduct additional studies, and, even after the creation of the correct description of the analyzed process in the form of sets of differential equations having known coefficients, it does not guarantee success during modeling. A very important thing is the proper choice of parameters of the modeling process itself: the solver algorithm and the calculation step.

3.2.2 *Integration Methods and Integration Step Selection for Simulation*

The created analytical description of the process allows modeling dynamic reactions, but in order to implement this, one has to develop numerical integration of proper differential equations. Numerical integration methods are an important problem (Spriet and Vansteenkiste, 1983).

The numerical integration problem consists in finding the values of the function $y(t)$ that is defined as $\frac{d}{dt}y(t) = f[u(t),u(t),t]$, $f,y \in R^n$, $u \in R^m$, $t \geq 0$ with the known initial condition $y_0 = y(t_0)$. The relation in this form defines a priori the character of changes of the function $y(t)$. It is a continuous function at least of class C. The easiest way to calculate the function in a numerical way is integration:

$$y(t) = \int_{t_0} f[u(\tau),y(\tau),\tau]\,d\tau, \qquad y_0 = y(t_0) \in R^n. \tag{3.32}$$

In practice, in order to numerically calculate the task (3.32), Runge–Kutta's integration algorithms are very often applied (Spriet and Vansteenkiste, 1983). They are non-extrapolation methods, i.e., they are not based on the estimation of the integrand $y(\cdot)$ beyond the integration range and are methods having the repeatable step. Depending on the multiplication factor of the division of the integration range and the number of repeats, there exist several versions of Runge–Kutta's method. Four of these algorithms are implemented in the *PExSim* package.

In order to explain the principle of operation, we present the description of the 4-th order Runge–Kutta method algorithm. The method is applied to most packages for dynamic system modeling. In the description, discrete time symbolism will be used, i.e., the integral $y(t)$ or the integrand $f[u(t),y(t),t]$ value will be calculated for the moment $t = n * \Delta$, where Δ denotes the integration step, and it will be denoted as $y(n)$ or $f[u(n),y(n),n]$, respectively.

The integrated function estimated value is defined according to the rule

$$y(n+1) = y(n) + \frac{\Delta}{6}\left[f(n) + 2f_{p1}\left(n+\frac{1}{2}\right) + 2f_{p2}\left(n+\frac{1}{2}\right) + f_p(n+1)\right], \tag{3.33}$$

where the first estimation of the function value in the middle of the range $f_{p1}\left(n+\frac{1}{2}\right)$ is defined as follows:

$$f_{p1}\left(n+\frac{1}{2}\right) = f\left[u\left(n+\frac{1}{2}\right), \left[y(n) + \frac{\Delta}{2}f[u(n),y(n),n]\right], \left(n+\frac{1}{2}\right)\Delta\right]. \tag{3.34}$$

The second estimation of the function value $f_{p2}\left(n+\frac{1}{2}\right)$ in the middle of the integration range is created on the grounds of the first one:

$$f_{p2}\left(n+\frac{1}{2}\right) = f\left[u\left(n+\frac{1}{2}\right), \left[y(n) + \frac{\Delta}{2}f_{p1}\left(n+\frac{1}{2}\right)\right], \left(n+\frac{1}{2}\right)\Delta\right], \tag{3.35}$$

and the initial estimation at the end of the range is equal to

$$f_p(n+1) = f\left[u(n+1), y(n) + \Delta f_{p2}\left(n+\frac{1}{2}\right), n\right]. \tag{3.36}$$

The above algorithm shows us the course of the numerical integration process. Recurrences repeated inside the integration range definitely improve calculation accuracy and allow a significant increase in the integration range (from a dozen to tens of times over in comparison with the ordinary Euler procedure).

The modeling of processes that are described in an analytical way must be carried out with correct selection of the integration step. When the step is not correctly chosen, the solution may show oscillatory character (when the step is too long) or may last a long time (when the step is too short).

The selection of the step depends on the dynamic properties and the presence of active stimulations of, e.g., control or regulation, and—since the differences may be very large—Table 3.2 below presents suggested integration step values for different kinds of processes and physical quantities (Isermann, 1988).

Table 3.2 Suggested ranges of integration step selection

Kind of quantity (process)	Integration step [s]
Controlled hydraulic flows	10^{-5} to 10^{-3}
Controlled pneumatic flows, electric drives	10^{-4} to 10^{-2}
Pressures in hydraulic and pneumatic circuits	10^{-3} to 10^{-1}
Forces, accelerations in mechanical systems	10^{-3} to 10^{-1}
Hydraulic and pneumatic free flows	10^{-2} to 10^{-1}
Temperatures	10^{1} to 10^{3}
Concentrations of gases and fluids	10^{1} to 10^{3}
Chemical processes, redox type reactions	10^{1} to 10^{4}
Biological processes, population growth	10^{3} to 10^{6}

3.2.3 *Pneumatic Cylinder Controlled by a Servo-Valve: A Balance Model of the System*

The method of analytical model construction will be presented with the use of a seemingly simple system we know almost everything about—a servo-valve controlled pneumatic cylinder. The system is described by both balance equations of flux flows and thermodynamic processes resulting from the phenomena of rapid air compression and de-compression. The existence of friction phenomena that have non-linear and discontinuous character makes the modeling process complex and non-trivial.

The system considered has been analyzed in the literature for over sixty years (Shearer, 1960), but it still does not have a complete description that would allow precise modeling of all of its reactions. The main reason is the lack of a good, easy-to-define (concerning its parameters) and numerically stable model of friction forces in the pneumatic cylinder.

With the help of an example, we will discuss different approaches to obtain the model of the system together with its specifics, advantages and limitations:

- the analysis of phenomena existing in the process, as well as the creation of a non-linear set of differential equations;

- the quickest and easiest modeling method is the use of ready templates of elements from the library adapted to the modeling of such systems;
- a simplification of the differential equation set to the linear form, and an attempt to define the resulting transfer function, and then the estimation of this transfer function's parameters by the way of optimization through the comparison of the modeling results by measured sequences of input-output data;
- the last method is the use of statistic identification methods based on known, sufficiently representative sequences of measurements, and the creation of the parametric model of the process as well as a fuzzy TSK type model.

The structure of the system considered is presented in Fig. 3.1. The electro-pneumatic five-way valve controls the air flow to the cylinder chambers and causes the movement of the piston. Dynamic and static behavior is defined by several quantities. Overall dimensions—cylinder sections A_1 and A_2, chamber volumes V_1 and V_2, dead volumes V_{m1} and V_{m2}, the piston middle position x, the piston thickness d, air pressure values in chambers p_1 and p_2, as well as air temperature in chambers T_1 and T_2, the mass of the piston and the moving unit M, supply pressure p_z, pressure at the run-off p_{atm}—all exert an obvious effect on piston behavior. Also the dynamic reactions such as friction forces F_t and an external force F_e will have a direct effect on the piston movement. Figure 3.1 presents the situation in which the chamber 1 is filled and the piston moves right with the velocity v.

The method of cylinder chambers filling depends on the relation between the pressures p_1, p_2, p_z and p_{atm} and the flowing properties of the valve (Olszewski, 2007). Assuming that the chamber 1, (Fig. 3.1) is filled, the servo-valve slider opens the sections 1 and 3, then the chamber 2 is emptied, and air flow conditions are defined by the equations of the air flow q_i to the chambers:

$$q_1 = a_1\sqrt{2\rho_1 p_z}\sqrt{\frac{\kappa}{\kappa-1}\left[\left(\frac{p_1}{p_{zas}}\right)^{2/\kappa} - \left(\frac{p_1}{p_{zas}}\right)^{(\kappa+1)/\kappa}\right]},$$

$$q_2 = a_2\sqrt{2\rho_2 p_2}\sqrt{\frac{\kappa}{\kappa-1}\left[\left(\frac{p_{atm}}{p_2}\right)^{2/\kappa} - \left(\frac{p_{atm}}{p_2}\right)^{(\kappa+1)/\kappa}\right]}, \tag{3.37}$$

where a_1, a_2 denote averaged sections of the inflow to chamber 1 and the outflow from the chamber 2, ρ_1 and ρ_2 are air densities at the inlet and outlet from the chambers, and $\kappa = 1.4$ is the polytrophic process coefficient (3.28). Values of a_1 and a_2 are not usually given by manufactures, values of ρ_1 and ρ_2 are not known either—they correspond with a specified air temperature T and the pressure p value.

For the needs of system dynamics modeling, air flow q conditions may be expressed in a simplified form that assumes the lack of heat exchange by air flows:

$$q = \alpha\sqrt{\Delta p}, \tag{3.38}$$

where α denotes the actual ability of the flow through the valve, and Δp is the pressure difference in appropriate chambers ($p_z - p_1$ or $p_2 - p_{atm}$).

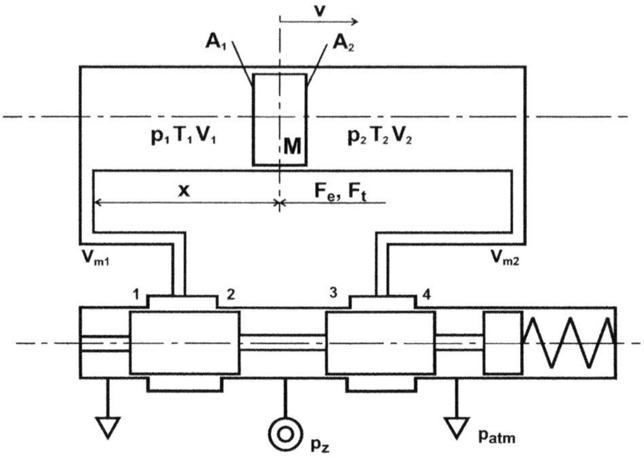

Fig. 3.1 Block diagram of the modeled process: piston position control with the use of the flow servo-valve

The actual value of α in static conditions (and the set control signal for the valve) depends on the level of the valve opening μ and the flow nominal value q_{nom} given by the valve manufacturer:

$$\alpha = \mu \frac{q_{nom}}{\sqrt{p_{znom} - p_{atm}}} = \mu\beta, \qquad \beta = \frac{q_{nom}}{\sqrt{p_{znom} - p_{atm}}}, \tag{3.39}$$

where p_{znom} and p_{atm} most often equal 6 and 1 bar, respectively.

The servo-valve has its own dynamics (it is a proportional magnet having a small but non-negligible mass), and its opening μ should be modeled as a reaction to the level of the control signal u/u_{max}:

$$\frac{d}{dt}\mu(t) = \frac{1}{T_z}\left[\frac{u(t-T_0)}{u_{max}} - \mu(t)\right], \tag{3.40}$$

where T_z denotes the time constant of the valve (usually given by the manufacturer) and T_0 is the time delay. The valve opening value has obvious limits $\mu(t) \in \langle \mu_{min}, \mu_{max} \rangle$. Using the description of the air flow to chambers, one can define the increase of the air mass in each of the chambers. The dependencies will be given under the assumption that the chamber 1 is filled and the chamber 2 emptied at the positive level $\mu > 0$:

$$\begin{aligned} \frac{d}{dt}m_1(t) = q_1 &= \begin{cases} \mu\sqrt{p_z - p_1} & \mu > 0 \\ \mu\sqrt{p_1 - p_{atm}} & \mu < 0 \end{cases} \\ \frac{d}{dt}m_2(t) = q_2 &= \begin{cases} -\mu\sqrt{p_2 - p_{atm}} & \mu > 0 \\ -\mu\sqrt{p_z - p_2} & \mu < 0. \end{cases} \end{aligned} \tag{3.41}$$

Assuming the isothermal process $p_iV_i = const$, one can calculate the dependency for the air mass in each of the chambers,

$$m_i = c_w V_i \frac{p_i}{p_{atm}} = c' p_i V_i, \qquad c' = c_w / p_{atm}, \quad i = 1,2, \tag{3.42}$$

where c_w denotes the air specific gravity.

In more advanced calculation systems, air leak m_p between chambers is also modeled which depends on the pressure difference $\frac{d}{dt} m_p = \rho(p_1 - p_2)$, as well as air leak from chambers through leakages of air supply, connections, seals, etc., in the form $\frac{d}{dt} m_i = \rho(p_i - p_{atm})$. Unfortunately, the values of appropriate coefficients ρ are difficult to define. A correctly installed and assembled unit should have leaks more or less 10^{-8} times lower than the main flows, q_1, and q_2. Since these effects are difficult to be precisely estimated, in the subsequent description they are omitted. The dependency (3.42) is an approximation of a real description. In the investigation of gas processes in pneumatic systems, the polytrophic process $p^{\kappa} V = const$ (3.30) with $\kappa = 1.4$ is sometimes assumed but, as examinations show, such a description should be applied to long-time displacements, when temperature changes occur. In single displacements, this effect modeling is not justified (too short time periods and negligible energy changes).

One should remember that each of the masses m_1, m_2 has a changing volume (due to the piston movement), and the mass change dynamics should be therefore described as $\frac{d}{dt} m_i = c' \frac{d}{dt} (p_i V_i) = c' \left(p_i \frac{d}{dt} V_i + V \frac{d}{dt} p_i \right)$. The volume V_1 increases with the velocity v, and the volume V_2 decreases with the same velocity. The final expressions for mass changes will have the following form:

$$c' V_1 \frac{d}{dt} p_1 = q_1 - c' p_1 \frac{d}{dt} V_1 = q_1 - c' p_1 A_1 v, \tag{3.43}$$

$$c' V_2 \frac{d}{dt} p_2 = q_2 - c' p_2 \frac{d}{dt} V_2 = q_2 + c' p_2 A_2 v. \tag{3.44}$$

Using the equations (3.39) and simplifying the above dependencies, one can obtain relations for pressure changes in the chambers:

$$\frac{d}{dt} p_1 = \mu \frac{\sqrt{p_z - p_1}}{c' V_1} - \frac{p_1 v}{x}, \tag{3.45}$$

$$\frac{d}{dt} p_2 = \mu \frac{\sqrt{p_2 - p_{atm}}}{c' V_2} + \frac{p_2 v}{L - x}, \tag{3.46}$$

where L is the maximum piston stroke. Knowing the pressure values in the cylinder chambers, one can obtain the equation for the piston movement dynamics with the use of the force balance,

$$p_1 A_1 - p_2 A_2 = Ma + F_t + F_e. \tag{3.47}$$

The acceleration a value obtained from the above dependency defines the time-derivative of the piston velocity v,

$$\frac{d}{dt}v = \frac{1}{M}\left[p_1A_1 - p_2A_2 - (F_t + F_e)\right], \tag{3.48}$$

which allows us to write the difference equation for the piston position x:

$$\frac{d}{dt}x = v. \tag{3.49}$$

The differential equations (3.40), (3.41), (3.45), (3.46), (3.48) and (3.49), together with the set of initial conditions μ_0, p_{10}, p_{20}, v_0 and x_0, define the servo-drive state and allow us to model the one-, two- or rod-less system type. In the case of divided controls (two control valves) in the equations (3.45), (3.46), there will appear two different control signals, μ_1 and μ_2.

The external force F_e (3.48) is the stimulation that should be recognized as the input quantity for the model, beside the control signal μ (3.45), (3.46). The friction force F_t has a different character. This force depends on many construction factors, most of all on piston velocity (Janiszowski, 2004).

Its model should be therefore given together with the description of the cylinder itself. The friction measurement is difficult and, unfortunately, this force strongly depends on the cylinder conditions of operation: temperature, lubrication, the type and state of seals, the bearings of the moving part of the drive, humidity, cylinder idle time, pressure in the chambers, etc. The number of factors is so high that deterministic formulae that would allow us to find the force value have not been created yet. Figure 3.2 presents the measured course of the friction force after many displacements (warm cylinder). One should notice (Fig. 3.2) the very low velocity values for which the non-linear effect appears—lower than 75 mm/s. For higher velocity values, the friction force change practically has linear character.

The most often used approximation of friction force changes vs. velocity v is the so-called Stribeck curve with the shape described by functional dependencies corresponding with Fig. 3.3:

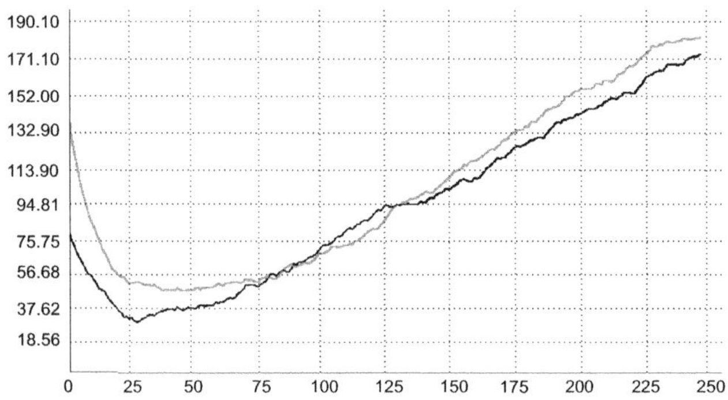

Fig. 3.2 Measured friction force during the acceleration (upper curve) and deceleration of the cylinder piston (lover curve)

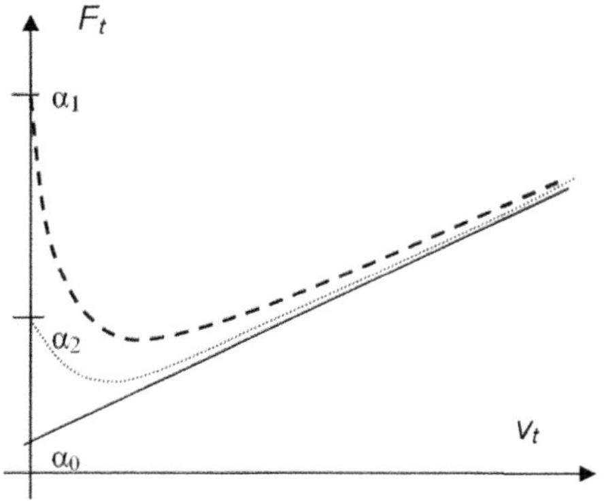

Fig. 3.3 Analytical approximation of friction force F_t changes vs. velocity: dot line—velocity decrease, solid line—linear approximation

$$F_t(v) = sign(v)\alpha_0 + \beta_0 v + sign(v) * \begin{cases} (\alpha_1 - \alpha_0)exp(-\beta_1|v|), & |v| > v_0 \cap sign(av) > 0 \\ (\alpha_2 - \alpha_0)exp(-\beta_2|v|), & |v| > v_0 \cap sign(av) < 0 \end{cases}. \tag{3.50}$$

The equation (3.50) is a precise description of changes presented in Figs. 3.2 and 3.3 but includes many *sign*() functions that are very inconvenient in modeling. It is difficult to ascertain precisely what velocity v_0 will be considered to be zero and in what range the values α_1, α_2 will be changed. Simultaneously, the backlash caused by the friction hysteresis (Fig. 3.3) forces us to apply very small integration steps in order to avoid oscillations during the modeling. A more advantageous formula is friction force description in the form of velocity functions:

$$F_t(v) = sign(v)\left[\alpha_0 + \beta_0|v|\right]\frac{exp(\beta|v|) + \left\{\frac{\alpha_1+\alpha_2}{2\alpha_0} - 1\right\}}{exp(\beta|v|)}. \tag{3.51}$$

The relation does not have the hysteresis observed in Figs. 3.2 and 3.3, but it is obviously easier in modeling.

Considering the above description of the system, one can state that practically all dependencies applied to the creation of the description of its operation are known: dimensions, masses, pressures, etc., and the only uncertainty element is introduced by friction force description, (3.50) or (3.51). Coefficient values appearing in the description (3.50) should be obtained by the way of response modeling and comparing it with the measured changes.

Concluding the investigation of the pneumatic cylinder controlled by the proportional valve, one can distinguish five variables in the presented description. These define in a unique way process behavior, i.e., the state variables including the μ position of the servo-valve throttle (3.40), two pressures p_1 and p_2 in the cylinder chambers (3.45), (3.46), the cylinder piston velocity v (3.48) and its position x (3.49). These variables, together with the algebraic equations (3.39) and (3.50) or (3.51), as well as adequate initial conditions, give a complete description of the examined process state.

The obtained model described the method for system analytical description that unites several elements: a servo-valve, pneumatic lines, a cylinder and the mechanical construction unit that is connected to the piston. Selected parameters in the model description integrate properties of different elements, e.g., the mass M represented the overall mass of the cylinder moving unit and the driven construction, the effective gain coefficient included the valve and conduit flow capacity, the friction characteristic included the overall friction of the cylinder and the driven unit, etc. In the above approach, the model actually represented the complete servo-drive together with the driven mechanism while the valve control signal acted as the input, and the dynamic parameters of the driven unit—position, velocity and acceleration—acted as the output.

3.2.4 *Pneumatic Cylinder Controlled by a Servo-Valve: A Block Model of the System*

When attempting to model a system with the application of special blocks that represent particular components and the construction of their shell versions, one should very carefully consider the way of description. The components should be connected together in desirable structures, i.e., the outputs of a particular type of component have to be compatible with the inputs of a component that is connected to it as an energy receiver.

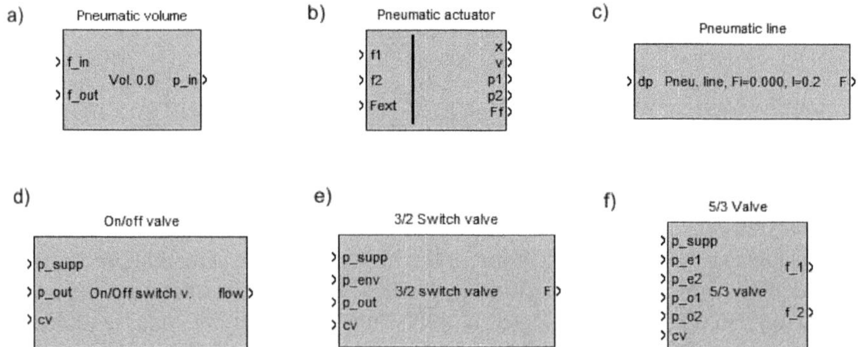

Fig. 3.4 *Pneumatic Elements* library components: pneumatic capacitance (a), pneumatic cylinder (b), pneumatic line (c), cut-off valve (d), three-way switching valve (e), five-way proportional valve (f)

As an example, we can point out the *Pneumatic Elements* library included in the *PExSim* package. Some selected components of the library are presented in Fig. 3.4. The library contains models of pneumatic capacity, the cylinder, the supply line, the cut-off valve, the three-way switching valve and the five-way proportional valve.

In order to model a system composed of these elements, one should select one out of two rules of phenomena modeling in pneumatic elements: flows create pressure differences or—what is applied more often—pressure differences decide on the flow values (3.38), (3.39). A combined description is prohibited since it does not allow us to create equation sets that would univocally define initial conditions that are required for the modeling.

In the case of the mentioned library, the first method has been accepted: flows create pressure differences. This allows modeling in a simple way, e.g., the states that appear in pneumatic capacitance, cylinder, etc. As a rule, we have information on the nominal supply and run-off pressure values, the nominal flow through the valve, valve reaction time delay, inertia (or the limit frequency), etc. The introduced models of components can be parameterized in such an approach (Fig. 3.5).

The second method of description would require giving information on pneumatic resistances as parameters of particular components. As a rule, they are not given by manufacturers. The approach applied does not have universal character—while describing electric circuits, one more often uses a description in which voltages or voltage drops are the initiating factors since electric element characterization is based more often on information such as resistance, inductance or capacitance.

The acceptance of a definite method of initiation requires us to model quantities that would allow calculating the pressure values and to give this information back to the structure elements that demand it. This situation is presented in Fig. 3.5, which shows the position control system with the use of the pneumatic cylinder and the supply unit (compressor with capacitance and relay). The outlet from the valve that

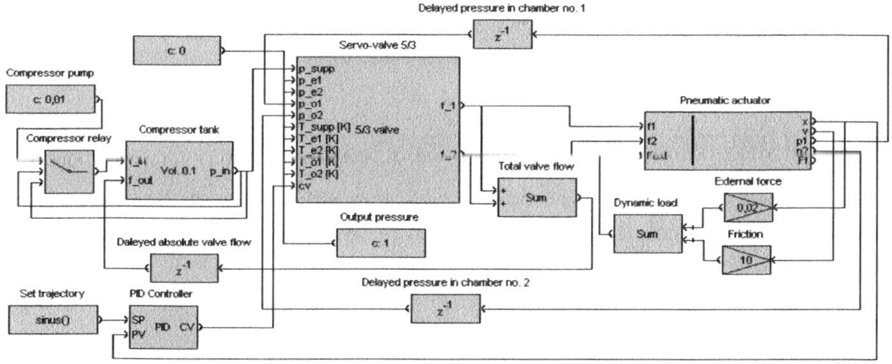

Fig. 3.5 Model of the structure of the position control system with the PID controller, supply unit, proportional valve, and pneumatic cylinder

supplies the cylinder is necessary for the calculation of the state of the tank that supplies the valve. In the case of too high air consumption and a pressure drop in the tank, the compressor must load up the tank to the required value. In the next stage of the calculations, the state of the flow through the valve depends on the pressure in the cylinder chambers; therefore, these values should be calculated in the cylinder and turned back as inputs *p_o1* and *p_o2* to the valve. The next turning back is, for instance, the modeling of the force value that loads the cylinder depending on the piston velocity that defines the kinetic friction.

In order to avoid problems with initial states modeling (the simulation in *PexSim* is carried out from signal sources to successive dynamic units), in loops of the turn-back one should introduce one-step delays of modeling (symbol z^{-1}), (Fig. 3.5). The model created in this way allows us to simulate the reaction of the pneumatic servo-drive system but, despite the introduction of all known dimensions, the value of the mass, the external force F_e, one should not expect to achieve full conformity between the simulated changes and the measured ones. The difference will result from accepted friction force parametrization by the descriptions (3.50) or (3.51), as well as the acceptance of unknown resistance values, e.g., at pneumatic terminals of the cylinder. There exist many such parameters in the presented description: two in the valve, six in the cylinder, two in the supply tank and one in lines—together 11 inaccurately defined parameters. One may expect that joined tuning of the parameter values will be very time-consuming even with the use of very fast computers. As an alternative, one can apply the following popular approach: to simplify the model to basic interactions and to find values of several parameters only, with the use of the optimization method.

3.3 Linear Models: Local Approximation of Dynamic Properties

All dynamic processes examined in the full range of changes have non-linear character, but when they are analyzed in a sufficiently limited range, they can be approximated by linear description with satisfactory accuracy.

3.3.1 Dynamic Model Linearization

Linearization should be carried out in a stable point $\mathbf{x} \in R^n$ of the process state space. In such a point, the description is given by the non-linear differential state space equation

$$\frac{d}{dt}\mathbf{x} = F(\mathbf{x},\mathbf{u},t), \qquad \mathbf{x} \in R^n, \mathbf{u} \in R^q, \tag{3.52}$$

and the algebraic output equation

$$\mathbf{y} = G(\mathbf{x},\mathbf{u},t) \qquad \mathbf{y} \in R^m, \tag{3.53}$$

where the vector $\mathbf{y}$ is measurable in the equilibrium state. If the input $\mathbf{u}_0$ exists the state vector $\mathbf{x}_0$, then the following equations are satisfied:

$$\frac{d}{dt}\mathbf{x} = F(\mathbf{x}_0, \mathbf{u}_0, t_0) = 0,$$

$$\mathbf{y}_0 = G(\mathbf{x}_0, \mathbf{u}_0, t_0). \tag{3.54}$$

If, in the point $\mathbf{p}_0 = \{\mathbf{x}_0, \mathbf{u}_0\}$, functions $F(\cdot)$ and $G(\cdot)$ are continuous, one can use Taylor's formula and expand the set of differential equations and the output equations:

$$\frac{d}{dt}\mathbf{x} = F(\mathbf{x}_0, \mathbf{u}_0, t_0) + \frac{\partial}{\partial \mathbf{x}} F(\mathbf{x}, \mathbf{u}, t)\bigg|_{\mathbf{p}_0} \Delta\mathbf{x} + \frac{\partial}{\partial \mathbf{u}} F(\mathbf{x}, \mathbf{u}, t)\bigg|_{\mathbf{p}_0} \Delta\mathbf{u} + R_F(\mathbf{x}_0, \mathbf{u}_0, t_0), \tag{3.55}$$

$$\mathbf{y} = G(\mathbf{x}_0, \mathbf{u}_0, t_0) + \frac{\partial}{\partial \mathbf{x}} G(\mathbf{x}, \mathbf{u}, t)\bigg|_{\mathbf{p}_0} \Delta\mathbf{x} + \frac{\partial}{\partial \mathbf{u}} G(\mathbf{x}, \mathbf{u}, t)\bigg|_{\mathbf{p}_0} \Delta\mathbf{u} + R_G(\mathbf{x}_0, \mathbf{u}_0, t_0), \tag{3.56}$$

where

$$\Delta\mathbf{x} = \mathbf{x} - \mathbf{x}_0, \qquad \mathbf{x}_0 = F(\mathbf{x}_0, \mathbf{u}_0, t_0), \tag{3.57}$$

and

$$\Delta\mathbf{y} = \mathbf{y} - \mathbf{y}_0, \qquad \mathbf{y}_0 = F(\mathbf{x}_0, \mathbf{u}_0, t_0). \tag{3.58}$$

The expansion range of these two functions, F and G, should be selected in such a way that the assumption that the factors R_F and R_G in the expansions (3.55), (3.56) can be omitted was justifiable. Remembering the properties of the point $\mathbf{p}_0$ (3.54) and introducing new symbols for the partial derivatives calculated in (3.55), (3.56), one may reach the following form of description:

$$\frac{d}{dt}\mathbf{x} = \mathbf{A}\Delta\mathbf{x} + \mathbf{B}\Delta\mathbf{u} \tag{3.59}$$

and

$$\Delta\mathbf{y} = \mathbf{C}\Delta\mathbf{x} + \mathbf{D}\Delta\mathbf{u}, \tag{3.60}$$

where

$$\mathbf{A} = \frac{\partial}{\partial \mathbf{x}} F(\mathbf{x}, \mathbf{u}, t)\bigg|_{\mathbf{p}_0}, \quad \mathbf{B} = \frac{\partial}{\partial \mathbf{u}} F(\mathbf{x}, \mathbf{u}, t)\bigg|_{\mathbf{p}_0},$$
$$\mathbf{C} = \frac{\partial}{\partial \mathbf{x}} G(\mathbf{x}, \mathbf{u}, t)\bigg|_{\mathbf{p}_0}, \quad \mathbf{D} = \frac{\partial}{\partial \mathbf{u}} G(\mathbf{x}, \mathbf{u}, t)\bigg|_{\mathbf{p}_0}, \tag{3.61}$$

which presents local properties of the dynamic system. The symbols $\mathbf{x}$ and $\mathbf{y}$ stress the local character of the description. For compactness of presentation, a simplified form of the linear description of the approximation of the equations (3.59), (3.60) is most often used:

$$\begin{aligned} \dot{\mathbf{x}} &= \mathbf{A}\mathbf{x} + \mathbf{B}\mathbf{u}, \\ \mathbf{y} &= \mathbf{C}\mathbf{x} + \mathbf{D}\mathbf{u}. \end{aligned} \tag{3.62}$$

The above relations are simple and the calculation of derivatives of analytical functions should pose no problems if they are continuous functions. The state variables system (3.62) allows us to obtain a description in the form of the transfer function matrix:

$$\mathbf{G}(s) = \mathbf{C}\left[s\mathbf{I} - \mathbf{A}\right]^{-1}\mathbf{B} + \mathbf{D}. \tag{3.63}$$

While introducing linear description, future applications should also be considered, e.g., to controller algorithm design. In such a case, one should think which state variables are measurable or which transformations should be carried out in order to obtain a description in a form more suitable for the intended application. This approach is presented for the above description of the pneumatic drive.

3.3.2 *Pneumatic Cylinder Controlled by a Servo-Valve: A Linear Model of the System*

The final set of differential equations (3.45) to (3.49), under the assumption of the following state variables: μ—position of the servo-valve, x—position of the cylinder piston, v—the piston velocity, p_1 and p_2—pressures in chambers, as well as two external stimulations: the voltage control u of the valve and the external force F_e, will have two forms dependent on the servo-valve throttle position (3.41). For $\mu > 0$, the description will have the form

$$\begin{cases} \dfrac{d}{dt}\mu &= \dfrac{1}{T_z}\left[\dfrac{u(t-T_0)}{u_{max}} - \mu\right], \\ \dfrac{d}{dt}x &= v, \\ \dfrac{d}{dt}v &= \dfrac{1}{M}\left[p_1A_1 - p_2A_2 - F_t(v) - F_e\right], \\ \dfrac{d}{dt}p_1 &= \mu\dfrac{\sqrt{p_z - p_1}}{c'A_1(x)} - \dfrac{p_1v}{x}, \\ \dfrac{d}{dt}p_2 &= -\mu\dfrac{\sqrt{p_2 - p_{atm}}}{c'A_2(L-x)} + \dfrac{p_2v}{L-x} \end{cases} \tag{3.64}$$

and, for $\mu < 0$, the form

$$\begin{cases} \frac{d}{dt}\mu &= \frac{1}{T_z}\left[\frac{u(t-T_0)}{u_{max}} - \mu\right], \\ \frac{d}{dt}x &= v, \\ \frac{d}{dt}v &= \frac{1}{M}\left[p_1A_1 - p_2A_2 - F_t(v) - F_e\right], \\ \frac{d}{dt}p_1 &= \mu\frac{\sqrt{p_1 - p_{atm}}}{c'A_1(x)} - \frac{p_1 v}{x}, \\ \frac{d}{dt}p_2 &= -\mu\frac{\sqrt{p_z - p_2}}{c'A_2(L-x)} + \frac{p_2 v}{L-x}. \end{cases} \tag{3.65}$$

The description is a relatively complex one and we can try to introduce simplifications in order to obtain a linear equations set suitable for, e.g., the design of a state variables controller.

One should begin by analyzing which state variables change so rapidly that their variations may be considered to be without inertia in comparison with the remaining quantities. The servo-valve (FESTO MPY 5/3 type) with the proportional magnet has a limit frequency of about 180 Hz, so the time constant T_z can be estimated as having 3 to 4 milliseconds. Time delays τ_{op} measured in a laboratory were equal to 2.5 and 2.8 milliseconds, respectively (for the control signal $u > 0$ and $u < 0$), and the air flow through lines having approximately 1 m required also the time τ_{dop} of about 3 milliseconds. Since free vibrations of the pneumatic cylinder that were observed in the laboratory had a frequency of about 2 to 8 Hz, it was assumed that the servo-valve operation would be represented by a proportional element having the time delay of $T_0 \cong T_z + \tau_{op} + \tau_{dop} = 7$ to 10 milliseconds. Hence, the first differential equation out of the set (3.64) or (3.65) is replaced by the algebraic equation and the description order is lowered by one. The control signal u will be included in appropriate equations with a specified gain factor and the time argument delayed by T_0.

The friction force is a non-continuous function for $v = 0$ (3.50). In the range $|v| \approx 0$, the friction jump appears and the linearization cannot be carried out, but for $|v| > 0.05$ m/s (see Fig. 3.3), one can assume that the equation (3.50) has linear character, and it can be expressed in the following form:

$$F_t(v) = \beta_0 v. \tag{3.66}$$

One may assume that such an approximation will be sufficiently good for fast piston movement, but during the piston's start and stop, some phenomena can appear that depart from this description.

The next step consists in taking the effect of the controls u into account. Depending on the control signal sign, two different processes (3.41) exist: for $\mu > 0$, the chamber 1 is filled up (to the maximum pressure value of 6 bar), and for $\mu < 0$, the chamber is emptied down to the minimum pressure value of 1 bar. The air flows practically through passage channels having the same construction, therefore it is possible to recognize air flow resistance in both cases as comparable. In this case, the value of the expressions $\sqrt{p_z - p_1}$ and $\sqrt{p_2 - p_{atm}}$ has a decisive effect. It was discovered during the experiments that the steady value of pressures in the symmetric cylinder chambers (with two piston rods or rod-less) is equal to approximately 3.4 to 3.6 bar. Taking into account the values $p_2 = 6$ bar and $p_{atm} = 1$ bar, it is easy to discover that the expressions $\sqrt{p_z - p_1}$ and $\sqrt{p_2 - p_{atm}}$ will have similar values and, as a result, directional gain values ($\mu > 0$ and $\mu < 0$) in (3.45), (3.46) can be considered comparable. One can therefore give up two sets of equations (3.43) and (3.45), and create a joint description in the form

$$\begin{cases} \frac{d}{dt}x &= v \\ \frac{d}{dt}v &= \frac{1}{M}\left[p_1A_1 - p_2A_2 - \beta_0 v - F_e\right] \\ \frac{d}{dt}p_1 &= k_{1u}u(t-T_0) - k_{1v}v \\ \frac{d}{dt}p_2 &= -k_{2u}(t-T_0) + k_{2v}v \end{cases}, \tag{3.67}$$

where $k_{1u} \cong \frac{\sqrt{2,5}}{c'A_1x}$, $k_{1v} = \frac{p_1}{x}$, $k_{2u} \cong \frac{\sqrt{2,5}}{c'A_2(L-x)}$ and $k_{2v} = \frac{p_2}{L-x}$.

The gain coefficient values k_{1u}, k_{1v}, k_{2u} and k_{2v} that appear in the above equations depend on the position and pressure values, but these values change relatively slowly (in comparison with the velocity v) and, in a selected position, they can be considered to be constant in (3.67).

Assuming the possibility of measurements of the cylinder piston position x and the pressures p_1 and p_2 in the chambers (some proportional valves are equipped with the measurement of the pressure in the chamber), the description of the existing phenomena depends on four state variables: the piston position x, its velocity v, and pressures p_1 and p_2. The control signal u and the external force F_e act as stimulations. Therefore, linear description may be introduced:

$$\dot{\mathbf{x}} = \mathbf{Ax} + \mathbf{Bu}, \tag{3.68a}$$

$$\mathbf{y} = \mathbf{Cx} + \mathbf{Du}, \tag{3.68b}$$

where

$$\mathbf{x} = \begin{bmatrix} x(t) \\ v(t) \\ p_1(t) \\ p_2(t) \end{bmatrix}, \quad \mathbf{u} = \begin{bmatrix} u(t-T_0) \\ F_e(t) \end{bmatrix}, \quad \mathbf{y} = \begin{bmatrix} x(t) \\ p_1(t) \\ p_2(t) \end{bmatrix} \tag{3.69}$$

and

$$\mathbf{A} = \begin{bmatrix} 0 & 1 & 0 & 0 \\ 0 & \frac{-\beta_0}{M} & \frac{A_1}{M} & \frac{-A_2}{M} \\ 0 & -k_{1v} & 0 & 0 \\ 0 & k_{2v} & 0 & 0 \end{bmatrix}, \quad \mathbf{B} = \begin{bmatrix} 0 & 0 \\ 0 & \frac{1}{M} \\ k_{1u} & 0 \\ -k_{2u} & 0 \end{bmatrix}, \quad \mathbf{C} = \begin{bmatrix} 1 & 0 & 0 & 0 \\ 0 & 1 & 0 & 0 \\ 0 & 0 & 1 & 0 \\ 0 & 0 & 0 & 1 \end{bmatrix}, \quad \mathbf{D} = \begin{bmatrix} 0 & 0 \\ 0 & 0 \\ 0 & 0 \\ 0 & 0 \end{bmatrix}. \tag{3.70}$$

When the measurement of the pressures p_1 and p_2 is not possible (as in the case of the employed MPY 5/3 valve) and the position x measurement is available only, one should try to obtain parametrization that makes the above description independent of the pressure values. To simplify the discussion, let us assume that the external resistance force $F_e = 0$. The piston acceleration may be expressed in the form

$$a = \frac{1}{M}\left[p_1 A_1 - p_2 A_2\right] - \frac{1}{M}\beta_0 v, \tag{3.71}$$

and its derivative may be defined from the equations (3.67):

$$\begin{aligned}\frac{d}{dt}a &= \frac{1}{M}\left[\dot{p}_1 A_1 - \dot{p}_2 A_2\right] - \frac{1}{M}\beta_0 \dot{v}\\ &= \frac{A_1 k_{1u}}{M}u(t-T_0) - \frac{A_1 k_{1u}}{M}v + \frac{A_2 k_{2u}}{M}u(t-T_0) - \frac{A_2 k_{2v}}{M}v - \frac{1}{M}\beta_0 a\\ &= k_u u(t-T_0) - k_{av}v - k_{aa}a,\end{aligned} \tag{3.72}$$

where $k_u = \frac{A_1 k_{1u}}{M} + \frac{A_2 k_{2u}}{M}$, $k_{av} = \frac{A_1 k_{1v}}{M} + \frac{A_2 k_{2v}}{M}$, $k_{aa} = \frac{\beta_0}{M}$.

The system will be described by three so-called phase state variables: x, v, a (position, velocity and acceleration), and the whole description will have the form (3.68), where $\mathbf{x} = \begin{bmatrix} x(t) \\ v(t) \\ a(t) \end{bmatrix}$, $\mathbf{u} = [u(t-T_0)]$ and $\mathbf{y} = [x(t)] \in R^1$, and $\mathbf{A} = \begin{bmatrix} 0 & 1 & 0 \\ 0 & 0 & 1 \\ 0 & -k_{av} & -k_{aa} \end{bmatrix}$,
$\mathbf{B} = \begin{bmatrix} 0 \\ 0 \\ -k_{aa} \end{bmatrix}$, $\mathbf{C} = \begin{bmatrix} 1\,0\,0 \end{bmatrix}$, $\mathbf{D} = [0]$.
This form is definitely simpler than the equations (3.64), (3.65), but it is an approximation for which the coefficients k_u, k_{av} and k_{uv} are clearly dependent on the point of operation.

The introduced description may be presented in the form of transfer function with the use of the transformation (3.63). Section 3.4 and 3.5 deal with looking for parametric models of the discussed pneumatic drive in the form

$$G_{vu}(s) = \frac{k_u}{s^2 + k_{aa}s + k_{av}} e^{-sT_0} \tag{3.73}$$

obtained with the use of the statistic evaluation method.

The observed piston velocity reactions to the control signal u on the input of the proportional valve have typical oscillatory character, therefore the transfer function $G_{vu}(s)$ should contain information on the velocity gain C, radial frequency of the proper vibrations ω_0 and the damping factor ξ,

$$G_{vu}(s) = \frac{C\omega_0^2}{s^2 + 2\xi\omega_0 s + \omega_0^2} e^{-sT_0}. \tag{3.74}$$

Comparing (3.72) and (3.74), one can obtain dependencies for the parameters C, ω_0 and ξ:

$$\omega_0^2 = \frac{A_1 p_{10}}{Mx} + \frac{A_2 p_{20}}{M(L-x)},$$
$$C\omega_0^2 = \frac{\sqrt{2,5}}{c'M}\left(\frac{1}{x} + \frac{1}{L-x}\right),$$
$$2\xi\omega_0 = \frac{\beta_0}{M}. \tag{3.75}$$

As is easy to observe, the frequency of proper vibrations ω_0 of the system will depend mainly on the mass M, pressure values in the chambers, the piston position and the cylinder diameter. The remaining parameters strongly depend on ω_0, but they are more independent of the position and air pressures in the chambers. In the case of a symmetric cylinder ($A_1 = A_2 = A$), the pressures ($p_{10} = p_{20} = p_0$) in the chambers in initial states are equal, and the parameters C, ω_0 and ξ are described by simple dependencies:

$$\omega_0^2 = \frac{Ap_0}{M}\left(\frac{1}{x} + \frac{1}{L-x}\right),$$
$$C = \frac{\sqrt{2,5}}{c'Ap_0},$$
$$\xi = \frac{\beta_0 x(L-x)}{2ALp_0}, \tag{3.76}$$

which means that the controlled cylinder will have the lowest frequency of proper vibrations ω_0 and the highest vibration damping ξ in the middle position of the piston, and the gain should be constant and dependent on the servo-valve flow capacity, as well as inversely proportional to the piston surface A. This simple model will be examined in the chapter on parametric models of the system.

3.3.3 Pneumatic Cylinder Controlled by a Servo-Valve: An Optimized Linear Model of the System

While designing the analytical model, we usually know a part of parameters resulting from geometry, physical constants and laws that describe particular phenomena. The remaining parameters are either difficult to be measured or they reflect some approximate relations used for the derivation of the final description. Modeling the process, we usually have additional knowledge in the form of recorded measurements. One can ask if both sources of information may be used, and if a model could be derived that has a partially defined description and will fit the gathered measurements.

A typical example of such a situation is the state variables description (3.72), in which values of the mass M, the piston surfaces A_1, A_2 are relatively easy to be determined, but the coefficients β_0, k_{1u}, k_{2u}, k_{1v}, k_{2v}, which depend on the friction conditions, flow resistances and servo-valve flow capacity, are not known.

In such a situation, the user can come to the conclusion that the usefulness of shell models, e.g., the above-described pneumatic drive system having a description based on analytical equations, is limited due to the lack of a method that would allow us to define coefficients necessary for the modeling. One can then use the option of parameter optimization (*PExSim Optimizer*) available in the *PExSim* package.

The option allows us to optimize selected coefficients of the model in the sense of the minimization of the required evaluation index. The use of this approach will be presented with the help of the above-described pneumatic cylinder model. Figure 3.6 presents transients measured during the experiment: changes of the valve control signal, piston velocity and pressures in the cylinder chambers.

Let us assume that the simplest model of the whole system is required in the form of the velocity transfer function (3.74), we do not know any of the examined transfer function coefficients, and we assume that the structure is known and we look for coefficients of both the numerator and the denominator:

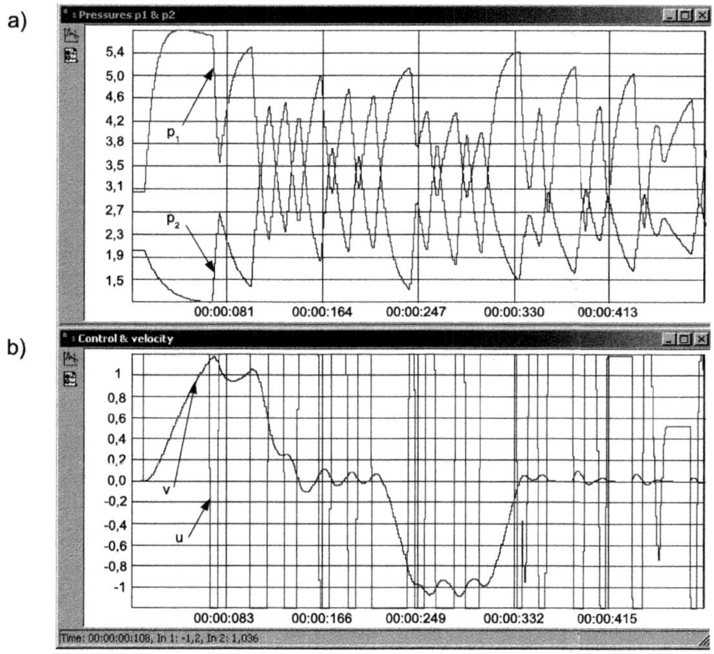

Fig. 3.6 Transients of registered values of pressures in the cylinder chambers (a), servo-valve control signal u and cylinder piston velocity v (b)

$$G(s) = \frac{x_0 + x_1 s}{1 + a_1 s + a_2 s^2} e^{-s0{,}005}. \tag{3.77}$$

The time delay value should be evaluated earlier and included in the examined model (3.77). Next, one should create the system structure in the *PExSim* package for the optimization purposes (Fig. 3.7). One should also define the type of the matched element, e.g., the transfer function, the pattern model, etc., and the form of the evaluation index that will be minimized (e.g., the sum of absolute values of the difference between the output from the model and the measured input), as well as introduce the reference signals, e.g., the sequence of registered inputs and outputs ("File with recorded input & output data" in Fig. 3.7).

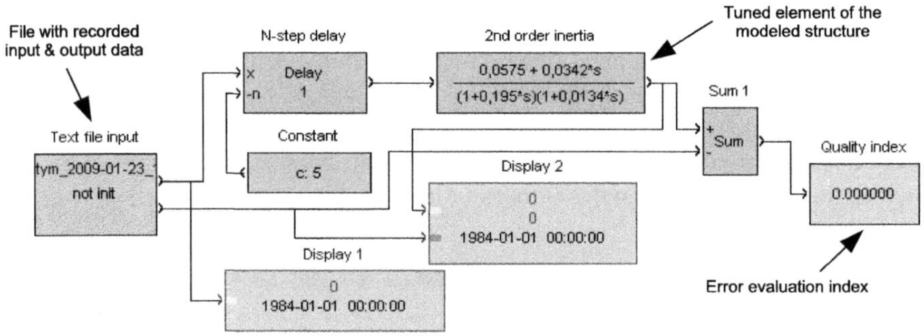

Fig. 3.7 Diagram of the modeling structure for the optimization of transfer function coefficients in the *PExSim* package

Then, the coefficient tuning process is set in motion. At the first stage of the tuning, global optimization algorithms are used, e.g., the Particle Swarm Optimization (PSO) algorithm, which allows us to select at first one (or more) points around of which optimum solutions will be looked for (Wnuk, 2006).

The optimization mechanism is an external thread for the *PexSim* package. It consists in sequential setting into motion the simulation module of the package. When the complete simulation cycle is achieved, the obtained results are evaluated. When the index value is obtained, the next point is analyzed (in the model coefficient space) according to the searching procedure applied. The range of parameter searching may be limited according to a priori knowledge of the person that implements the optimization. One can also choose the mode of the free choice of unknown values out of a very broad range, as in the presented example. Such a process is relatively time consuming but it allows us to look for investigated parameters in the sense of freely defined evaluation indices, e.g., the sum of absolute values of errors of the model or the introduction of multi-criteria indices (Kreisselmeier and Steinhauser, 1979).

The point chosen by the global optimization algorithm is then recognized as the initial one for solution searching by the local searching methods. The optimization procedure is again set in motion, and selected parameters are tuned according to the procedure described earlier. Figure 3.9 presents changes of the tuned coefficient x_0 of the transfer function (3.77). As one can see, the value of the coefficient x_0 at the initial time instant was very distant from the final result, and the evaluation index was (for a relatively long time, approximately 100 objective function evaluations) practically constant, but the final result was quite satisfactory. The comparison of changes obtained from the tuned model and results introduced as the reference data is presented in Fig. 3.8.

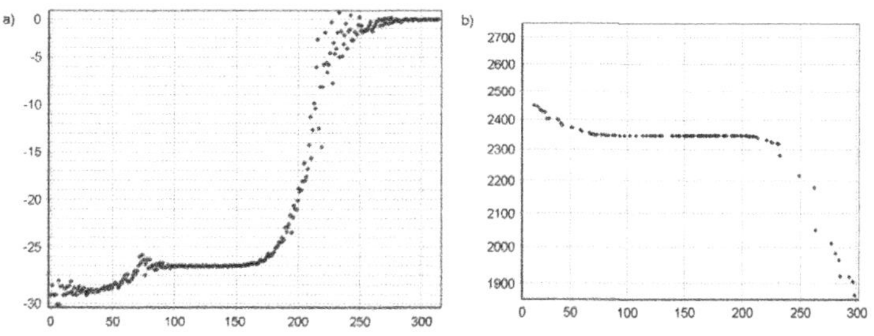

Fig. 3.8 Changes of the values of the coefficient x_0 of the transfer function (3.77) during optimization (a), changes of the evaluation index during optimization (b)

The transfer function obtained in the above example had the form

$$G(s) = \frac{0,0575 + 0,0342s}{1 + 0,195s + 0,0134s^2} e^{-s0,005} = \frac{C\omega_0^2}{s^2 + 2\xi\omega_0 s + \omega_0^2} e^{-s0,005s}, \tag{3.78}$$

where $\omega_0 = 8,63$ rad/s, $\xi = 0,84$, $C = 0,0575$(m/s)/V.

While comparing the possibilities of recovering the recorded input-output data by the model (3.78), Fig. 3.9, it can be considered quite satisfactory. Of course, the system described by the 2-nd order transfer function will show more rapid changes than the real system that is at least of the 4-th or the 5-th order (3.70), but dynamics reproduction is correct.

Considering the relations (3.76), it is easy to notice that the value of the proper vibrations ω_0 is relatively easy to be estimated since it depends merely on the dimensions and mass of the moving assembly. In the case considered, the cylinder had the overall length L equal to 0.5 m and the experiment was conducted in the cylinder center. Knowing the piston active surface $A \approx 0.0008$ m^2 and assuming that the pressure value in the state of equilibrium is equal to $p0 \approx 3,5$ bar, one can calculate the value of ω_0 as 8,48 rad/s (3.76), which agrees quite well with the value obtained during the optimization experiment.

The above example presents the possibilities of the analytical modeling technique—one can know the structural description of the investigated process but the accuracy or knowledge of some parameters may be limited. The presented optimization technique allows us then to effectively match these parameters to available measurements.

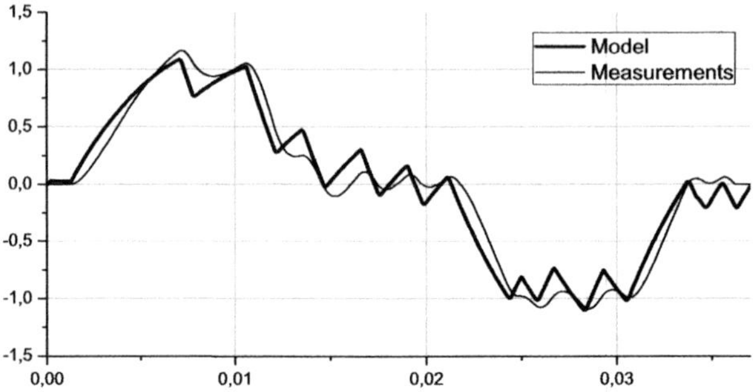

Fig. 3.9 Changes of the output of the optimized transfer function (curve having sharp summits) and the measured transients having smooth value changes

3.4 Parametric Models

Parametric models, as well as neural ones, belong to the group of models determined through identification, i.e., numerical processing of data collected during measurements. Such models may recreate static or dynamic relations between the measured signals. Unlike models in the form of artificial neural networks, parametric models better correspond with traditional forms of description, static characteristics or transfer functions that describe the examined process dynamics. Since coefficients of these models can be transformed into values of physical parameters that describe the process dynamic properties: gain, time constants, time delays, etc., the models are often called parametric. The *DiaSter* platform contains the *MITforRD* package, which helps us to create parametric models in a broad range. As plug-ins, such models may cooperate with the *PExSim* package.

Unlike analytical models, parametric ones depend on the data sampling interval Δ. The description of the model will be calculated with the discrete time $t = k\Delta$, i.e., both the modeled quantity signal $y(k)$ and the signals of inputs $u_i(k)$, as well as disturbances $z_i(k)$, will be introduced to the model in the form of sampled values.

The parametric models structure may be presented in a general form:

$$\hat{y}(k) = M(\mathbf{u}, \mathbf{z}, \mathbf{y}, \theta), \tag{3.79}$$

where $\mathbf{u} = [u_1(\tilde{k}), u_2(\tilde{k}), \ldots, u_q(\tilde{k})]^T$ is the control vector at different discrete time instants $\tilde{k} \in \langle k - l_1, k - l_2 \rangle$, $l_2 > l_1$, $\mathbf{y}$ is the output vector, $\mathbf{z} = [z_1(\tilde{k}), z_2(\tilde{k}), \ldots, z_r(\tilde{k})]^T$ is the disturbance vector and θ denotes the vector of unknown coefficients. The structure of the model is very vital and depends on the planned application.

Models for controller design and automatic control applications will require the exposition of the effect of stimulating factors—the output quantity $y(k)$ should depend on the control signals u and the disturbances z that do not depend on the control signals. In the structure of these models, it is better not to include information gathered in measured signals that depend on the controls u_j. For example, investigating the pneumatic drive dynamics model for the synthesis of the control algorithm it would be undesirable to include the measurements of the pressures p_1, p_2 (in the chambers) as the input information. In order to obtain the compact form of description and to take into account the proper dynamics of the process considered, it is reasonable to include into the model structure M an evaluation of the output signal $\hat{y}$, i.e., $\hat{y}(k) = M(\mathbf{u}, \mathbf{z}, \hat{y}, \theta, k)$. Such models must have a form that allows us to evaluate the future effect of the decisions made (therefore, $l_1, l_2 \in N$) and to design a control or an algorithm of automatic control.

Models for diagnostic purposes will include information carried by all of the signals u, z that are reliable measurements, in general, without the signal y that is modeled, i.e., $\hat{y}(k) = M(\mathbf{u}, \mathbf{z}, \theta, k)$. On the other hand, they can use future signal values in relation to the time instant k, if it improves the evaluation quality—therefore, $l_1, l_2 \in C$. The aim of this model is the most reliable calculation of the signal value at the current time instant k. Such a model may be also successfully used for the so-called "soft measurements", but then it is usually obtained during a special experiment, where the measurement is implemented by the reference system and the recorded pattern data allow us to calculate a precise model that may replace the pattern measurement with given accuracy. All forms of models may be used in this application, including models based on artificial neural networks, very popular recently.

The third group consists of parametric models for prediction purposes. In this case, the only limitation is the prediction horizon $l_2 \in N$. As the input information, one can use all measurable signals that are suspected to have an effect on the output at time instants $k - l_2, k - l_2 - 1$, i.e., $\hat{y}(k) = M(u, z, y, \theta, k)$. The form of the models is not limited.

The quality index of the estimated model is most often calculated in the form of the sum of squares of the model errors:

$$I_{LS} = \sum_{k=1}^{N} [y(k) - \hat{y}(k)]^2 , \tag{3.80}$$

which is used very readily due to its analytical form and advantageous behavior during minimization. The index of the sum of absolute errors

$$I_{SAE} = \sum_{k=1}^{N} |y(k) - \hat{y}(k)| \tag{3.81}$$

has no analytical formula but can be more suitable for the comparison of the determined models.

3.4.1 *Discrete Linear Parametric Models*

Parametric dynamic models are often a representation of dynamic dependencies between signals, expressed in the form of the state variables diagram (3.62) or the transfer function (3.63). Measurement data are sampled with the sampling interval Δ, hence appropriate descriptions are also expressed in a form parameterized by Δ. For example, equations of discrete state variables may be used:

$$\mathbf{x}(k+1) = \tilde{\mathbf{A}}\mathbf{x}(k) + \tilde{\mathbf{B}}\mathbf{u}(k), \tag{3.82a}$$

$$\mathbf{y}(k) = \mathbf{C}\mathbf{x}(k) + \mathbf{D}\mathbf{u}(k), \tag{3.82b}$$

where $\mathbf{x} \in R^n, \mathbf{u} \in R^q, \mathbf{y} \in R^m, \tilde{\mathbf{A}} \in R^{n\times n}, \tilde{\mathbf{B}} \in R^{n\times q}, \mathbf{C} \in R^{m\times n}, \mathbf{D} \in R^{m\times q}$, in which the time argument k denotes the appropriate signal value at the time $t = k\Delta$. The matrices $\mathbf{C}$ and $\mathbf{D}$ appearing in this description are identical with those appearing in the description (3.62), and the matrices $\tilde{\mathbf{A}}$ and $\tilde{\mathbf{B}}$ may be obtained by the transformation of the matrices $\mathbf{A}$ and $\mathbf{B}$ in the following way:

$$\begin{aligned}\tilde{\mathbf{A}} &= exp(\mathbf{A}\Delta),\\ \tilde{\mathbf{B}} &= \mathbf{A}^{-1}\big(exp(\mathbf{A}\Delta) - \mathbf{I}\big)\mathbf{B} = \big(exp(\mathbf{A}\Delta) - \mathbf{I}\big)\mathbf{A}^{-1}\mathbf{B},\end{aligned} \tag{3.83}$$

where $\tilde{\mathbf{A}} \in R^{n\times n}, \tilde{\mathbf{B}} \in R^{n\times q}$, $exp(\mathbf{A}\Delta)$ is the matrix function and $\mathbf{I}$ is the unitary matrix having the appropriate dimension. In the case of small (in comparison with the described process dynamics) value of the sampling interval Δ, one may use approximate description being the result of the approximation of the expansion of the function $exp(\mathbf{A}\Delta)$ to the first term: $\tilde{\mathbf{A}} = \mathbf{I} - \mathbf{A}\Delta$, $\tilde{\mathbf{B}} = \Delta\mathbf{B}$, where $\tilde{\mathbf{A}} \in R^{n\times n}$, $\tilde{\mathbf{B}} \in R^{n\times q}$.

The description with the use of the transfer function (3.63) may also be expressed in a form parametrized by the sampling interval Δ. It is usually convenient to apply the description of a single modeled output y. One may try to use tables with transforms of particular types of continuous transfer functions, but for more complex descriptions (of the 3-rd order and higher), their application requires us to run many calculations. Approximated formulae are applied.

Let us consider a continuous process of the description (3.63) that has j output signals and is simulated by the signals $u_i(t), i = 1,\ldots,q$,

$$y_j(s) = \sum_i G_{ij}(s)u_i(s), \tag{3.84}$$

having values sampled with the step Δ. Let us assume that the simulations change so slowly that they can be considered constant in the time intervals $\langle k\Delta, (k+1)\Delta)$. The particular output signals $y_j(t)$ may be then described by the following dependency:

$$y_j(z^{-1}) = \sum_i \tilde{G}_{ij}(z^{-1})u_i(z^{-1}) = \sum_i \frac{B_{ij}(z^{-1})}{A_j(z^{-1})}u_i(z^{-1}), \tag{3.85}$$

in which z^{-1} is the shift back operator, $x[k\Delta]z^{-p} = x[(k-p)\Delta]$. In order to simplify the notation, let us denote the discrete time argument simply as k in our further discussion. The expressions $B_{ij}(z^{-1})$ and $A_j(z^{-1})$ are polynomials of the variable z^{-1}, and the polynomial $A_j(z^{-1})$ always has $a_0 = 1$ for the term z^0.

The discrete transfer function forms $\tilde{G}(z^{-1})$ may be calculated directly from the description (3.83), applying different approximation formulae in which proper dependency with the operator z^{-1} is introduced instead of the operator s (Spriet and Vansteenkiste, 1983; Janiszowski, 1993).

Both ways of description, (3.82) and (3.85), lead us to the acquisition of the selected output signal $y_j(k)$ in the form of the weighted sum of the stimulation signal values $u_i(k)$. In order to simplify the notation, let us assume that the process described by the dependency (3.85) has one output signal, and the stimulation signals $u_i(k)$ have time delays $T_i = d_i\Delta$. The description can be presented in the form

$$\begin{aligned}
&y(k)A\left(z^{-1}\right) = \sum_{i=1}^{q} B_i\left(z^{-1}\right) u_i\left(k-d_i\right) \Leftrightarrow \\
&y(k)\left(1+a_1 z^{-1}+\cdots+a_n z^{-n}\right) \\
&\quad = \sum_{i=1}^{q}\left(b_{i0} z^{-d_i}+b_{i1} z^{-1-d_i}+\cdots+b_{in} z^{-n-d_i}\right) u_i(k).
\end{aligned} \tag{3.86}$$

The above equation leads to a simple recursive relation that describes the modeled output value at $t = k\Delta$ as the effect of the measured signal values at previous time instants:

$$\begin{aligned}
y(k) &= y(k)\left(-a_1 z^{-1}+\cdots+a_n z^{-n}\right) \\
&\quad +\sum_{i=1}^{q}\left(b_{i0} z^{-d_i}+b_{i1} z^{-1-d_i}+\cdots+b_{in} z^{-n-d_i}\right) u_i(k) \Leftrightarrow \\
y(k) &= -a_1 y(k-1)-\cdots-a_n y(k-n) \\
&\quad +\sum_{i=1}^{q}\left(b_{i0} u_i\left(k-d_i\right)+\cdots+b_{in} u_i\left(k-n-d_i\right)\right).
\end{aligned} \tag{3.87}$$

Relations of this kind are the basis for the creation of parametric models of the process dynamics. The modeled output signal $y(k)$ is determined by the vector $\mathbf{v}(k)$, composed of measured signal values, as well as the vector of coefficients θ,

$$y(k) = \mathbf{v}(k)\theta, \tag{3.88}$$

where $\mathbf{v}(k) = [-y(k-1),\ldots,-y(k-n),u_1(k-d_1),\ldots,u_1(k-d_1-n),\ldots,u_q(k-d_q),\ldots,u_q(k-d_q-n)] \Leftrightarrow \mathbf{v}(k) = [v_1(k),v_2(k),\ldots,v_m(k)]$ and $\theta = [a_1,\ldots,a_n,b_{10},b_{11},\ldots,b_{1n},\ldots,b_{q0},b_{q1},\ldots,b_{qn}]^T$.

In further deliberations, the vector $\mathbf{v}(k)$ will be called the vector of model inputs (one must not confuse it with inputs to the modeled process!), and the vector θ the vector of the model coefficients. The vector $\mathbf{v}(k)$ of inputs of the model can have various forms.

When looking for the transfer function model (3.86), the vector of the model inputs $\mathbf{v}(k)$ will have the form described by the equation (3.88) and will be called the Auto-Regressive Moving Average (ARMA) structure. The name shows that the vector θ contains coefficients of the a_i type, which express the dependency of the output $y(k)$ on previous values of the same signal (auto-regressive), as well as coefficients of the b_{ij} type, which show the relation having Moving Average (MA) character with respect to the input values u_i.

The vector $\mathbf{v}(k)$ may also have the signal values of inputs u_i to the process only, and such a structure of the model will be called MA or the Finite Impulse Response (FIR).

The problem of the modeling of the output signal $y(k)$ can be expressed, as the above relation shows, in a form much simpler than the description (3.59), which requires us to apply complex integration techniques. If the vector of coefficients θ of the model is known, the vector of the model inputs $\mathbf{v}(k)$ is measurable, then modeling with the use of an equation in the form of (3.87) is evidently simpler.

One should remember, however, that the representation (3.87) is local linear approximation of dynamic properties only, calculated for the given sampling interval Δ, and it cannot be used for simulation with the step (of integration) $\Delta' < \Delta$. In fact, the sampling interval change in the parametric model (3.88) is a difficult task and requires us to implement very complex re-calculation of the coefficients θ of the model.

For modeling purposes, the form (3.88) is very often completely sufficient. However, its acquisition may pose certain problems. In the case when the accurate description (3.82) or (3.85) is known, obtaining the formula (3.88) is simple. When the description (3.88) should be calculated directly based on the knowledge of the signal measurements that appear in the vector $\mathbf{v}(k)$, as well as the knowledge of the output signal values $u(k)$, one should apply one of the available methods for coefficient vector θ value estimation, and the final result of the approach depends on many factors.

3.4.2 Identification of the Coefficients of Parametric Models

The problem can be statistically defined in the following way: we are looking for a reasonable approach to estimate the vector of the coefficients $\hat{\theta}$ in the model presented in the general form (3.88). The measurement sequences are given and their values appear in the vector $\mathbf{v}(k)$, while the value of the output signal $y(k)$ is subjected to the effect of non-measurable disturbances $\eta(k)$,

$$y(k) = \mathbf{v}(k)\theta + \eta(k) \qquad k = 1, \ldots, N. \tag{3.89}$$

An intuitive way is to define a measure $I(\hat{\theta})$ of model errors:

$$\varepsilon(k) = y(k) - \mathbf{v}(k)\theta \qquad k = 1, \ldots, N, \tag{3.90}$$

and calculate the estimation $\hat{\theta}$ that will result in the minimum value of the measure $I(\tilde{\theta})$. The simplest way is to apply the Least sum of Square (LS) errors method. In this method, the vector of coefficients $\hat{\theta}_{LS}$ should be selected in such a way that the sum of square errors of the process output reaches its minimum value:

$$I_{LS}(\theta) = \sum_{k=1}^{N} [y(k) - \mathbf{v}(k)\theta]^2 \Rightarrow \min_{\theta} I(\theta) = I(\hat{\theta}_{LS}). \tag{3.91}$$

The solution is a formula called the LS estimator:

$$\hat{\theta}_{LS} = \left[\mathbf{V}^T\mathbf{V}\right]^{-1}\mathbf{V}^T\mathbf{y}, \tag{3.92}$$

where

$$\mathbf{V} = \begin{bmatrix} \mathbf{v}(1) \\ \mathbf{v}(2) \\ \vdots \\ \mathbf{v}(N) \end{bmatrix} \in R^{N\times M}, \qquad \mathbf{y} = \begin{bmatrix} y(1) \\ y(2) \\ \vdots \\ y(N) \end{bmatrix} \in R^{N}. \tag{3.93}$$

The LS estimator will exist only when a determinant of the matrix $\mathbf{V}^T\mathbf{V}$ exists and is not equal to zero. The estimator's properties depend on the structure of the model and signal values (Soderstrom and Stoica, 1988).

The LS estimator (3.92) is most often biased, i.e., $E[\theta_{LS}] \neq \theta$. This effect is caused by the introduction of the value of the signal $y(k-i)$ to the vector $\mathbf{v}(k)$ of the model inputs (3.89). Such a case appears when the ARMA-structured model is estimated. This property does not have to have a decisive effect on the reproductive properties of the model (3.92).

If there are some indications to suspect that the disturbance $\eta(k)$ in the model (3.89) is not a white noise signal, it is reasonable to investigate the parametric model in an extended structure. LS estimator generalization, the so-called estimator of the Generalized Least sum of Square (GLS) errors, allows us to estimate the vector of coefficients of the ARMAX-structured model in which (beside stimulation signals) also the estimation of the non-measurable disturbance effect exists:

$$\begin{aligned} y(k) = \mathbf{v}(k)\theta_{GLS} = [&-y(k-1),\ldots,-y(k-n),u_1(k-d_1),\ldots,u_{qn}(k-d_q-n), \\ &\eta(k-1),\ldots,\eta(k-n)]\,[a_1,\ldots,a_n,b_{10},\ldots,b_{qn},c_1,\ldots,c_n]^T. \end{aligned} \tag{3.94}$$

The GLS estimator (Soderstrom and Stoica, 1988) allows us to obtain asymptotically unbiased estimations:

$$\lim_{N\to\infty} E\left[\theta_{GLS}\right] = \theta, \tag{3.95}$$

under similar assumptions as in the case of the LS estimator but for the ARMAX model, which can be more useful than the ARMA structured model.

The Instrumental Variable (IV) estimator has statistically better properties than the LS estimator. Here, instead of the matrix $\mathbf{V}$ (3.92) consisting of the vectors of the model inputs $\mathbf{v}(k)$, the matrix $\mathbf{W}$ is introduced which contains signals that should not be correlated with the signal $\eta(k)$, e.g., instead of the signal $y(k-i)$ value, its estimation $\hat{y}(k-i)$ is introduced that should show no correlation with $\eta(k)$:

$$\theta_{IV} = \left[\mathbf{W}^T\mathbf{V}\right]^{-1}\mathbf{W}^T\mathbf{y}, \tag{3.96}$$

where

$$\mathbf{V}=\begin{bmatrix}\mathbf{v}(1)\\ \mathbf{v}(2)\\ \vdots\\ \mathbf{v}(N)\end{bmatrix}\in R^{N\times M},\qquad \mathbf{W}=\begin{bmatrix}\mathbf{w}(1)\\ \mathbf{w}(2)\\ \vdots\\ \mathbf{w}(N)\end{bmatrix}\in R^{N\times M},\qquad \mathbf{y}=\begin{bmatrix}y(1)\\ y(2)\\ \vdots\\ y(N)\end{bmatrix}\in R^{N}. \tag{3.97}$$

The IV estimator (3.96) can lead to calculations of convergent and unbiased estimations also for models having the ARMA structure.

Certainly, the better the statistic properties of the estimator, the higher the comfort of the investigator, but, as the experience shows, success in model creation for diagnostic purposes or in the design of the control system depends on the data quality, i.e., stimulation properties of signals during the experiment.

In order to evaluate the quality of the obtained process parametric model, one may use the index I_{LS} (3.88) or I_{SAE}, the Sum of Absolute Errors (SAE):

$$I_{SAE}(\theta)=\sum_{k=1}^{N}|y(k)-\mathbf{v}(k)\theta|. \tag{3.98}$$

The universal application of LS, GLS or IV estimators results exclusively from the possibility of easier introduction of calculation algorithms and the possibility of proving estimation convergence with the index (3.91), whose derivative with respect to the coefficients vector θ may be easily calculated. It does not mean, however, that the index always has physical sense. For instance, when the error of the model means the inaccuracy of the stock exchange rate, then the losses due to such an error should be calculated in the currency of a given stock, e.g., in \$ and not in $\$^2$.

While discussing the forms of indices and their usefulness, one should remember the way in which modeling is implemented. When values of the output signals from the process are included in the model inputs vector $\mathbf{v}(k)$, modeling can be implemented in two ways. The first one allows us to evaluate the prediction properties and then, in both of the indices (3.91) and (3.98), the formula for the calculation of the output value of the model may contain measured values of the output signal $y(k)$. The second one consists in the replacement of the estimate value $\hat{y}(k)$ instead of the measured output signal $y(k)$ in the vector $\mathbf{v}(k)$. The calculated index value will be denoted by the MRO symbol:

$$I_{MRO}=\sum_{k=1}^{N}\left[y(k)-M\left(\mathbf{u},\mathbf{z},\hat{y},\theta,k\right)\right]^{2}. \tag{3.99}$$

It evaluates the model better with respect to the reaction of the model output to stimulation (u and z) changes. In the MRO mode, the index SAE (3.98) may also be applied.

While calculating the dynamics of the model with the use of identification, one should make use of all available information about the examined process. One should not approach the examinations with the assumption that we do not know anything about the model, it is a black box for us, but the identification algorithm is very exquisite, it will find the model. The initial information that may be used during identification implementation consists in the evaluation of the number of accumulators of energies that appear in the examined process. It allows us to make an initial guess regarding the order of the model. Another very important information is the expected delay time in interactions of particular stimulations. One can relatively easily evaluate delays having transport type character. Both of the premises allow initial fitting of the structure of the examined model. Very valuable is also information from process operators since it can improve the structure or allow us to verify the obtained results.

3.4.3 *Pneumatic Cylinder Controlled by a Servo-Valve: A Parametric Linear Model of the System*

Experiments and investigation of pneumatic servo-drive systems were performed at the Laboratory of Servo-drives of the Institute of Automatic Control and Robotics of the Warsaw University of Technology. The stand for pneumatic servo-drive investigation was equipped with a very precise position measurement device optical linear unit of the resolution of 1 micrometer, and a controller from *dSPACE*. Many experiments were conducted and the obtained results were relatively broadly discussed (Janiszowski, 2002). A single case will be discussed below: the long jump of the cylinder piston, by 600 mm (Fig. 3.10). During the displacement, the piston achieved a relatively high negative velocity—approximately 1.5 m/s, and then the braking began. The movement time was 800 milliseconds, and the sampling interval was 1 millisecond.

In order to define the optimum state space feedback algorithm of the state variable algorithm for the positioning drive, the model having the form (3.74) was looked for. Due to that, the SAE index (3.98) in the MRO version (3.99) was applied to the evaluation of the quality of the model of the reaction of the piston velocity v to the control signal level u. At the first stage of examinations, the delay time was relatively accurately defined. The implemented tests with the use of the

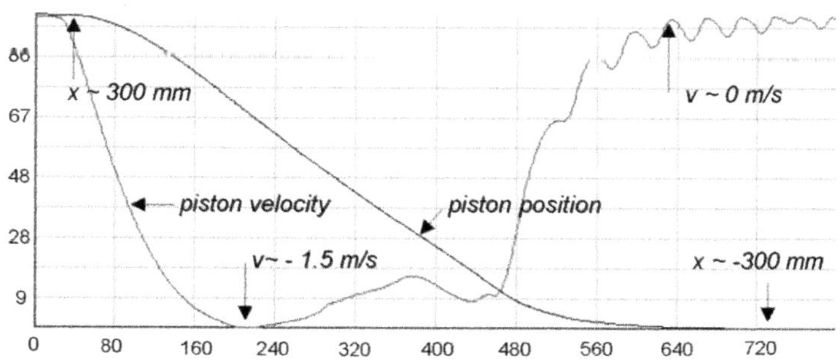

Fig. 3.10 Changes of the cylinder piston position x and the velocity v at the stroke of 600 mm

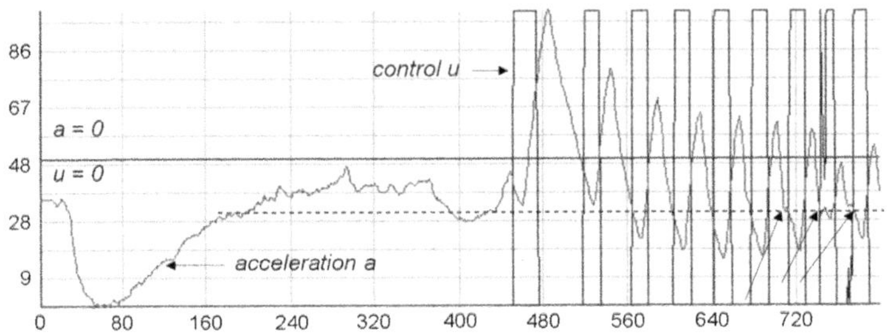

Fig. 3.11 Changes of the signal controlling the valve u and the acceleration a of the cylinder piston at the stroke of 600 mm

proportional servo-valve showed that the minimum value of the delay introduced by the valve was approximately 2.5 to 3 ms. Due to the length of connections, being approximately of 1 meter, one should expect the delay of about 5 to 6 ms. At the time instant of the start, the cylinder piston must overcome the adhesive friction effect that introduces further delay. After many experiments (14 different displacements were tested), the optimum delay value was established at 9 ms. The I_{MROSAE} value for transients presented in Figs. 3.10 and 3.11 had the minimum for the order $n = 6$ of the model. The modeling results in the MRO mode are presented in Fig. 3.12.

Comparing both transients, one may acknowledge that in the range of high velocities the model correctly reproduces the process gain but presents rather too high a damping value—the measured transients in Fig. 3.10 had a visible tendency for oscillations. Simultaneously, during the piston braking, the model over-estimated the velocity gain value for lower velocities—therefore one may conclude that the servomotor valve-piston assembly had different properties in this range of velocity than at the acceleration stage.

The discrete transfer function of the velocity v [mm/s] as a reaction to the valve control signal u [V] for the optimum structure ($n = 6, d = 9$ milliseconds) had the form

$$\begin{aligned} v(k) = {} & 1.94v(k-1) - 0.5338v(k-2) - 0.4789v(k-3) - 0.5804v(k-4) \\ & +1.008v(k-5) - 0.3553v(k-6) + 0.4124u(k-9) - 0.2717u(k-10) \\ & +0.3074u(k-11) - 0.1638u(k-12) - 0.1887u(k-13) - 0.07809u(k-14). \end{aligned} \tag{3.100}$$

The above form was inconvenient for model properties evaluation, but the step function response and the pulse response calculated for it (Fig. 3.13) suggested that the lower-order description is possible.

The implemented reduction to the model having the order of n = 2, d = 9 led to the form

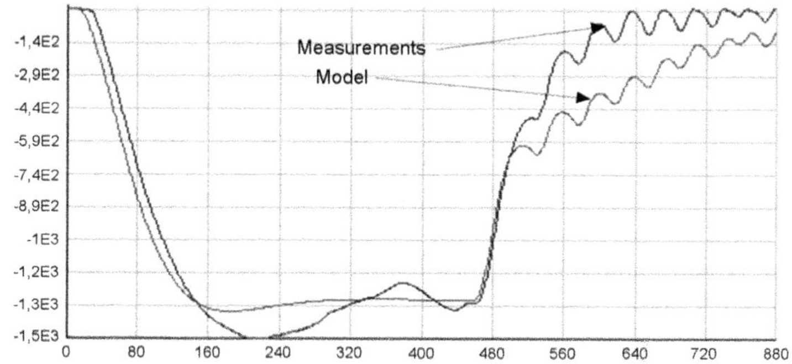

Fig. 3.12 Changes of the measured and the modeled velocity (in the MRO mode)

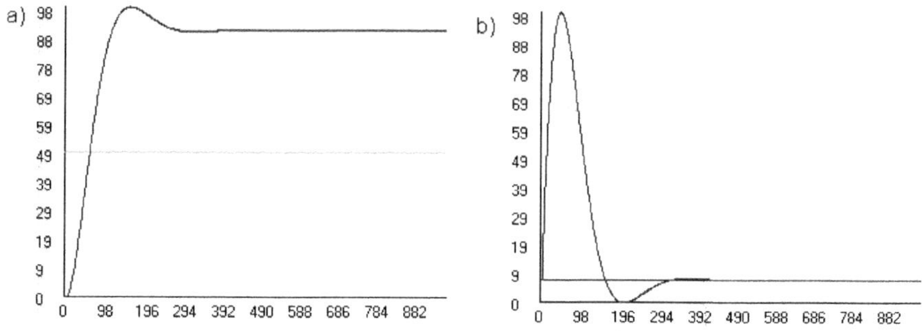

Fig. 3.13 Time responses of the examined cylinder model: step function response (a), impulse response (b)

$$v(k) = 1.9465v(k-1) - 0.9655v(k-2) + 0.4661u(k-9) + 0.3991u(k-10), \tag{3.101}$$

and approximation to the continuous domain allowed us to obtain the following description:

$$G_{vu}(s) = \frac{8.62 \cdot 10^5}{s^2 + 33.17s + 634.4} e^{s0.008}. \tag{3.102}$$

The reduction quality was very good—time responses of the model before, and after, the reduction calculated for the model (3.102) differed by less than 2% (this is why they were not shown). Comparing the above description with the form (3.74), one can calculate the transfer function parameters for the examined cylinder load and conditions of displacement:

$$C = 1359[(mm/s)/V], \qquad \omega_0 = 25.18[rad/s], \qquad \xi = 0.66. \tag{3.103}$$

In the above example, a different cylinder was examined than the one described by the model (3.78). The cylinder described by (3.102) had the length of 800 mm, a higher section $A = 0.0019\ \mathrm{m}^2$, and it was weighted by the mass $M = 10$ kg, which had an effect on the increase of the proper vibrations frequency ω_0 and the gain C.

3.5 Fuzzy Parametric Models

The methodology of the fuzzy set approach finds newer and newer applications in technology (Yager and Filev, 1994). The *MITforRD* package integrated with *PExSim* helps us to calculate parametric fuzzy models. This approach is especially useful for modeling—it yields the investigation of process operation regions in which the process dynamics may have different local properties in a consistent way. It allows us to integrate the model description and apply it to, e.g., diagnosing algorithm synthesis, controller selection, etc., and sometimes also to detect new properties of the process.

Fuzzy description is based on intuitive presentation of the expert's knowledge in the form of the creation of premises and conclusions. The most often used method of premise definition is based on the definition of the set X_i to which there belongs the event or state x describing the characteristic situation, followed by the conclusion K_j. The mechanism for fuzzy inference system creation consists in the creation of the set of rules having the form $R_{ij} : if(x \in X_i) then K_j$, while the essence of the approach is the fact that rules ranges X_i partially overlap one another. The sets X_i do not divide the ranges of changes of the quantities $x_i \in \sum X_i$ in an acute way. The application of the fuzzy set method should begin with the definition of ranges X_i and the degree of affiliation x to this set, i.e., the membership function $\varphi_i(x)$. This function belongs to the range $\langle 0,1\rangle$, while $\varphi_i(x) = 0 \Leftrightarrow x \notin X_i, \varphi_i(x) = 1 \Leftrightarrow x \in X_i$ and $x \notin X_{j \neq i}$.

It is possible to define the membership function in different ways, e.g., one may apply analytical functions (Gauss' distribution or spline functions), or continuous but non-analytical functions, e.g., trapezoidal distribution. The basic assumption applied in the *MITforRD* package for the membership function is as follows:

$$\forall i \qquad \varphi_i(x) \in \langle 0,1\rangle \quad \text{and} \quad \forall x \subset X \qquad \sum_{i=1}^{s} \varphi_i(x) = 1, \tag{3.104}$$

where s denotes the number of selected subsets of X_i and is called the number of partitions. The second part of the definition is important since it introduces the condition of necessary normalization of the membership function before the inference. Such a mechanism protects the search against the introduction of numerically unstable models.

3.5.1 Fuzzy Parametric TSK Models

Fuzzy inference ends with a specific formula that usually results from taking into account several premises $P_s, s = 1,\ldots,k$ for which $\varphi_s(x) > 0$. The formula may have the form of an algebraic dependency:

$$F(x) = \sum_{s=1}^{k} \varphi_s(x) K_s, \tag{3.105}$$

where K_s are conclusions for the premises P_s. When the conclusions K_s are algebraic formulae, the model (3.102) is called the Takagi–Sugeno–Kang model (Yager and Filev, 1994). In the work of Wnuk (2006), the generalization of the above formula in the field of parametric model identification was very successfully applied by the introduction of the parametric model fuzzy form:

$$\theta(x) = \sum_{s=1}^{k} \varphi_s(x)\theta_S \Leftrightarrow \hat{y}(k) = v(k) \sum_{s=1}^{k} \varphi_s(x)\theta_S, \tag{3.106}$$

where θ_S denotes the local model calculated for data belonging to the subset X_s according, e.g., to (3.92) or (3.96). All models θ_S must have the same structure (inputs u_i, order n, time delays d_i). The above model is the linear form of coefficients of the models θ_S but, since they depend on the state of $\mathbf{x}$, it introduces in general non-linear dependencies between the stimulation signals u_i and the reaction of the output y that is modeled. This is especially visible after, e.g., static gains for the partial models θ_S have been calculated.

The basic advantage of the fuzzy model is the possibility of good matching of the partial models θ_S to the ranges X_s in which a defined process character dominates. Such a linear form (3.106) is well understood by users. Local descriptions of dynamic properties allow us to compare differences between particular partitions and to consider if the accepted division is not too dense or too sparse. Having the partial models θ_S one may think of a separate choice of controllers, models for soft-measurement, prediction or diagnostics, etc. The algorithm applied in the *MITforRD* package allows us to calculate fuzzy models with the use of the LS estimator (3.92). LS estimation allows obtaining asymptotically un biased estimations exclusively for a high number of measurements applied to calculation. In practice, the number of measurements should be at least five to ten times higher than the number m of the coefficients vector θ of the model.

The main disadvantage of TSK models is the necessity of a sensible definition of borders of the sets X_i as well as the membership function φ_i. It is possible to define the sets with the use of experts' knowledge, or one can try various methods: clusterization, GMDH, or running automatic optimization. In each one of the cases, much effort is required, but it may considerably improve the model quality.

The number of fuzzy variables x_j may be higher, $j = 1, \ldots, p$. In such a case the number of local models may be large and equal to

$$M = \prod_{j=1}^{p} \kappa_j, \tag{3.107}$$

where κ_j denotes the number of partitions (a set of x where $\varphi_i(x) \neq 0$) of membership functions for the successive fuzzy variables. During the calculation of the fuzzy model of the m coefficients, the validation level of the models is very important:

$$\rho_i(k) = \sum_{k=1}^{N} \varphi_i\big(x(k)\big) \geq 5m, \tag{3.108}$$

where the argument k points summation after all gathered time instants. In many cases, persons that run the identification experiment tend to choose a very complicated structure of the model (too many partitions) in order to obtain seemingly better matching of the model. One should always check the dependency (3.108), especially for a high number of fuzzy variables.

3.5.2 Estimation of Fuzzy TSK Model Coefficients

The problem of a numerical run of estimation can cause some doubts—in many cases, the point x considered may belong simultaneously to several partitions and then the question is if it should be used for calculations, and if so, in what partition it should be included.

The problem has a simple solution: since all vectors of inputs to the model in the LS estimator (3.92), (3.93) are summed in the matrices $\mathbf{V}^T\mathbf{V}$ and $\mathbf{V}^T\mathbf{y}$, then the sequence of the introduction of the vectors $\mathbf{v}(k)$ to the matrix $\mathbf{V}$ is not important. If the vector of model inputs $\mathbf{v}(k)$ to the model belongs to the interval P_i (corresponding to the model θ_i) to a degree $\varphi_i < 1$ only, then it may be included in the LS method, together with the requirement that it recreate the part equal to $\varphi_i y(k)$ of the measured output signal only. With such an approach to the problem of the estimation of a fuzzy model composed of L partitions, one can decompose the solution of the problem to the definition of models $\varphi_i, i = 1, \ldots, L$. Using an index in the form of the least sum of error squares, the definition of each one of the partial models θ_i is described by a relatively simple dependency:

$$\sum_{k=1}^{N} \Big[\varphi_i\big(P(k)\big)y(k) - \varphi_i\big(P(k)\big)\mathbf{v}(k)\theta_i\Big]^2 = I(\theta_i) \Rightarrow \min_{\theta_i} I(\tilde{\theta}_i). \tag{3.109}$$

The above shows the filtration of the sequence of the values of the output $\mathbf{y}(k)$, as well as the vector of the model inputs v(k) by the function of membership values $\varphi_i\big(P(k)\big)$, where $P(k)$ denotes the fuzzy variable value at the discrete time instant k. LS estimation of the vector of coefficients of the partial model θ_i is defined by the following rule:

$$\tilde{\theta}_i = \big[\mathbf{V}_i^T\mathbf{V}\big]^{-1}\mathbf{V}_i^T\mathbf{y}_i, \tag{3.110}$$

where

$$\mathbf{V}_i = \begin{bmatrix} \varphi_i\big(P(1)\big)\mathbf{v}(1) \\ \varphi_i\big(P(2)\big)\mathbf{v}(2) \\ \vdots \\ \varphi_i\big(P(N)\big)\mathbf{v}(N) \end{bmatrix} \in R^{N\times M} \qquad \mathbf{y}_i = \begin{bmatrix} \varphi_i\big(P(1)\big)y(1) \\ \varphi_i\big(P(2)\big)y(2) \\ \vdots \\ \varphi_i\big(P(N)\big)y(N) \end{bmatrix} \in R^N. \tag{3.111}$$

After (3.110) has been calculated, the dependency (3.108) is verified. Among the advantages of the estimation (3.110), one can list its simple algorithm and limited calculation effort.

One should notice, however, that the condition (3.109) for the indices $I(\theta_i)$ to reach their minimal values is not equivalent to global minimization of the sum of error squares. The latter can be obtained by simultaneous optimization of the coefficient values for all models described by the vectors θ_i. In the case of global minimization, one looks simultaneously for the vector $\theta^T = \left[\theta_1^T, \theta_2^T, \ldots, \theta_L^T\right]$ composed of L vectors θ_i that ensures the minimum value of the index

$$I(\Theta) = \sum_{k=1}^{N} \left[y(k) - \sum_{l=1}^{L} \varphi_l\big(P(k)\big)\left[\mathbf{v}(k)\theta_l\right] \right]^2 \Rightarrow \min_{\Theta} I(\hat{\Theta}). \tag{3.112}$$

The above problem has a solution in another form:

$$\hat{\Theta} = \left[\mathbf{V}^T\mathbf{V}\right]^{-1}\mathbf{V}^T\mathbf{y}, \tag{3.113}$$

where

$$\mathbf{V} = \begin{bmatrix} \varphi_1\big(P(1)\big)\mathbf{v}(1) & \ldots & \varphi_L\big(P(1)\big)\mathbf{v}(1) \\ \varphi_2\big(P(2)\big)\mathbf{v}(2) & \ldots & \varphi_L\big(P(2)\big)\mathbf{v}(2) \\ & \vdots & \\ \varphi_N\big(P(N)\big)\mathbf{v}(N) & \ldots & \varphi_L\big(P(N)\big)\mathbf{v}(N) \end{bmatrix} \in R^{N\times(M\times L)}$$

$$\mathbf{y} = \left[\underbrace{y(1),\ldots,y(N)}, \ldots, \underbrace{y(1),\ldots,y(N)}\right]^T \in R^N, \tag{3.114}$$

and it requires converting visibly larger matrices. The results of such a solution may be better, but numerical problems with insufficient numeric conditioning due to poor stimulation signals will also be higher. The higher calculation effort in the case of the estimator (3.113) is caused by the inversion of a matrix having the order of L times higher in comparison with the estimation (3.110). It will not be remarkable in the case of a single estimator calculation, but if the estimator is applied to recursive optimization calculations, then the calculation effort may limit application possibilities.

Both of the estimators (3.110) and (3.113) may be used after the membership function has been precisely defined. When one defines parametric fuzzy models, the proper structure of fuzzyfication takes most of the time: the number of fuzzy signals, partitions and the form of each function. In the determination of the fuzzyfication structure one may and should take into account the observations of process operators, but such information is not always available. In this case, one should test successive variants and verify values of selected indices for the calculated models.

3.5.3 Pneumatic Cylinder Controlled by a Servo-Valve: A TSK Fuzzy Model

Examinations of TSK fuzzy models for the pneumatic drive system were based on laboratory experiment results gathered for a symmetric (two-piston rod) cylinder having the length of 800 mm and the diameter of 50 mm. The results described in Section 3.4 yield the conclusion that the parametric model of the pneumatic drive can be improved when one relates the estimation results to cylinder piston velocity. Observations made during parametric model calculations justified such an approach due to strong dependency of the dynamic properties on the friction force, which is, in turn, strongly dependent on piston velocity (Fig. 3.2). The second quantity that may have an effect on the estimation of coefficients of the dynamic model is the piston position x in the cylinder.

After different tests, conducted to find a proper choice of the membership function in the velocity range, final models were calculated with the division into particles presented in Fig. 3.16 (negative range of v).

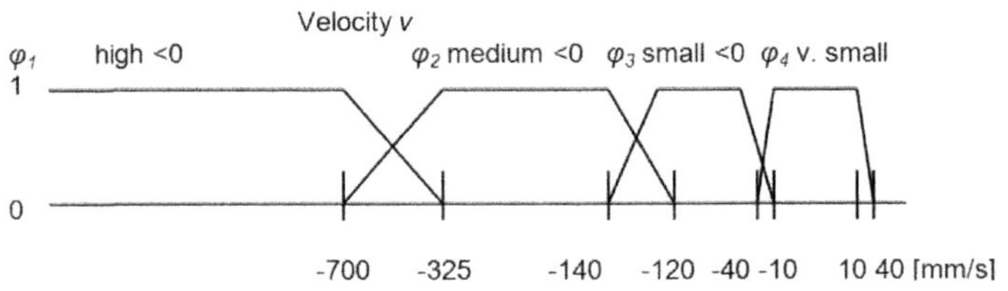

Fig. 3.14 Selection of partitions for membership functions with respect to the piston velocity v at the calculations of the TSK fuzzy model for the displacement presented in Figs. 3.10 and 3.11

The calculations were conducted in a similar mode as in the case of the parametric model (Section 3.4). TSK fuzzy models were calculated which had the structure $n = 6$, $d = 9$ for each one of the partitions (Fig. 3.14), and then the reduction with approximation to 2-nd order models was implemented. The results of the estimation of the parameters C, ω_0, ξ are presented in Table 3.3.

Table 3.3 Changes of approximated parameters of partial models θ_i as a function of the piston velocity v

Velocity range	C [mm/s/V]	ω_0 [rad/s]	ξ [-]
$\lvert v \rvert > 700$ mm/s	1514	25,2	0,49
140 mm/s $< \lvert v \rvert < 325$ mm/s	667	43,1	0,57
40 mm/s $< \lvert v \rvert < 120$ mm/s	197	84,2	0,96
$\lvert v \rvert < 10$ mm/s	6,8	198	0,23

The above results should be commented on. The decrease of the value of the gain C and the increase of the damping when the piston velocity decreases (for velocities $|v| > 40$ mm/s) can be expected (the higher participation of the friction force should exert such an effect), but the range for the lowest velocities $|v| < 10$ mm/s was completely incomprehensible. The observed oscillations were initially interpreted as an inaccuracy of the measuring technique, but the optic measurement device used was not disturbed.

After other test sequences were examined more thoroughly, the same effects were observed—micro-vibrations having a small amplitude and very low damping. This effect resulted from vibrations of the piston suspended on elastic seals which, due to the adhesive friction, were motionless in relation to the cylinder walls, but the piston itself could vibrate with a high frequency. This effect was practically for the first time observed and explained.

The estimated TSK fuzzy model was better than the single parametric model (3.100). The quality of the output signal for this model, examined in the same way, yielded the modeling result presented in Fig. 3.15. Successive tests were conducted for two fuzzy quantities: the piston velocity v and the piston position x in relation to the middle of the cylinder.

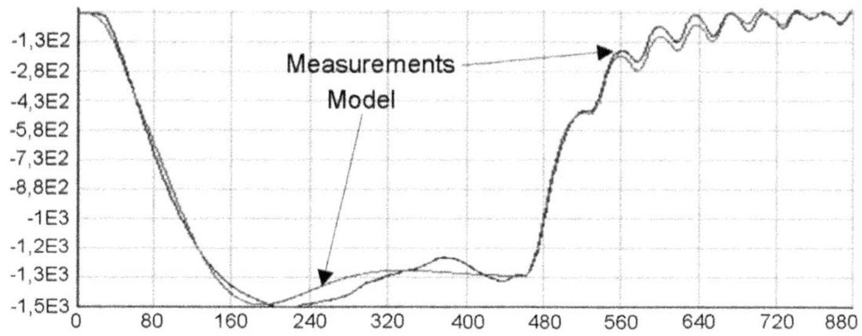

Fig. 3.15 Velocity signal generated by the TSK fuzzy model

In this case, we gave up the partition with micro-vibration (measurement series without micro-vibration sequences were used), but models were calculated for 15 partitions. Three partitions parameterized changes of the models in relation to the piston position x, and five partitions in relation to the velocity in the way presented in Figs. 3.16 and 3.17.

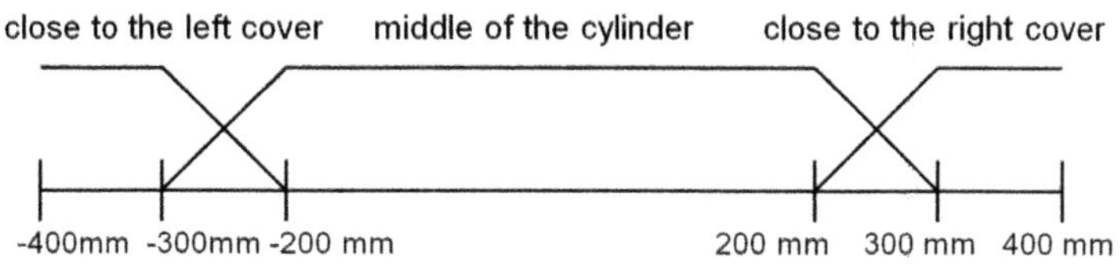

Fig. 3.16 Parametrization of trapezoid partitions with respect to the piston position x

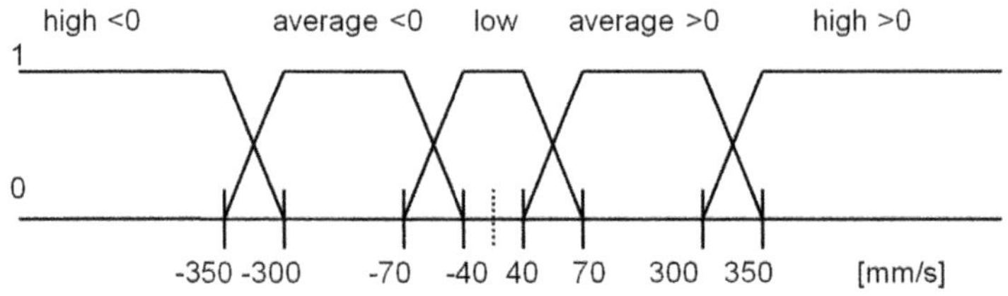

Fig. 3.17 Parametrization of trapezoid partitions with respect to the piston velocity v

We expected, according to (3.76), that the pulsation of the proper vibrations ω_0 would change in a symmetrical way together with the gain C by approximately 70 to 80% close to the covers of the cylinder with respect to the piston position x, and the damping ξ would increase for low velocities. The obtained results have shapes definitely different than we expected (Figs. 3.18 and 3.19).

The expected symmetry with respect to the velocity v and positive values of x could have been observed, but the range for $x < -250$ mm was definitely different than the expectations. Changes of properties appeared regularly (for negative values of x, abnormalities appeared) but up to a certain extent. In successive tests, we decided to replace the measurement system, and with a new system put in place, we observed a very strong jam of the cylinder piston in the left cover ($x = -400$ mm), caused by inaccurate tightening up of the mounts of two shears the mass load was moved over. This fact completely explained the shown abnormalities that did not appear during successive tests.

The above results show the main advantage of TSK fuzzy models. Locally they can describe the dynamic properties very precisely. The quality of the TSK fuzzy

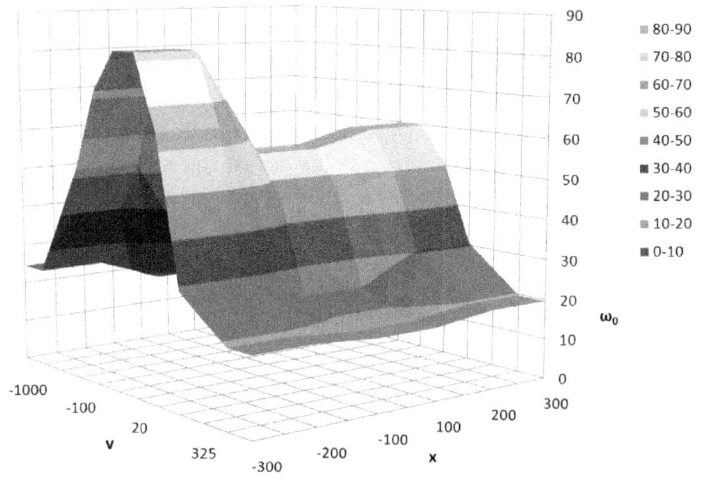

Fig. 3.18 Changes of the pulsation ω_0 of the proper vibrations for parametrization with respect to the position x and the velocity v

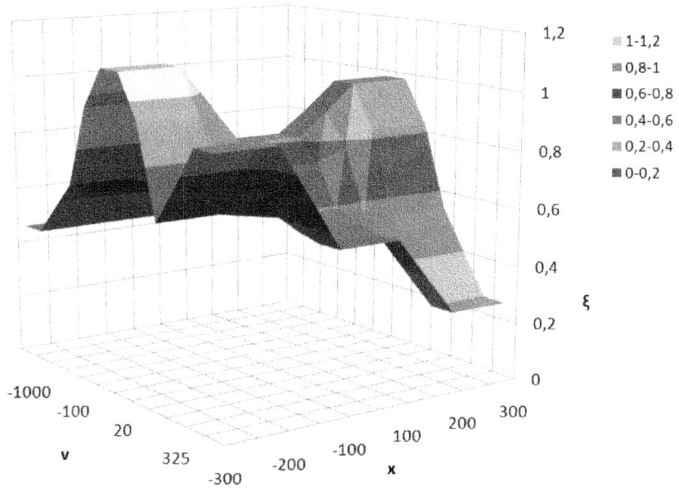

Fig. 3.19 Changes of the damping factor ξ for parametrization with respect to the position x and the velocity v

model can be so high that it can detect quite unexpected, new technical effects. The close correspondence of the TSK fuzzy models to a linear representation yields an easy explanation of the observed phenomena.

3.6 Neural Models

Artificial neural networks provide an excellent mathematical tool for dealing with non-linear problems (Abdessemed et al., 2006; Nelles, 2001; Haykin, 1999; Gupta et al., 2003; Deng et al., 2009). They have an important property according to which any continuous non-linear relation can be approximated with arbitrary accuracy using a neural network with a suitable architecture and weight parameters. Their another attractive property is the self-learning ability. A neural network can extract the system features from historical training data using the learning algorithm, requiring little or no a priori knowledge on the process. This provides the modeling of non-linear systems with great flexibility (Nelles, 2001; Norgard et al., 2000). However, the application of neural networks to the modeling or fault diagnosis of control systems requires taking into account the dynamics of the processes or systems considered (Korbicz et al., 2004; Calado et al., 2001). A neural network, to be dynamic, must contain a memory. The memory can be divided into a short-term memory and a long-term memory, depending on the retention time (Arbib, 1989; Elman, 1990; Mozer, 1994; Haykin, 1999). The short-term memory refers to a compilation of knowledge representing the current state of the environment. In turn, the long-term memory refers to knowledge stored for a long time or permanently. One simple way of incorporating a memory into the structure of a neural network is the use of time delays, which can be implemented at the synaptic level or in the input layer of the network. Another important way in which the dynamics can be built into the operation of a neural network in a implicit manner is through the use of feedbacks.

3.7 Neural Networks with External Dynamics

The neural network commonly and willingly used for modeling processes is a multi-layer perceptron. Neural models of this class, however, are of a static type and can be used to approximate any continuous non-linear, although static, function (Cybenko, 1989; Hornik et al., 1989). Therefore, neural network modeling of control systems should take into account the dynamics of the processes or systems considered. Two main methods exist to provide a static neural network with dynamic properties: the insertion of an external memory to the network or the use of feedbacks. The strategy most frequently applied to model dynamic non-linear mapping is the external dynamics approach (Narendra and Parthasarathy, 1990; Hunt et al., 1992; Haykin, 1999; Norgard et al., 2000; Nelles, 2001). It is based on the non-linear input/output model:

$$y_m(k+1) = f\big(y(k),\ldots,y(k-m),u(k),\ldots,u(k-m)\big), \tag{3.115}$$

where $f(\cdot)$ is a non-linear function, $u(k)$ is the input, $y(k)$ and $y_m(k)$ are outputs of the process and the model, respectively, m is the order of the process. The non-linear model is clearly separated into two parts: a non-linear static approximator (multi-layer perceptron) and an external dynamic filter bank (tapped delay lines) (Fig. 3.20). As a result, a model known as a multi-layer perceptron with tapped delay lines (time delay neural network) is obtained. Time delay neural networks can describe a large class of systems but are not as general as non-linear state-space models. Limitations are observed for processes with non-unique non-linearities, e.g., hysteresis or backslash, where internal unmeasurable states play a decisive

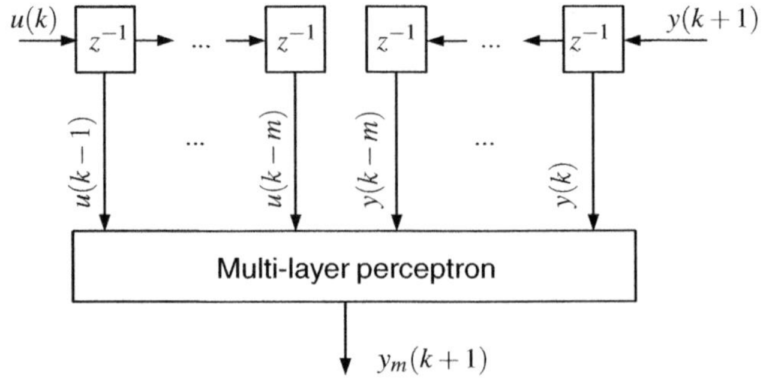

Fig. 3.20 Realization of the external dynamics approach

role, and partly for processes with non-invertible non-linearities (Zamarreno and Vega, 1998; Nelles, 2001). Moreover, the problem of order selection is not satisfactorily solved yet. This issue is equivalent to the determination of relevant inputs for the function $f(\cdot)$. If the order of a process is known, all necessary past inputs and outputs should be fed to the network. In this way, the input space of the network becomes large. In many practical cases, there is no possibility to learn the order of the modeled process, and the number of suitable delays has to be selected experimentally by using the trial and error procedure (Norgard et al., 2000).

Many papers show that the multi-layer perceptron is able to predict the outputs of various dynamic processes with high precision, but its inherent non-linearity makes assuring stability a hard task, especially when the output of the network is fed back to the network input, as in the case of the parallel model (Narendra and Parthasarathy, 1990; Norgard et al., 2000):

$$y_m(k+1) = f\big(y_m(k),\ldots,y_m(k-m),u(k),\ldots,u(k-m)\big). \tag{3.116}$$

Indeed, the architecture (3.116) uses its own delayed outputs as part of the input space, and the neural network becomes a recurrent model. There are also situations in which this type of network is not capable of capturing the whole plant state information of the modeled process (Williams, 1990; Zamarreno and Vega, 1998). The use of real plant outputs avoids many of the encountered analytical difficulties, assures stability and simplifies the identification procedure. This type of feedforward network is known as a series-parallel model (3.115), introduced by Narendra and Parthasarathy (1990). Such networks are capable of modeling systems if they have a weakly visible state, i.e., if there is an input-output equivalent to a system whose state is a function or a fixed set of finitely many past values of its inputs and outputs (Williams, 1990). Otherwise, the model has a strongly hidden state and its identification requires recurrent networks of a fairly general type.

3.7.1 Recurrent Networks

Standard feed-forward networks are capable to represent static mappings only. However, neural network modeling of control systems requires taking into account the dynamics of the processes or systems considered. This can be achieved by introducing tapped delay lines into the system model. Unfortunately, neural networks with external dynamics have a number of drawbacks, listed in the previous section. These shortcomings resulted in the fact that other neural models of a dynamic type have been proposed. Such neural networks are called recurrent networks.

Recurrent networks are neural networks with one or more feedback loops. As a result of feedback introduced to the network structure, it is possible to accumulate the information and use it later. Feedback can be either of a local or a global type. Taking into account the possible location of feedback, recurrent networks can be divided as follows (Tsoi and Back, 1994; Haykin, 1999; Campolucci et al., 1999):

- *globally recurrent networks*: there is feedback allowed between neurons of different layers or between neurons of the same layer. Such networks incorporate a static multi-layer perceptron or parts of it. Moreover, they exploit the non-linear mapping capability of the multi-layer perceptron;
- *locally recurrent networks*: there is feedback only inside neuron models. This means that there are neither feedback connections between neurons of successive layers nor lateral links between neurons of the same layer. These networks have a structure similar to that of static feedforward ones but consist of the so-called dynamic neuron models.

The most general architecture of recurrent neural networks was proposed by Williams and Zipser (1989a; 1989b). This structure is often called the Real Time Recurrent Network (RTRN), because it was designed for real time signal processing. The network consists of m neurons, and each of them creates a feedback. Any connections between neurons are allowed. Thus, a fully connected neural architecture is obtained. Only M from m neurons are established as the output neurons. The remaining $H = m - M$ units are the hidden ones. The fundamental advantage of such networks is the possibility of approximating a wide class of dynamic relations. Such a kind of network, however, exhibits some well-known disadvantages. One of them is large structural complexity: $O(n^2)$ weights needed for n neurons. Also, the training of the network is usually complex and slowly convergent (Hertz et al., 1991; Haykin, 1999; Campolucci et al., 1999). Moreover, there are problems with keeping network stability.

Partially recurrent networks have less general character (Stornetta et al., 1988; Mozer, 1989; Elman, 1990; Jordan and Jacobs, 1990). Contrary to the case of the fully recurrent network, the architecture of partially recurrent networks is based on a feedforward multi-layer perceptron consisting of an additional layer of units called the context layer. The Elman network is the best-known example of a partially recurrent neural network. The realization of such networks is considerably less expensive than in the case of a multi-layer perceptron with tapped delay lines. The Elman network consists of four layers of units: the input layer, the context layer, the hidden layer and the output layer. The input and output units interact with the outside environment, whereas the hidden and context units do not. The context units are used only to memorize the previous activations of the hidden neurons. A very important assumption is that in the Elman structure the number of context units is equal to that of hidden units. All the feedforward connections are adjustable; the recurrent connections are fixed. Theoretically, this kind of network is able to model the s-th order dynamic system, if it can be trained to do so (Haykin, 1999; Pham and Liu, 1996). At some specific time k, the previous activation of the hidden units (at time $k - 1$) and the current inputs (at time k) is used as inputs to the network. In this case, the Elman network's behavior is analogous to that of a feedforward network. Therefore, the standard back-propagation algorithm can be applied to train the network parameters. However, it should be kept in mind that such simplifications limit the application of the Elman structure to the modeling of dynamic processes (Hertz et al., 1991). Partially recurrent networks possess over fully recurrent networks the advantage that their recurrent links are more structured, which leads to faster training and fewer stability problems (Haykin, 1999; Nelles, 2001). Nevertheless, the number of states is still strongly related to that of hidden neurons, which severely restricts their flexibility.

Another architecture can be found in the recurrent network elaborated by Parlos (1994). A Recurrent Multi-Layer Perceptron (RMLP) is designed based on the multi-layer perceptron network, and by adding delayed links between neighboring units of the same hidden layer (cross-talk links), including the unit feedback itself (recurrent links). Empirical evidence indicates that by using delayed recurrent and cross-talk weights the RMLP network is able to emulate a large class of non-linear dynamic systems. The feedforward part of the network still maintains the well-known curve-fitting properties of the multi-layer perceptron, while the feedback part provides its dynamic character. Moreover, the usage of past process observations is not necessary, because their effect is captured by internal network states. The RMLP network has been successfully used as a model for dynamic system identification (Parlos et al., 1994). However, a drawback of this dynamic structure is increased network complexity strictly dependent on the number of hidden neurons and the resulting long training time. For the network containing one input, one output and only one hidden layer with v neurons, the number of the network parameters is equal to v^2+3v.

3.7.2 State Space Neural Networks

Figure 3.21 shows another type of recurrent neural network known as the state-space neural network (Zamarreno and Vega, 1998; Haykin, 1999; Nelles, 2001). The output of the hidden layer is fed back to the input layer through a bank of unit delays. The number of unit delays used here determines the order of the system. The user can choose how many neurons are used to produce feedback.

Let $\mathbf{u}(k) \in \mathbb{R}^n$ be the input vector, $\mathbf{x}(k) \in \mathbb{R}^q$ the output of the hidden layer at time k, and $\mathbf{y}(k) \in \mathbb{R}^m$ the output vector. Then the state-space representation of the neural model presented in Fig. 3.21 is described by the equations

$$\mathbf{x}(k+1) = \mathbf{f}(\mathbf{x}(k), \mathbf{u}(k)), \tag{3.117}$$

$$\mathbf{y}(k) = \mathbf{C}\mathbf{x}(k), \tag{3.118}$$

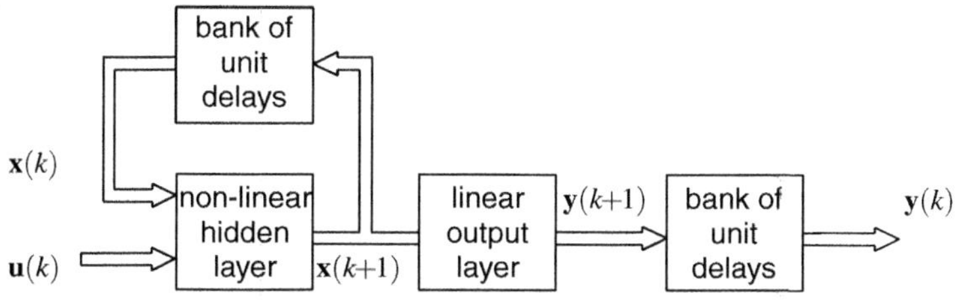

Fig. 3.21 Block scheme of the state-space neural network with one hidden layer

where $\mathbf{f}(\cdot)$ is a non-linear function characterizing the hidden layer, and $\mathbf{C}$ is a matrix of synaptic weights between hidden and output neurons. This model looks similar to the external dynamic approach presented in Fig. 3.20, but the main difference is that for the external dynamics the outputs which are fed back are known during training, while for the state-space model the outputs which are fed back are unknown during training. As a result, state-space models can be trained only by minimizing the simulation error. State-space models possess a number of advantages, contrary to fully and partially recurrent networks (Haykin, 1999; Nelles, 2001):

- the number of states (model order) can be selected independently of the number of hidden neurons. In this way only those neurons that feed their outputs back to the input layer through delays are responsible for defining the state of the network. As a consequence, the output neurons are excluded from the definition of the state;
- since model states feed the input of the network, they are easily accessible from the outside environment. This property can be useful when state measurements are available at some time instants (e.g., initial conditions).

The state-space model includes several recurrent structures as special cases. The previously analyzed Elman network has an architecture similar to that presented in Fig. 3.21, except for the fact that the output layer can be non-linear and the bank of unit delays at the output is omitted.

In spite of the fact that state-space neural networks seem to be more promising than fully or partially neural networks, in practice a lot of difficulties can be encountered (Nelles, 2001):

- model states do not approach true process states;
- wrong initial conditions can deteriorate the performance, especially when short data sets are used for training;
- training can become unstable;
- the model after training can be unstable.

In particular, these drawbacks appear when neither state measurements nor initial conditions are available.

On the other hand, a very important property of the state-space neural network is that it can approximate a wide class of non-linear dynamic systems (Zamarreno and Vega, 1998). There are, however, some restrictions. Approximation is only valid on compact subsets of the state-space and for finite time intervals, thus interesting dynamic characteristics are not reflected (Sontag, 1992; Haykin, 1999).

3.7.3 Locally Recurrent Networks

A biological neural cell not only contains a non-linear mapping operation on the weighted sum of its inputs, but it also has some dynamic properties such as state feedback, time delays hysteresis or limit cycles. In order to cope with such dynamic behavior, a special kind of neuron model has been proposed (Gori et al., 1989; Back and Tsoi, 1991; Frasconi et al., 1992; Poddar and Unnikrishnan, 1991; Gupta and

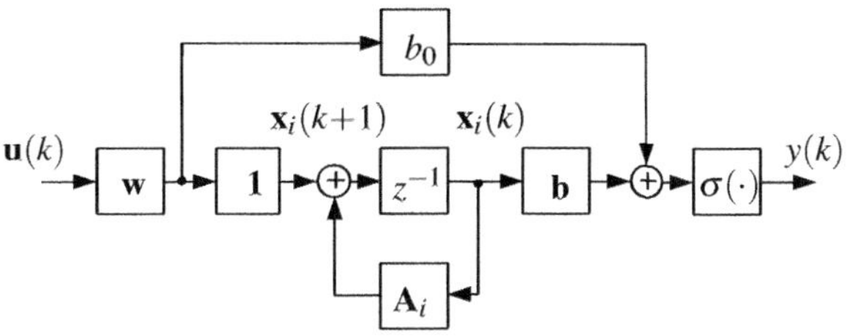

Fig. 3.22 State-space form of the i-th neuron with the IIR filter

Rao, 1993). Such neuron models constitute a basic building block for designing a complex dynamic neural network. Locally Recurrent Globally Feed-forward (LRGF) networks (Tsoi and Back, 1994; Campolucci et al., 1999) have an architecture that is somewhere inbetween a feedforward and a globally recurrent one. The topology of such neural networks is analogous to that of the multi-layered feedforward ones, and the dynamics are reproduced by the so-called dynamic neuron models. Based on the well-known McCulloch–Pitts neuron model, different dynamic neuron models can be designed. In general, differences between these depend on the localization of internal feedback (Patan, 2008b).

One possible solution is to introduce an Infinite Impulse Response (IIR) filter into the neuron structure. In this way, the neuron reproduces its own past inputs and activations using two signals: the input $u_i(k)$, for $i = 1, 2, \ldots, n$, and the output $y(k)$. The states of the neuron can be described by the following state equation:

$$\mathbf{x}(k+1) = \mathbf{A}\mathbf{x}(k) + \mathbf{W}\mathbf{u}(k), \tag{3.119}$$

where $\mathbf{x}(k) \in \mathbb{R}^r$ is the state vector, $\mathbf{W} = \mathbf{1}\mathbf{w}^T$ is the weight matrix ($\mathbf{w} \in \mathbb{R}^n$, $\mathbf{1} \in \mathbb{R}^r$ is the vector with one in the first place and zeros elsewhere), $\mathbf{u}(k) \in \mathbb{R}^n$ is the input vector, n is the number of inputs, and the state matrix $\mathbf{A}$ has the form

$$\mathbf{A} = \begin{bmatrix} -a_1 & -a_2 & \ldots & -a_{r-1} & -a_r \\ 1 & 0 & \ldots & 0 & 0 \\ 0 & 1 & \ldots & 0 & 0 \\ \vdots & \vdots & \ddots & \vdots & \vdots \\ 0 & 0 & \ldots & 1 & 0 \end{bmatrix}. \tag{3.120}$$

Finally, the neuron output is described by

$$y(k) = \sigma\big(g_2(\mathbf{b}\mathbf{x}(k) + \mathbf{d}\mathbf{u}(k) - g_1)\big), \tag{3.121}$$

where $\sigma(\cdot)$ is a non-linear activation function, $\mathbf{b} = [b_1 - b_0 a_1, \ldots, b_r - b_0 a_r]$ is the vector of feedforward filter parameters, $\mathbf{d} = [b_0 w_1, \ldots, b_0 w_n]$. The block structure of the state-space representation of the neuron considered is presented in Fig. 3.22. Much more powerful modeling properties are exhibited by a neural network consisting of a number of dynamic neurons.

3.7.3.1 Network with one Hidden Layer

A neural model with one hidden layer is described by the following equations:

$$\mathbf{x}(k+1) = \mathbf{A}\mathbf{x}(k) + \mathbf{W}\mathbf{u}(k), \tag{3.122a}$$

$$\mathbf{y}(k) = \mathbf{C}\sigma(\mathbf{G}_2(\mathbf{B}\mathbf{x}(k) + \mathbf{D}\mathbf{u}(k) - \mathbf{g}_1))^T, \tag{3.122b}$$

where $N = v \times r$ represents the number of model states, $\mathbf{x} \in \mathbb{R}^N$ is the state vector, $\mathbf{u} \in \mathbb{R}^n$, $\mathbf{y} \in \mathbb{R}^m$ are input and output vectors, respectively, $\mathbf{A} \in \mathbb{R}^{N\times N}$ is the block diagonal state matrix ($\text{diag}(\mathbf{A}) = [\mathbf{A}_1, \ldots, \mathbf{A}_v]$), $\mathbf{W} \in \mathbb{R}^{N\times n}$ ($\mathbf{W} = [\mathbf{w}_1\mathbf{1}^T, \ldots, \mathbf{w}_v\mathbf{1}^T]^T$, where $\mathbf{w}_i$ is the input weight vector of the i-th hidden neuron), and $\mathbf{C} \in \mathbb{R}^{m\times v}$ are the input and output matrices, respectively, $\mathbf{B} \in \mathbb{R}^{v\times N}$ is a block diagonal matrix of feedforward filter parameters ($\text{diag}(\mathbf{B}) = [\mathbf{b}_1, \ldots, \mathbf{b}_v]$), $\mathbf{D} \in \mathbb{R}^{v\times n}$ is the transfer matrix ($\mathbf{D} = [b_{0_1}\mathbf{w}_1^T, \ldots b_{0_v}\mathbf{w}_v^T]^T$), $\mathbf{g}_1 = [g_{1_1} \cdots g_{1_v}]^T$ denotes the vector of biases, $\mathbf{G}_2 \in \mathbb{R}^{v\times v}$ is the diagonal matrix of slope parameters ($\text{diag}(\mathbf{G}_2) = [g_{2_1} \cdots g_{2_v}]$), and $\sigma : \mathbb{R}^v \to \mathbb{R}^v$ is the non-linear vector-valued function.

A locally recurrent network with only one hidden layer is represented by the linear state equation (Patan, 2004; Patan, 2007b). Thus, its ability to approximate non-linear mappings is limited. Therefore, in order to model a wider class of dynamic systems, a network with two hidden layers should be taken into account.

3.7.3.2 Network with Two Hidden Layers

A neural model composed of two hidden layers with v_1 neurons in the first layer and v_2 neurons in the second layer is represented as follows:

$$\mathbf{x}(k+1) = \mathbf{g}(\mathbf{x}(k), \mathbf{u}(k)), \tag{3.123a}$$

$$\mathbf{y}(k) = \mathbf{h}(\mathbf{x}(k), \mathbf{u}(k)), \tag{3.123b}$$

where $\mathbf{g}$, $\mathbf{h}$ are non-linear functions. Taking into account the layered topology of the network, one can decompose the state vector as follows: $\mathbf{x}(k) = [\mathbf{x}^1(k) \quad \mathbf{x}^2(k)]^T$, where $\mathbf{x}^1(k) \in \mathbb{R}^{N_1}$ ($N_1 = v_1 \times r$) represents the states of the first layer, and $\mathbf{x}^2(k) \in \mathbb{R}^{N_2}$ ($N_2 = v_2 \times r$) represents the states of the second layer. Then the state equation can be rewritten in the following form:

$$\mathbf{x}^1(k+1) = \mathbf{A}^1\mathbf{x}^1(k) + \mathbf{W}^1\mathbf{u}(k), \tag{3.124a}$$

$$\mathbf{x}^2(k+1) = \mathbf{A}^2\mathbf{x}^2(k) + \mathbf{W}^2\sigma\big(\mathbf{G}_2^1(\mathbf{B}^1\mathbf{x}^1(k) + \mathbf{D}^1\mathbf{u}(k) - \mathbf{g}_1^1)\big), \tag{3.124b}$$

where $\mathbf{u} \in \mathbb{R}^n$, $\mathbf{y} \in \mathbb{R}^m$ are inputs and outputs, respectively, the matrices $\mathbf{A}^1 \in \mathbb{R}^{N_1\times N_1}$, $\mathbf{B}^1 \in \mathbb{R}^{v_1\times N_1}$, $\mathbf{W}^1 \in \mathbb{R}^{N_1\times n}$, $\mathbf{D}^1 \in \mathbb{R}^{v_2\times n}$, $\mathbf{g}_1^1 \in \mathbb{R}^{v_1}$, $\mathbf{G}_2^1 \in \mathbb{R}^{v_1\times v_1}$ have a form analogous to that of the matrices describing the network with one hidden layer, $\mathbf{A}^2 \in \mathbb{R}^{N_2\times N_2}$ is the block diagonal state matrix of the second layer ($\text{diag}(\mathbf{A}^2) = [\mathbf{A}_1^2, \ldots, \mathbf{A}_{v_2}^2]$), $\mathbf{W}^2 \in \mathbb{R}^{N_2\times v_1}$ is the weight matrix between the first and second hidden layers defined in a similar manner as $\mathbf{W}^1$. Finally, the output of the model is represented by the equation

$$\mathbf{y}(k) = \mathbf{C}^2\sigma\big(\mathbf{G}_2^2(\mathbf{B}^2\mathbf{x}^2(k) + \mathbf{D}^2\sigma\big(\mathbf{G}_2^1(\mathbf{B}^1\mathbf{x}^1(k) + \mathbf{D}^1\mathbf{u}(k) - \mathbf{g}_1^1)\big) - \mathbf{g}_1^2)\big), \quad (3.125)$$

where $\mathbf{C}^2 \in \mathbb{R}^{m \times v_2}$ is the output matrix, $\mathbf{B}^2 \in \mathbb{R}^{v_2 \times N_2}$ is the block diagonal matrix of the second layer feedforward filter parameters, $\mathbf{D}^2 \in \mathbb{R}^{v_2 \times v_1}$ is the transfer matrix of the second layer, $\mathbf{g}_1^2 \in \mathbb{R}^{v_2}$ is the vector of second layer biases, $\mathbf{G}_2^2 \in \mathbb{R}^{v_2 \times v_2}$ represents the diagonal matrix of second layer activation function slope parameters. The matrices $\mathbf{B}^2$, $\mathbf{D}^2$, $\mathbf{g}_1^2$ and $\mathbf{G}_2^2$ have a form analogous to that of the matrices of the first hidden layer.

The presented neural network with two hidden layers possesses pretty good approximation abilities. In the work of Patan (2008a) it was proved that the network with a suitably large number of neurons is able to approximate a state-space trajectory produced by any Lipschitz continuous function with arbitrary accuracy.

3.7.3.3 Cascade Network

The universal approximation theorem for the locally recurrent network with two hidden layers makes it possible to design a less complex neural network. Modifications are as follows:

1. to use in the second layer of the network linear neurons with finite impulse response filters instead of non-linear neurons with IIR filters,
2. to introduce additional synaptic connections between the input and the second layer of the network.

A cascade neural network obtained in this way (Patan, 2010) is presented in Fig. 3.23. Let us consider a discrete-time neural network with n inputs and m outputs. The cascade locally recurrent network is composed of two processing layers consisting of v_1 and v_2 neurons, respectively. Neurons of the second layer receive excitation not only from the neurons of the previous layer but also from the external inputs (Fig. 3.23) (Patan, 2008a; Patan et al., 2008). The first layer includes neu-

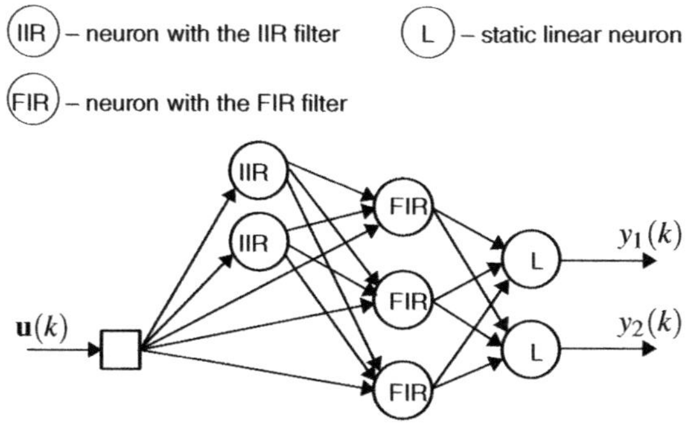

Fig. 3.23 Cascade structure of the locally recurrent network

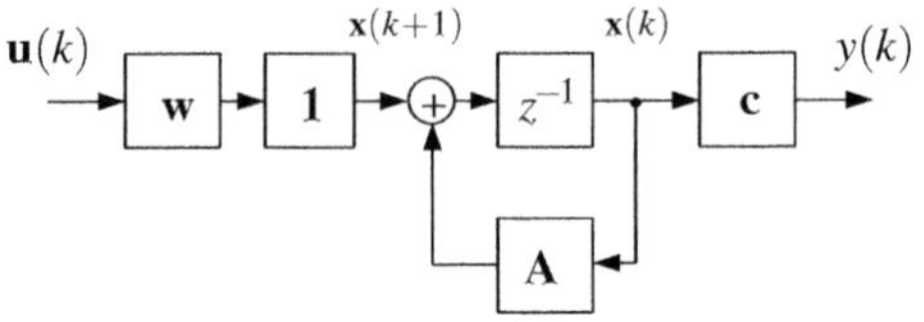

Fig. 3.24 State space form of the neuron with the FIR filter

rons with IIR filters while the second one consists of neurons with finite impulse response filters. The state-space representation block scheme of the neuron with the FIR filter is presented in Fig. 3.24. The states of the i-th neuron with the FIR filter are represented by (3.119), that is, in the same way as for the neuron with the IIR filter. The difference is in the representation of the observation equation. The neuron output is described as follows:

$$y(k) = \mathbf{c}^T \mathbf{x}(k), \tag{3.126}$$

where $\mathbf{c} \in \mathbb{R}^r$ is the output vector. Next, let us consider a cascade locally recurrent network with n inputs and m outputs. The state of the cascade network is represented as follows:

$$\mathbf{x}^1(k+1) = \mathbf{A}^1\mathbf{x}^1(k) + \mathbf{W}^1\mathbf{u}(k), \tag{3.127a}$$

$$\begin{aligned} \mathbf{x}^2(k+1) = \mathbf{A}^2\mathbf{x}^2(k) + \mathbf{W}^2\sigma\big(\mathbf{G}_2^1(\mathbf{B}^1\mathbf{x}^1(k) \\ +\mathbf{D}^1\mathbf{u}(k) - \mathbf{g}_1^1)\big) + \mathbf{W}^u\mathbf{u}(k), \end{aligned} \tag{3.127b}$$

where $\mathbf{x}^1(k) \in \mathbb{R}^{N_1}$ ($N_1 = v_1 \times r$) represents the states of the first layer and $\mathbf{x}^2(k) \in \mathbb{R}^{N_2}$ ($N_2 = v_2 \times r$) represents the states of the second layer, $\mathbf{A}^1 \in \mathbb{R}^{N_1 \times N_1}$ and $\mathbf{A}^1 \in \mathbb{R}^{N_2 \times N_2}$ are the block diagonal state matrices of the first and second layers, respectively, $\mathbf{W}^1 \in \mathbb{R}^{N_1 \times n}$ is the input weight matrix, $\mathbf{W}^2 \in \mathbb{R}^{N_2 \times v_1}$ is the weight matrix between the first and second layers, $\mathbf{W}^u \in \mathbb{R}^{N_2 \times n}$ is the weight matrix between the input and the second layer, $\mathbf{B}^1 \in \mathbb{R}^{v_1 \times N_1}$ is the block diagonal matrix of feedforward filter parameters of the first layer, $\mathbf{D}^1 \in \mathbb{R}^{v_1 \times n}$ is the transfer matrix, $\mathbf{g}_1^1 \in \mathbb{R}^{v_1}$ denotes the vector of biases of the first layer, $\mathbf{G}_2^1 \in \mathbb{R}^{v_1 \times v_1}$ is the diagonal matrix of slope parameters of the first layer, and $\sigma : \mathbb{R}^{v_1} \rightarrow \mathbb{R}^{v_1}$ is the non-linear vector-valued function.

The presented cascade neural network possesses pretty good approximation abilities. In the work of Patan (2008b) it was proved that the cascade network (3.127) with a suitably large number of neurons with IIR filters in the first layer and a suitably large number of neurons with FIR filters in the second layer is able to approximate a state-space trajectory produced by any Lipschitz continuous function with arbitrary accuracy. Additionally, the cascade network has a less complex structure (lower number of parameters), contrary to the locally recurrent network with two hidden layers (Patan, 2008a; 2008b).

3.7.3.4 Training of the Network

The neural network composed of dynamic neuron models does not include any recurrent link between processing units, thus its training procedure is relatively simpler than in the case of globally recurrent networks. To derive values of the network parameters, the back propagation idea can be applied. The Extended Dynamic Back Propagation (EDBP) algorithm, developed for locally recurrent networks, can work in both off-line and on-line modes. The selection of the proper training mode is dependent on the requirements and specificity of the problem to be solved. However, taking into account the fact that a cost function to be minimized usually has a non-linear multimodal form, gradient-based algorithms often get stuck in local minima and training results are not satisfactory. Even the multi-start technique does not improve the quality of the neural model. To avoid such problems, one can use global optimization methods, e.g., the Adaptive Random Search (ARS) or Simultaneous Perturbation Stochastic Approximation (SPSA) (Patan and Parisini, 2002; Patan, 2008a).

3.7.3.5 Neural Model of a DC Motor

In this section, neural modeling of the AMIRA DR300 laboratory system is presented. The laboratory system shown in Fig. 3.25 is used to control the rotational speed of a DC motor with a changing load. The laboratory object considered consists of five main elements: a DC motor M1, a DC motor M2, two digital incremental encoders and a clutch K. The input signal of the engine M1 is an armature current and the output signal is the angular velocity. The available sensors for the output are an analog tachometer on an optical sensor, and a digital incremental encoder. The available measurements of the plant are as follows: I_m is the motor current of the DC motor M1, I_g is the motor current of the DC motor M2, T is the tachometer signal, and the control signals include C_m: the input of the motor M1, C_g: the input of the motor M2. The separately excited DC motor is composed of two subsystems:

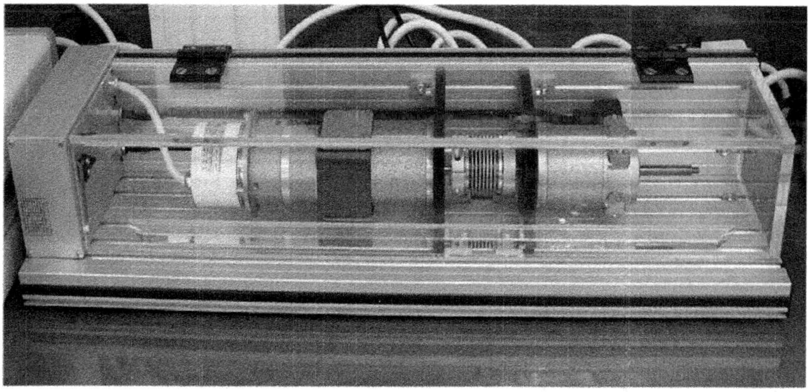

Fig. 3.25 AMIRA DR300 laboratory stand

electrical and mechanical. The electrical subsystem is governed by a linear differential equation. In turn, the mechanical subsystem is in general non-linear. One of the viscous components is the Stribeck friction, which is a non-linear component occurring at low angular velocities. Therefore, to model a DC motor, a non-linear modeling technique, e.g., the locally recurrent network, should be employed.

A separately excited DC motor was modeled by using the dynamic neural network (3.127). The model of the motor was selected as follows:

$$T = f(C_m). \tag{3.128}$$

The following input signal was used in the experiments:

$$C_m(k) = 3\sin(2\pi 1.7k) + 3\sin(2\pi 1.1k - \pi/7) + 3\sin(2\pi 0.3k + \pi/3). \tag{3.129}$$

Using (3.129), a learning set containing 1000 samples was formed. The neural network model (3.127) and (3.126) had the following structure: one input, three IIR neurons with 1-st order filters and hyperbolic tangent activation functions, six FIR neurons with 1-st order filters and linear activation functions, and one linear output neuron (Patan, 2007a; Patan et al., 2007). The neural model structure was selected using the trial and error method. The quality of each model was determined using the Akaike Information Criterion (AIC) (Ljung, 1999). This criterion contains a penalty term and makes it possible to discard too complex models. The training process was carried out for 100 steps using the ARS algorithm (Walter and Pronzato, 1997; Patan and Parisini, 2002) with the initial variance $v_0 = 0.1$. The outputs of the neural model and the separately excited motor generated for another 1000 testing samples are depicted in Fig. 3.26. The efficiency of the neural model was also checked during the work of the motor in closed loop control. The results are presented in Fig. 3.27. After transitional oscillations (Fig. 3.27(a)), the neural model settled at a proper value.

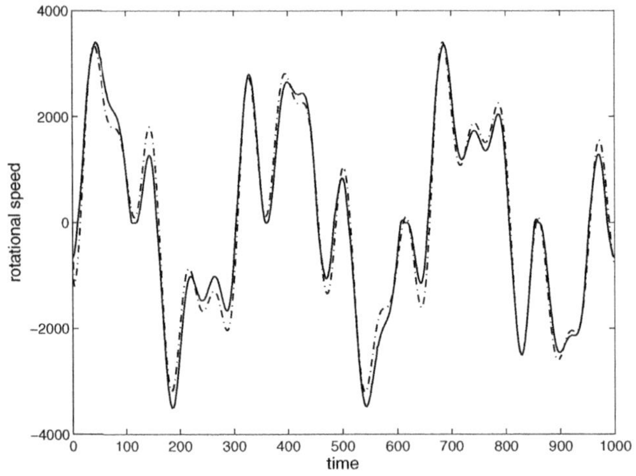

Fig. 3.26 Outputs of the motor (solid) and the neural model (dash-dot): open loop control

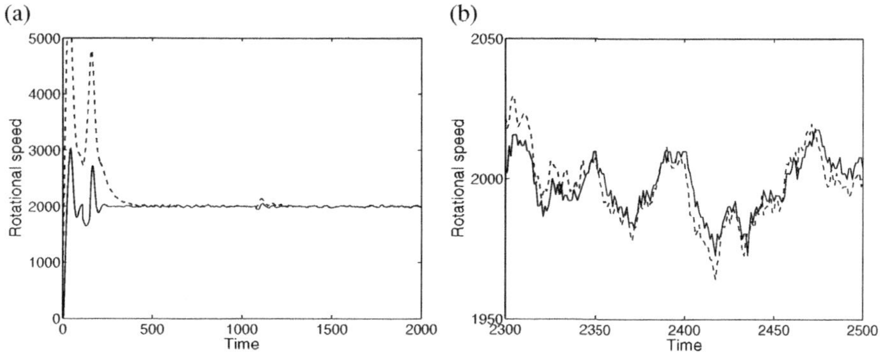

Fig. 3.27 Outputs of the motor (solid) and neural model (dash-dot): closed loop control

For clarity of presentation of the modeling results, in Fig. 3.27(b) the outputs of the process and the neural model for 200 time steps are illustrated only. The above results give a strong argument that the neural model mimics the behavior of the DC motor pretty well and confirm its good generalization abilities.

3.7.4 GMDH Neural Networks

The group method of data handling (Ivakhnenko, 1971) enables automatic structure selection of a neural model. Initially, the method was designed for the identification of static Multi-Input-Single-Output (MISO) systems but, in practice, most systems possess a dynamic structure. Thus, it is necessary to extend the GMDH algorithm for the identification of dynamic systems. Unfortunately, there are only few works in the literature dealing with this problem. All these techniques rely on the introduction of external tapped delay lines into the model structure and usually lead to unstable models due to permanent structure changes of the GMDH neural network during the identification procedure. This problem can be effectively solved with the introduction of the so-called dynamic neurons (Witczak et al., 2006; Witczak, 2007). Another problem is to modify the GMDH identification strategy in such a way that it can be applied for Multi-Input-Multi-Output (MIMO) systems (Mrugalski, 2004; Witczak et al., 2006; Mrugalski and Korbicz, 2007).

3.7.4.1 Synthesis of GMDH Models

The general idea of the GMDH consists in replacing the problem of designing the entire model with the hierarchical structure of polynomial models. The synthesis of the GMDH model (Fig. 3.28) consists in parameter estimation of the partial models and combining them with appropriate selection methods in such a way that the resulting structure evolves towards the best replica of the system being identified (in

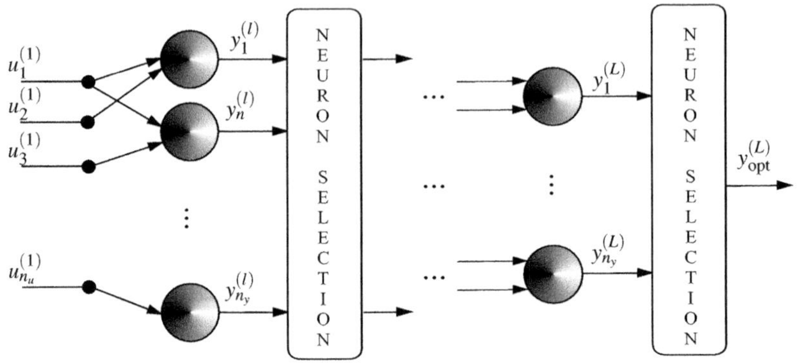

Fig. 3.28 Synthesis of the GMDH model

the sense of the identification criterion used). In a general case, the system can be described by

$$y(k) = f(u_1(k), \ldots, u_{n_u}(k)), \tag{3.130}$$

where $y \in \mathbb{R}$ is the system output, $u = [u_1(k), \ldots, u_{n_u}(k)]^T \in \mathbb{R}^{n_u}$ is the system input, and $f(\cdot)$ is the system structure. The most important feature of the GMDH algorithm is that the partial models are designed separately before introduction to the entire model structure. The parameters of the partial models are estimated in such a way as to achieve the best modeling quality of (3.130), which means that the response of the partial model should be as close as possible to that of the system. The synthesis procedure is stopped when the optimal structure of the entire model is attained. It is assumed that the input vector of the partial models is composed of two inputs of the system. The output of the partial model (cf. Fig. 3.28) can be described, in a general way, by

$$y_n^{(l)} = f(u) = f(u_1^{(l)}, \ldots, u_{n_u}^{(l)}), \tag{3.131}$$

where $n = 1, \ldots, n_y$, while n_y stands for the number of partial models in a layer, $l = 1, \ldots, L$ and L is the number of layers.

In the case of a GMDH neural network, there is a large degree of freedom in defining (3.131), which describes the partial model. On the other hand, it cannot be too complex because, due to the large number of partial models, it may lead to a large computational burden. The general description of the GMDH model is as follows:

$$y = f(u) = p_0 + \sum_{i=1}^{n_u} p_i u_i + \sum_{i=1}^{n_u} \sum_{j=1}^{n_u} p_{ij} u_i u_j +, \ldots, \tag{3.132}$$

where p stands for the parameter (weight) vector of the network.

In the classical approach (Ivakhnenko, 1971), it is assumed that the function partial model is a polynomial of the second degree. Based on the set of n_u available inputs, the first layer of n_y neurons is constructed for all combinations of input couples. In the case of the subsequent l-th layer, the number of newly created neurons depends on the number of neurons in the preceding layer $n_y{}^{(l-1)}$ and the selected number of neuron inputs n_p:

$$n_y{}^{(l)} = \binom{n_y{}^{(l-1)}}{n_p} = \frac{n_y{}^{(l-1)}!}{n_p!(n_y{}^{(l-1)} - n_p)!}. \tag{3.133}$$

The first layer of the GMDH network, for $n_p = 2$, is presented in Fig. 3.29. The

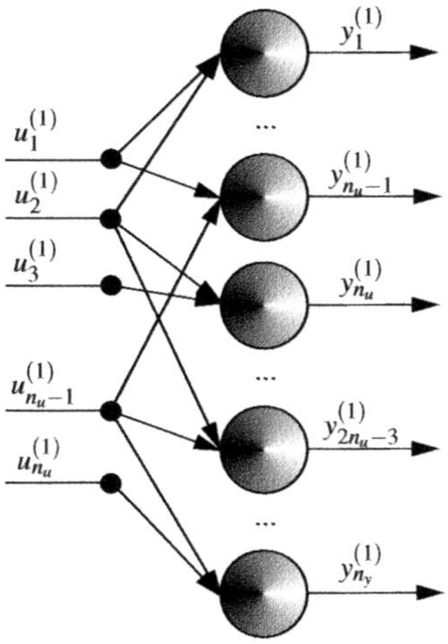

Fig. 3.29 First layer of the GMDH model

responses provided by the particular neurons can be described by

$$\begin{cases} y_1^{(1)} &= f(u_1^{(1)}, u_2^{(1)}), \\ & \dots \\ y_{n_u-1}^{(1)} &= f(u_1^{(1)}, u_{n_u}^{(1)}), \\ y_{n_u}^{(1)} &= f(u_2^{(1)}, u_3^{(1)}), \\ & \dots \\ y_{2n_u-3}^{(1)} &= f(u_2^{(1)}, u_{n_u}^{(1)}), \\ & \dots \\ y_{n_y}^{(1)} &= f(u_{n_u-1}^{(1)}, u_{n_u}^{(1)}). \end{cases} \tag{3.134}$$

The characteristic feature of the GMDH is the quality index $Q(\hat{y}_n^{(l)})$ (performance error), which makes it possible to describe the performance of each partial model. In order to achieve good generalization abilities, this quality index is evaluated with the so-called validation data set T. This quality index is utilized for partial model selection, i.e., a decision is made which of the neurons is introduced to the entire structure of the network. The parameters of the neurons (partial models) of the newly created layer remain unchanged during the remaining identification procedure of the

GMDH neural network. The outputs of the selected neurons stand for the inputs for the subsequent layer:

$$\begin{cases} u_1^{(l+1)} = y_1^{(l)}, \\ u_2^{(l+1)} = y_2^{(l)}, \\ \dots \\ u_{n_u}^{(l+1)} = y_{n_y}^{(l)}. \end{cases} \tag{3.135}$$

The remaining layers are created in an analogous way. The synthesis process is continued until the entire structure achieves an optimal form in the sense of the identification criterion used. Thus, $Q_{\min}^{(l)}$ stands for the quality index for the best performing partial model in the l-th layer:

$$Q_{\min}^{(l)} = \min_{n=1,\dots,n_y} Q(\hat{y}_n^{(l)}). \tag{3.136}$$

$Q_{\min}^{(l)}$ are calculated for each layer and the synthesis process is stopped when

$$Q_{\text{opt}}^{(L)} \geq \min_{l=1,\dots,L} Q_{\min}^{(l)}. \tag{3.137}$$

Thus, the synthesis process is continued until the quality index for the best performing neuron is decreasing (Fig. 3.30) while introducing the subsequent layers. In order to achieve the final structure of the GMDH model (Fig. 3.31), all but the best performing neuron of the last layer are removed. In a similar way, the neurons of the hidden layers are removed.

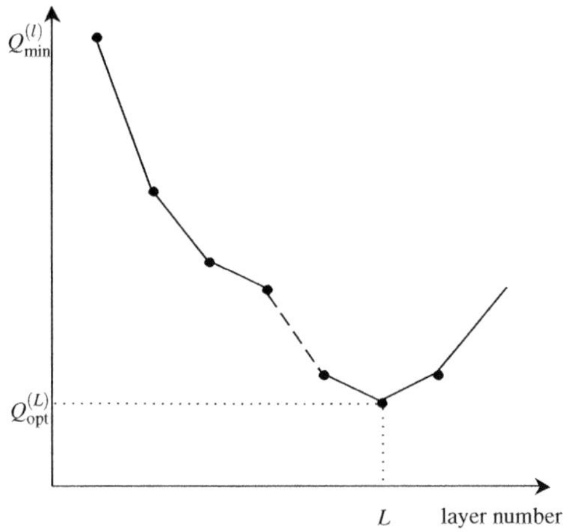

Fig. 3.30 Evolution of the quality index

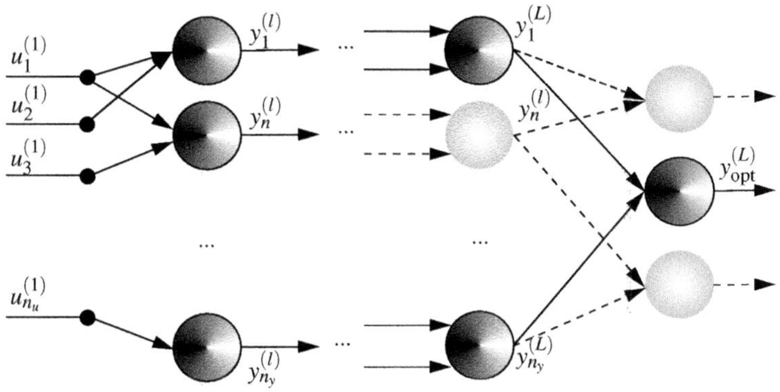

Fig. 3.31 Final structure of the GMDH model

3.7.4.2 Selection Criteria of the Partial Models

The selection criterion of the partial models plays an important role in the synthesis of the GMDH neural network. This choice may have a strong influence on various properties of the resulting model, such as its generalization abilities. In the work of Mrugalski (2004), a number of selection criteria were proposed, which can be divided into three groups:

- *criteria without the partitioning of the data set*: the whole data set is used for parameter estimation and validation of the partial models. Following Mrugalski (2004), it can be observed that the models obtained with such criteria have good prediction abilities. The quality measure of the partial model can be expressed in the form of a variance error:

$$s_\tau^2 = \frac{1}{\tau} \sum_{k=n_{\mathscr{D}}+1}^{n_{\mathscr{D}}+\tau} (y(k) - \hat{y}_n^{(l)}(k))^2 = \frac{1}{\tau} \sum_{k=n_{\mathscr{D}}+1}^{n_{\mathscr{D}}+\tau} \varepsilon(k)^2, \tag{3.138}$$

where τ stands for the prediction horizon, while $n_{\mathscr{D}}$ denotes the number of observations. The expectation of the variance is

$$\mathscr{E}(s_\tau^2) = \sigma_\varepsilon^2 + \mathscr{E}\Big(\frac{1}{\tau} \sum_{k=n_{\mathscr{D}}+1}^{n_{\mathscr{D}}+\tau} (y(k) - \hat{y}_n^{(l)}(k))^2\Big), \tag{3.139}$$

where σ_ε^2 corresponds to an unknown ideal partial model. Among the most popular criteria without partitioning of the data set, three can be distinguished:

  - the FPE (*Final Prediction Error*):

$$J_{FPE} = \frac{n_{\mathscr{D}} + n_p}{n_{\mathscr{D}} - n_p} s_e^2, \tag{3.140}$$

– the AIC (*Akaike Information Criterion*):

$$J_{AIC} = n_{\mathscr{D}} \log s_e^2 + 2n_p + c, \tag{3.141}$$

– the convergence criterion $i^2(n_{\mathscr{D}})$:

$$J_{i^2} = \frac{\sum_{k=1}^{n_{\mathscr{D}}} (\hat{y}_n^{(l)}(k) - y(k))^2}{\sum_{k=1}^{n_{\mathscr{D}}} y(k)^2}, \tag{3.142}$$

where $s_e^2 = \frac{1}{n_{\mathscr{D}}} \sum_{k=1}^{n_{\mathscr{D}}} \varepsilon(k)^2 = \frac{1}{n_{\mathscr{D}}} \sum_{k=1}^{n_{\mathscr{D}}} (y(k) - \hat{y}_n^{(l)}(k))^2$, $\sigma_{\hat{p}}^2$ stands for an estimate of σ_e^2, n_p is the number of parameters, c is a constant, and $\hat{y}_n^{(l)}$ is the response of the partial model obtained with $n_{\mathscr{D}}$.

In practical applications, the convergence criterion (3.142) is most frequently used;

- *criteria with the partitioning of the data set*: in this case, the data are partitioned into the training $n_{\mathscr{U}}$ and validation $n_{\mathscr{T}}$ sets, respectively. The first one is utilized by the training algorithm, while the second one is employed by selection procedures. Thus, in this way, models with good generalization abilities can be obtained. The underlying criteria can be divided into two groups:

1. *Accuracy criteria:*
   - The regularity criterion
     – non-symmetric:

$$\Delta^2(\mathscr{U}) = \sum_{k \in \mathscr{U}} (y(k) - \hat{y}_n^{(l)}(k)|_{\mathscr{T}})^2, \tag{3.143}$$

$$\Delta^2(\mathscr{T}) = \sum_{k \in \mathscr{T}} (y(k) - \hat{y}_n^{(l)}(k)|_{\mathscr{U}})^2, \tag{3.144}$$

     – symmetric:

$$Q\Delta^2 = \Delta^2(\mathscr{U}) + \Delta^2(\mathscr{T}); \tag{3.145}$$

2. *The stability criterion:*
   – non-symmetric:

$$x^2 = \sum_{k \in \mathscr{U} \cup \mathscr{T}} (y(k) - \hat{y}_n^{(l)}(k)|_{\mathscr{U}})^2, \tag{3.146}$$

   – symmetric:

$$S^2 = \sum_{k \in \mathscr{U} \cup \mathscr{T}} [(y(k) - \hat{y}_n^{(l)}(k)|_{\mathscr{U}})^2 + (y(k) - \hat{y}_n^{(l)}(k)|_{\mathscr{T}})^2], \tag{3.147}$$

where $\hat{y}_n^{(l)}(k)|_{\mathscr{U}}$, $\hat{y}_n^{(l)}(k)|_{\mathscr{T}}$ stands for the models outputs, whose parameters were estimated with $\mathscr{U}$ i $\mathscr{T}$.
Consistency criteria express standard deviation of either parameters or outputs of the partial models obtained for the training and validation data sets, respectively. The main drawback of this approach is that the partial models have to be designed for both the training and validation data sets, which significantly increases the computational burden;

- *combined criteria:* these are developed as a result of the combination of selected criteria, e.g.,

$$C_{\text{comb}} = \sqrt{\beta C_1^2 + (1-\beta)C_2^2}, \qquad (3.148)$$

where C_1 and C_2 may represent two arbitrarily selected criteria from the above-described list, while $(0 < \beta < 1)$.

3.7.4.3 Selection Methods for GMDH Neural Networks

The selection mechanism of partial models of the GMDH neural network is responsible for automatic selection of the entire model structure. This process is realized by removing the worst performing partial models according to the selection criterion used. In the literature, the most frequently used selection method is the constant population one (Mrugalski, 2004). The method boils down to selecting g neurons with the smallest processing error $Q(\hat{y}_n^{(l)})$, where g is selected in an empirical way. The main advantage of this method is implementation simplicity, while the main

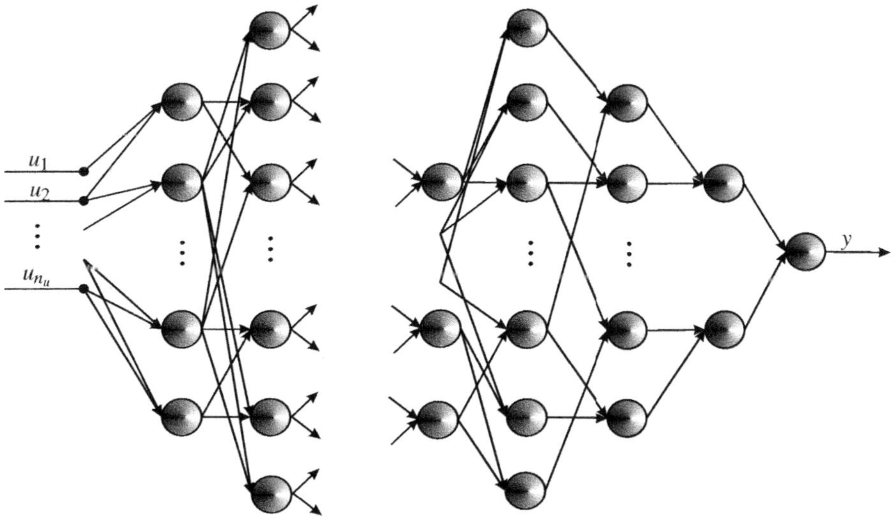

Fig. 3.32 Structure of the GMDH model obtained with the optimal population method

drawback is the fact that all layers have the same structure and hence their evolution is impossible. A similar situation is the case of the so-called decreasing population method. In this approach, a maximum number of neurons is assumed, which is decreasing while new layers are introduced. Contrary to the above-mentioned approaches, the so-called optimal population method (Fig. 3.32) possesses the ability of varying the number of neurons for the subsequent layers. The main principle of this approach is to select all neurons for which the performance error is smaller than an arbitrarily selected threshold, which can be different for different layers. The value of this threshold is usually selected in an intuitive way, which depends on a given example. This constitutes the main drawback of the presented approach.

3.7.5 Implementation of Neural Models in the DiaSter System

For the purposes of the *DiaSter* environment, two neural models were implemented: locally recurrent networks and neural networks of the GMDH type. Additionally, for both models, a robustness problem was considered in Chapter 5. In turn, implementation details of the plug-ins are presented in Chapter 7.

Chapter 4
Knowledge Discovery in Databases

Wojciech Moczulski, Robert Szulim,
Piotr Tomasik, and Dominik Wachla

4.1 Introduction

Contemporary SCADA systems allow collecting huge amounts of data that originates from diverse data sources. There are at least two groups of such sources:

- measuring systems installed on the object (machine or installation) that permit the acquisition of data that are instant values of sampled analogue signals or numeric/functional values of features of such signals;
- sources of messages which can be either automatic systems (e.g., systems for assessing residuals in a model-based diagnostic system, supervisory systems generating warnings and/or alarms etc.), or process operators and other managing personnel.

Data acquired from the enumerated data sources can be the carrier of important information on the process state. During the realization of the *DiaSter* project, particular meaning is assigned to the issue that these data can be the source of diagnostic knowledge that might be used for automatic detection, localization and diagnostics of faults in dynamic industrial processes.

The module described in this chapter includes several applications whose purpose consists in discovering useful knowledge in databases. The most fundamental concept that expresses a piece of raw (uninterpreted) knowledge is *regularity*. It is defined as a *pattern* and a *range* within which this pattern appears (Żytkow and

Wojciech Moczulski · Piotr Tomasik · Dominik Wachla
Department of Fundamentals of Machine Design, Silesian University of Technology,
ul. Konarskiego 18A, 44–100 Gliwice, Poland
e-mail: {wojciech.moczulski,piotr.tomasik,
dominik.wachla}@polsl.pl

Robert Szulim
Institute of Computer Engineering and Electronics, University of Zielona Góra,
ul. Podgórna 50, 65–246 Zielona Góra, Poland
e-mail: r.szulim@ime.uz.zgora.pl

Zembowicz, 1993). Examples of patterns are contingency tables, equations and logical equivalences. The range of appearance of regularity is defined as a subset of data satisfying some complex condition that is a conjunction of simple conditions (such as inequalities).

The task of knowledge acquisition from data is carried out by means of several methods that are classified as Knowledge Discovery in Databases (KDD). Rapid development of this subdomain of artificial intelligence started about the 1990s. A good overview of the methodology is contained in the book by Cichosz (2000). It is worth emphasizing, though, that the majority of work concerns the acquisition of knowledge of static character, while the diagnostics of industrial processes focuses our attention on knowledge that describes phenomena which develop with time. Therefore, knowledge of dynamical processes is of great value.

The methodology of discovering static knowledge is developed quite well. Nowadays, commercial software is available on the market, allowing carrying out analyses of data files with the general goal to find dependencies that might occur between data. As examples of KDD tools for discovering static knowledge one can enumerate *STATISTICA Data Miner*, *Oracle Data Mining* or the *SPSS Clementine* system. However, functional dependencies of static character play a very limited role in process diagnostics. This is because of the fact that the most crucial property of the process consists in the change, which for the purpose of description requires dependencies that either openly or even indirectly include the time variable. Further on, such dependencies will be called dynamic dependencies (or models).

In contrast to well-expanded and quite commonly used methodology and software for discovering static knowledge from databases, the methodology of discovering knowledge represented by different dynamic models is not so satisfactorily developed. Some attention should be paid to the results obtained by the authors of this chapter. The first work concerning applications of KDD methodology to the discovery of diagnostic knowledge was initiated by Moczulski and Żytkow (1997). Although the object of research was mainly static models, the need for discovering dynamic models was stressed.

A group managed by W. Moczulski subsequently carried out several research projects focused on the development of the methodology of acquiring diagnostic knowledge, with special attention paid to applications of KDD methodology. Some PhD theses were carried out, too. They concerned methods of knowledge discovery based on diverse ways of representing this knowledge. Particular attention should be paid to the works of

- Cholewa (2004), concerning the *representation of sequences of events* for reasoning in technical diagnostics;
- Szulim (2004), developing methods of building a knowledge base containing fuzzy dynamic models of processes, which allows inferring founded on Case-Based Reasoning (CBR);
- Wachla (2006), concerning the identification of dynamic diagnostic models using support vectors machines;
- Tomasik (2006), addressing methods of discovering models of processes with the usage of approximation methods.

Theoretic results of these theses (excluding the first one) have been the basis for elaborating algorithms and then modules of the package *Knowledge Discovery in Databases of the System DiaSter*. Several of them are the subject of this chapter, which is composed as follows. This section deals with the issues of knowledge discovery about dynamic processes. Section 4.2 concerns the fundamental task of the selection of process variables. Section 4.3 deals with methods of discovering qualitative knowledge that are of particular usefulness at the introductory stages of the KDD process. The next section (4.4) is devoted to three independent approaches to the discovery of quantitative dependencies. The first one is based upon the SVM algorithm and permits the discovery of models that can further be applied for predictions of outputs of the modeled system or process. The two remaining approaches allow building reasoning systems based on cases: the first one is founded on approximation models of dynamic processes, whereas the other one utilizes an original fuzzy description of process realizations. The chapter ends with conclusions.

4.2 Selection of Input Variables of Models

One of the phases of knowledge acquisition processes based on a learning data set is the feature selection phase. Knowledge acquisition with high-quality learning data including many features does not guarantee the acquisition of adequate models, i.e., those which are characterized by simplicity and a high degree of generalization. Therefore, the representativity of the data set used, together with properties of the algorithms applied, plays a decisive role in the above-mentioned models. Consequently, model identification should be preceded by the feature selection phase, which results in the necessity of equipping knowledge acquisition systems with suitable algorithms.

The definition of relevant feature selection (Hall, 1998) determines this task as the identification and elimination process of redundant features and those which do not carry any important information (Kohavi, 1995; Hall, 1998). Accordingly, the mentioned process results in the reduction of the space dimension of the problem considered. This, in turn, increases the productivity and efficiency of knowledge acquisition and, in some cases, enables obtaining models of higher exactness. Such disciplines as statistics, machine learning and pattern recognition provide two general solutions in the domain of feature selection (Kohavi, 1995).

The first one is called *filter feature selection* and is based on the use of some measures (such as correlation, entropy, etc.) or algorithms (e.g., RELIEF (Kira and Rendell, 1992)) which provide statistical information on relations between the purpose feature and the selected subset of descriptive features which have been chosen by means of a search algorithm.

The second approach is characterized by the fact that, for the evaluation of a given subset of descriptive features, a learning algorithm is used, e.g., the tree induction algorithm (Quinlan, 1986), the algorithm of the supporting vectors method (Schölkopf and Smola, 2001), etc. The evaluation of the feature subset which has been pointed out by the search algorithm is realized by creating a model on this set and by calculating chosen statistical evaluations. The determined statistics provide quantitative information on the model quality and, at the same time, on the quality of the feature subset being analyzed. This approach is known as *wrapper feature selection*.

4.2.1 Correlation-Based Feature Selection

Correlation-based Feature Selection (CFS) (Hall, 1998) is a simple algorithm which uses heuristic functions based on correlation for the evaluation of subset features. This function is defined in the following way (Hall, 1998):

$$H_s = \frac{k\bar{r}_{cf}}{\sqrt{k + k(k-1)\bar{r}_{ff}}}, \tag{4.1}$$

where H_s is heuristic *merit* of subset S including k features, $\bar{r}_{cf}$ is the mean value of correlation determined between class feature f_c and descriptive features f_i from the set S $(f \in S)$, $\bar{r}_{ff}$ is the mean value of correlation determined between descriptive features f_i from the set S $(f \in S)$.

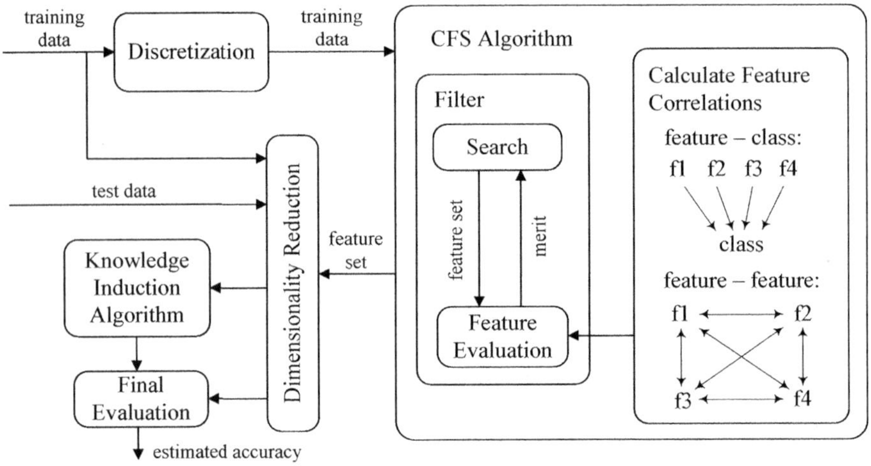

Fig. 4.1 Diagram of correlation-based feature selection (Hall, 1998)

Figure 4.1 shows particular phases of CFS algorithm operation as well as its correlation with the general scheme of feature selection. The algorithm operation can be described in the following manner: firstly, a copy of the training data set is created, and then it undergoes discretization and is handed over to the CFS algorithm, where the following values are calculated:

- the mean value of $\bar{r}_{fc}$ correlation between the class feature and other descriptive features,
- the mean value of $\bar{r}_{ff}$ correlation between descriptive features.

Based on these calculations, CFS determines values of the heuristic function H_s (4.1), which is further transferred to the search algorithm. Subsequently, the algorithm, based on the value of the heuristic function, generates a subset of descriptive features to be checked. Information on a new subset is redirected to the CFS algorithm, which calculates heuristic functions for that subset. Thus, the presented cycle lasts until the stop criterion is fulfilled. As a rule, the search algorithm finishes its operation when the feature subset of the biggest value of the heuristic function is found. Information carrying the finally determined feature subset is used for feature reduction in original sets of training and testing data. Consequently, both reduced data sets are then used for model identification.

4.2.2 Measures Based on Correlation

The determination of r correlations constitutes the basic problem with practical usage of the CFS algorithm. For a data set consisting of features of continuous values, the value of correlation r between two features f_i and f_j can be determined from correlations:

$$r_{f_i,f_j} = \frac{\sum\limits_{x \in T} \left(f_i(x) - \bar{f}_i(x)\right)\left(f_j(x) - \bar{f}_j(x)\right)}{\sqrt{\sum\limits_{x \in T} \left(f_i(x) - \bar{f}_i(x)\right)^2 \sum\limits_{x \in T} \left(f_j(x) - \bar{f}_j(x)\right)^2}}, \tag{4.2}$$

where

$$\bar{f}_i(x) = \frac{1}{\text{card}(T)} \sum_{x \in T} f_i(x), \tag{4.3}$$

$$\bar{f}_j(x) = \frac{1}{\text{card}(T)} \sum_{x \in T} f_j(x). \tag{4.4}$$

Both continuous and nominal features occur in learning data sets frequently at the same time. In such cases, r determination requires a different way of estimation from the correlations described by (4.2). For the set of data including only nominal features, r is determined in the form of a *symmetric uncertainty coefficient* (Press et al., 1992):

$$SUC = 2.0 \times \left[\frac{g_I}{H(f_i) + H(f_j)}\right], \tag{4.5}$$

where g_I is the information gain (Quinlan, 1986), and $H(f_i)$, $H(f_j)$ is information entropy in f_i, f_j, respectively.

Suitable elements appearing in the correlation (4.5) are calculated in the following way:

$$g_I = H(f_i) - H(f_i|f_j) = H(f_j) - H(f_j|f_i), \tag{4.6}$$

$$H(f) = -\sum_{x \in T} p(f(x)) \log_2(p(f(x))), \tag{4.7}$$

$$H(f_i|f_j) = -\sum_{x \in T} p(f_j(x)) \sum \log_2(p(f(x))). \tag{4.8}$$

For data sets including both continuous and nominal features or only continuous ones, the discretization of those sets has to take place before the application of the measure (4.5). Among other things, this can be done by means of the method developed by Fayyad and Irani (1993).

4.2.3 Searching through the Feature Space

In order to realize the CFS algorithm, it is necessary to use at least one method of searching through the state space. The simplest way to achieve this is to use the exhaust searching method, which is easy to implement. However, it requires checking all possible combinations of feature subsets in order to determine the subset with the biggest (or the smallest) value of the criteria function. This way of searching gives a global solution and is effective for a data set with a very small number of features. Even a small increase in the number of features means a drop of efficiency. It results in the fact that exhaust searching is practically used only for data sets with a small number of features. Therefore, only heuristic methods are used in the selection process. Within the elaborated system, the *Best First (BF) search* algorithm (Michalewicz and Fogel, 2004) has been implemented.

Practical usage of the above-mentioned search algorithm requires a suitable way of representing feature subsets. Binary representation seems to be the most natural one. Its essence consists in creating a binary sequence of the length corresponding to the number of features in a given space of F features. Next, one of the f_i features belonging to the F space is attributed to each sequence element. The information if a given feature f_i should create the analyzed feature subset or not is determined on the basis of the element value of the binary sequence which is connected with that feature. It is assumed that, if the value of a given element of the binary sequence is 1, then the feature connected with that sequence element is included in the analyzed set of features. Otherwise, i.e., when the sequence element is of the 0 value, the feature connected with that sequence element is not included. This concept of the way of representation is presented in Fig. 4.2.

$$\begin{matrix} f_1 & f_2 & f_3 & f_4 & f_5 & f_6 & f_7 & f_8 & f_9 & f_{10} \\ [\,1 & 0 & 0 & 1 & 0 & 0 & 1 & 1 & 0 & 0\,] \end{matrix}$$

Fig. 4.2 Scheme of coding solutions for the feature selection task

The chosen way of feature subset representation and the number of discussed features $N = card(F)$ determine the searched state space (Fig. 4.3). In the case of the best first algorithm, state space searching requires defining operators which would allow gradual moving in the space. For feature selection with binary representation, two operators are defined:

- the operator of adding a feature,
- the operator of deleting a feature.

Using these two operators leads to distinction between three categories of moving in the state space:

- *forward selection*: the process starts from an empty feature set, transition to the next state (feature subset) takes place when one of the inactive features is activated;
- *backward elimination*: the process starts from a full feature set, transition to the next state (feature subset) takes place when one of the active features is deactivated;
- *bi-directional*: the process starts simultaneously from two directions, i.e., from an empty and a full feature set.

The above-mentioned techniques of state space searching as well as the space itself are presented in Fig. 4.3 for the problem of three features selection.

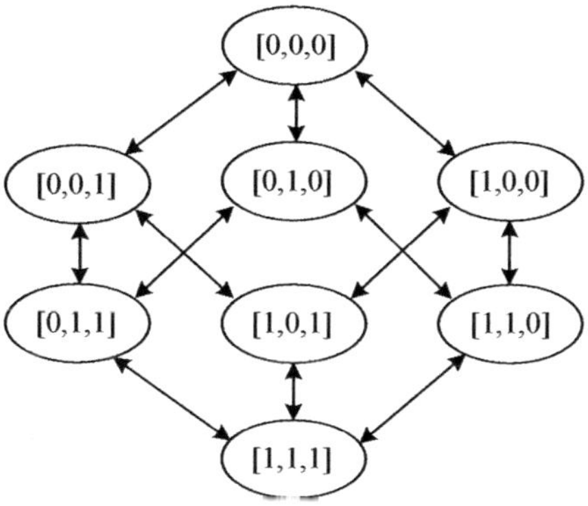

Fig. 4.3 Example of the state space for feature subset selection (Kohavi, 1995)

4.3 Discovery of Qualitative Dependencies

Contingency tables are used to represent qualitative dependencies between two random variables (Aczel, 2001). Most frequently they take the form of tables either of dimensions (2×2) to quickly determine whether there are relationships between attributes, or of tables of dimensions $(n \times m)$ in order to conduct an accurate search.

The discovered dependencies, represented by data tables, are evaluated using the confidence measure Q defined as the probability of random generation of a particular pattern for variables (features) which are independent. If in the contingency table there is an essential dependency, then the probability of event Q is equivalently low. It was concluded that it is necessary to use values of $Q < 10^{-5}$, which means there is a very low probability that patterns may have been derived in a random way. Higher Q values cause the discovery system to identify numerous regularities, amongst which very few are of interest on account of the low relevance value (and thus the probability of being accidental is very low). This makes it difficult for further selection of regularities and extends the operating time of the discovery system (Żytkow and Zembowicz, 1996).

To determine Q, the statistic χ^2 is used:

$$\chi^2 = \sum_{i,j} \frac{(A_{ij} - E_{ij})^2}{E_{ij}}, \tag{4.9}$$

where A_{ij} represents frequencies calculated from the sample and E_{ij} represents expected frequencies in case of truth of the null hypothesis of no relationship between two variables.

Distribution E_{ij} calculated from the sample is determined by the following relationship:

$$E_{ij} = \frac{n_{x_i} \cdot n_{y_j}}{N}, \tag{4.10}$$

where n_{x_i} is the sum of records for the i-th column of the contingency table, n_{y_j} is the sum of records for the j-th row of the contingency table, and N is the total number of records.

Most often the relevance of dependency represented by a given contingency table is evaluated by the following:

- Cramer's V measure (Żytkow and Zembowicz, 1993), defined as

$$V = \sqrt{\frac{\chi^2}{N \cdot \min\{(M_{row} - 1), (M_{col} - 1)\}}} \in [0,1]. \tag{4.11}$$

  The higher value of the measure V, the more unique dependencies may be acquired based upon the contingency table considered. An important feature of the measure V is its independence of the size of the table and the number of records. For $V > 0.9$, the dependency represented by the table can be considered an equivalence;

- Pearson's contingency coefficient C (Żytkow and Zembowicz, 1993):

$$C = \sqrt{\frac{\chi^2}{\chi^2 + N}} \in [0,1]. \tag{4.12}$$

The coefficient C may take values from the interval 0 to 1 (when the number of cells in the table increases to infinity).The value of C is equal to 0 if there is no correlation. The upper limit of the range of the coefficient C depends on the number of rows and columns in the contingency table. The more rows and columns, the higher the value of the coefficient C. Therefore, the value of the coefficient C obtained from the calculations should be considered in relation to its maximum value ($C_{\max}$) of the specified size of the contingency table. In the case of a square table, the maximum value of the coefficient C is

$$C_{\max} = \sqrt{\frac{\{M_{row}, M_{col}\} - 1}{\{M_{row}, M_{col}\}}}. \tag{4.13}$$

For rectangular tables, the approximate maximum value of the contingency coefficient C can be obtained from the formula

$$C_{max} = \frac{\sqrt{\frac{M_{row}-1}{M_{row}}} + \sqrt{\frac{M_{col}-1}{M_{col}}}}{2}. \tag{4.14}$$

The adjusted value of the coefficient C is calculated as

$$C_{kor} = \frac{C}{C_{\max}}; \tag{4.15}$$

- Czuprow's convergence coefficient, defined as (Sobczyk, 2000)

$$T = \sqrt{\frac{\chi^2}{N \cdot \sqrt{(M_{row} - 1) \cdot (M_{col} - 1)}}}. \tag{4.16}$$

  It takes values from the interval $[0, 1]$.When $T = 0$, tested attributes are stochastically independent, but when $T = 1$, the attributes are dependent functionally. When the convergence coefficient is closer to zero, the relationship between the attributes is weaker. In determining the convergence coefficient, it is not essential which attribute is treated as dependent and which as independent, the fact that is important in the study of dependency in the sense of correlation. This property of the convergence coefficient is called symmetry (Sobczyk, 2000). The advantage of this coefficient is that it can be used to measure correlation characteristics. Its disadvantage is the fact that it does not indicate the correlation direction (it is always non-negative).

Contingency tables can be used for pre-processing for methods of discovering knowledge in databases, and for the selection of attributes due to the presence of significant relationships between the attributes considered. Dependencies detected using contingency tables are then used to search for equations.

4.4 Discovery of Quantitative Dependencies

The previous section concerned methods of discovering qualitative knowledge. However, if one examines potential applications of discovered models, crucial meaning is born by methods of discovering regularities described by *quantitative dependencies*. Such dependencies may be used for predicting values of outputs of a process, so that they may be applied in process diagnostics, and even more—in aiding the control of the process course.

This section contains descriptions of three independent methodologies of building dynamic models of a given process by means of the methodology of knowledge discovery in process databases. The first one is based on the SVM algorithm, while the latter concern applications of knowledge discovery methodology to the development of systems of case-based reasoning.

4.4.1 Support Vector Machines

The method of support vector machines was developed on the basis of research concerning statistic learning theory by Vapnik and Chervonenkis (Vapnik, 1995). Initially, it was formulated for the problem of discrimination between two linear separable populations. The main purpose of the method is to determine an *optimal hyperplane* $h(\mathbf{x})$ for which the distance between ρ and $\mathbf{x}_i$ vectors closest to it is maximal (Fig. 4.4). $\mathbf{x}_i$ vectors, which are interpreted as free vectors, are data in a multi-dimensional space.

4.4.1.1 Linear Model

If we assume that

- $\mathcal{X} = \{(\mathbf{x}_1, y_1), \ldots, (\mathbf{x}_N, y_N)\}$ is a set of learning data, where $\mathbf{x}_i$ is the learning vector and $y_i \in \{\pm 1\}$ is the class label ($i = 1, \ldots, N$);
- $h(\mathbf{x}) = \mathbf{w}^T \mathbf{x} + w_0$ is the equation of a discriminative hyperplane,

as well as that

$$\rho = \frac{\tau}{||\mathbf{w}||}, \tag{4.17}$$

where $\tau > 0$ and

$$y_i \left(\mathbf{w}^T \mathbf{x}_i + w_0\right) \geq 0 \;\; i = 1, \ldots, N,$$

then the task of determining the optimal hyperplane $h(\mathbf{x})$ amounts to solving an optimization task (Schölkopf and Smola, 2001; Cristianini and Shawe-Taylor, 2000):

$$\left(\frac{1}{2}||\mathbf{w}||^2\right) \xrightarrow[\mathbf{w}, w_0]{} \min, \tag{4.18}$$

subject to

$$y_i \left[\mathbf{w}^{\mathrm{T}} \mathbf{x}_i + w_0\right] \geq 1 \;\; i = 1, \ldots, N. \tag{4.19}$$

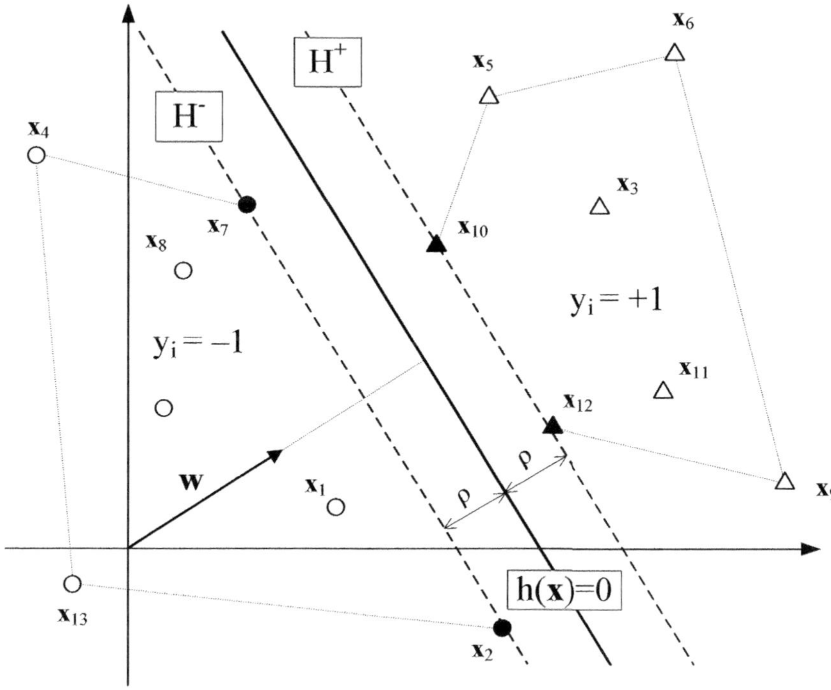

Fig. 4.4 Optimal hyperplane $h(\mathbf{x})$ (Schölkopf and Smola, 2001)

The above-presented problem of optimization may be solved only in the case of linear separability of two populations. For linear inseparable data, the problem is defined as

$$\left(\frac{1}{2}||\mathbf{w}||^2 + C\sum_{i=1}^{N}\xi_i\right) \xrightarrow[\mathbf{w},w_0,\xi]{} \min, \tag{4.20}$$

subject to

$$y_i\left[\mathbf{w}^{\mathrm{T}}\mathbf{x}_i + w_0\right] \geq 1 - \xi_i \quad i = 1, \ldots, N, \tag{4.21}$$

$$\xi_i \geq 0. \tag{4.22}$$

Entering ξ_i variables into the objective function (4.20) results in reducing the inequality (4.19). Owing to this, it is accepted that particular $\mathbf{x}_i$ vectors lie at incorrect sides of canonical planes H^+ and H^- which determine the margin ρ. The influence of ξ_i on (4.19) is controlled by using a C parameter. It is a regulator by means of which the stage of algorithm generalization (scope of the confidence margin) is controlled. At the same time, the number of errors in a training data set is controlled as well.

The optimization problems (4.18) and (4.20) are solved by means of Lagrange's multiplier method. As an example, for the objective function (4.20) and the constraints (4.21) and (4.22), the following Lagrangian is constructed (Cristianini and Shawe-Taylor, 2000; Schölkopf and Smola, 2001):

$$L(\mathbf{w}, w_0, \xi, \alpha, \beta) = \frac{1}{2}||\mathbf{w}||^2 + C\sum_{i=1}^{N} \xi_i \\ - \sum_{i=1}^{N} \alpha_i (y_i[\mathbf{x}_i^{\mathrm{T}} \mathbf{w} + w_0] - 1 + \xi_i) - \sum_{i=1}^{N} \beta_i \xi_i, \tag{4.23}$$

where $\alpha_i, \beta_i \geq 0$ are Lagrange's multipliers. Next, the Lagrangian (4.23) is minimized due to $\mathbf{w}$, w_0 and ξ, and maximized due to α_i. For this purpose, partial derivatives are determined:

$$\frac{\partial}{\partial \mathbf{w}} L(\mathbf{w}, w_0, \xi, \alpha, \beta) = 0, \tag{4.24}$$

$$\frac{\partial}{\partial w_0} L(\mathbf{w}, w_0, \xi, \alpha, \beta) = 0, \tag{4.25}$$

$$\frac{\partial}{\partial \xi} L(\mathbf{w}, b, \xi, \rho, \alpha, \beta, \delta) = 0. \tag{4.26}$$

The result of differentiating the Lagrangian (4.23) with respect to $\mathbf{w}$, w_0 and ξ is presented as the following set of equations:

$$\mathbf{w} = \sum_{i=1}^{N} \alpha_i y_i \mathbf{x}_i, \tag{4.27}$$

$$\sum_{i=1}^{N} \alpha_i y_i = 0, \tag{4.28}$$

$$\alpha_i + \beta_i = C, \tag{4.29}$$

Subsequently, the equations (4.27), (4.28) and (4.29) are inserted into the Lagrangian (4.23), which results in a dual formulation of the optimization problem (4.20)–(4.22):

$$\left(\sum_{i=1}^{N} \alpha_i - \frac{1}{2} \sum_{i,j=1}^{N} \alpha_i \alpha_j y_i y_j \mathbf{x}_i^T \mathbf{x}_j \right) \underset{\alpha}{\longrightarrow} \max, \tag{4.30}$$

subject to

$$0 \geq \alpha_i \geq C, \ i = 1, \ldots, N, \tag{4.31}$$

$$\sum_{i=1}^{N} \alpha_i y_i = 0. \tag{4.32}$$

According to the Karush–Kühn–Tucker theory (Schölkopf and Smola, 2001), a solution to the problem (4.30) subject to (4.31) and (4.32) is determined by the following equations:

$$\alpha_i \left[y_i(\mathbf{w}^T \mathbf{x}_i + w_0) - 1 + \xi_i \right] = 0, \tag{4.33}$$

$$\beta_i \xi_i = (C - \alpha_i)\xi_i = 0. \tag{4.34}$$

A set of Lagrange's multipliers α_i is the solution to (4.30)–(4.32). The $\mathbf{x}_i$ vectors, for which $\alpha_i > 0$, are called Support Vectors (SVs). Finally, the equation (4.35) gives a general structure of the SVM model for classification:

$$\hat{h}(\mathbf{x}) = \text{sign}\left(\sum_{i=1}^{N} \alpha_i y_i \mathbf{x}^T \mathbf{x}_i + w_0 \right). \tag{4.35}$$

The optimization problem (4.30), along with the constraints (4.31) and (4.32), is solved by means of Quadratic Programming (QP) methods. In practice, however, the Sequential Minimal Optimization (SMO) algorithm introduced by Platt (1998) is most frequently used.

4.4.1.2 Non-linear Model

In the SVM, the $h(\mathbf{x})$ hyperplane is not determined in the input space but in a particular high-dimensional space of features Φ. The Φ is usually a non-linear dot product of particular base functions $\phi_j(\mathbf{x})$ which are defined in the input space.

The application of base functions requires that the transformation of input space be explicitly provided. Such a task, including a two-, three- and four-dimensional space, is a fairly simple one. In the case of a bigger number of dimensions, explicit definition of input space transformation is more complicated.

The solution to the mentioned problem is based on the application of the dot product of the base functions $\phi_j(\mathbf{x})$, $(j = 1, 2, \ldots, m)$. This dot product is represented by a kernel function:

$$K(\mathbf{x}_i, \mathbf{x}) = \phi_j(\mathbf{x})^{\mathrm{T}} \phi_j(\mathbf{x}_i). \tag{4.36}$$

It is expected that the kernel functions K representing the dot product of base functions $\phi_j(x)$ will meet Mercer's condition (Vapnik, 1995; Schölkopf and Smola, 2001), i.e., that they will be symmetrical and positively defined:

$$\int\int K(\mathbf{x}_i, \mathbf{x}_j) f(\mathbf{x}_i) f(\mathbf{x}_j) d\mathbf{x}_i d\mathbf{x}_j > 0, \tag{4.37}$$

for

$$f \neq 0, \quad \int f^2(\mathbf{x}) d\mathbf{x} < \infty. \tag{4.38}$$

Among the most frequently applied functions of the kernel K one finds the following (Cristianini and Shawe-Taylor, 2000; Schölkopf and Smola, 2001):

- a linear kernel:
$$K(\mathbf{x}, \mathbf{x}_i) = \mathbf{x}^{\mathrm{T}}\mathbf{x}_i; \tag{4.39}$$
- a polynomial kernel:
$$K(\mathbf{x}, \mathbf{x}_i) = \left[\gamma \mathbf{x}^{\mathrm{T}}\mathbf{x}_i + \theta\right]^q, \tag{4.40}$$
(K is positively defined when $\theta \in \{0; 1\}$ and $q \in \mathbb{N}$ and $\gamma \geq 0$.);
- an RBF kernel:
$$K(\mathbf{x}, \mathbf{x}_i) = \exp\left(-\gamma ||\mathbf{x}^{\mathrm{T}} - \mathbf{x}_i||^2\right) = \exp\left(-\frac{||\mathbf{x} - \mathbf{x}_i||^2}{2\sigma^2}\right), \tag{4.41}$$
where $\gamma = 1/2\sigma^2 \geq 0$;
- a sigmoid kernel:
$$K(\mathbf{x}, \mathbf{x}_i) = \tanh\left(\gamma \mathbf{x}^{\mathrm{T}}\mathbf{x}_i + \theta\right). \tag{4.42}$$

Considering the kernel function K, the equation of the *optimal hyperplane* takes the following form:
$$h(\mathbf{x}) = \sum_{i=1}^{N} w_i K(\mathbf{x}_i, \mathbf{x}) + w_0. \tag{4.43}$$
In turn, the form of the SVM model for the decision problem involves the following functional dependency:
$$\hat{h}(\mathbf{x}) = \mathrm{sign}\left(\sum_{i=1}^{N} \alpha_i y_i K(\mathbf{x}_i, \mathbf{x}) + w_0\right). \tag{4.44}$$

In practice, introducing the kernel function K into the SVM allowed the acquisition of a method for any data analysis. Applying it to a particular problem is conditioned by the presence of an adequate function of the kernel K. Proper selection of the function and values of its parameters is essential for ensuring the quality of the acquired model (4.44). In practice, kernel functions presented here are most frequently used. It is possible to define given kernel functions adjusted to particular tasks; however, the functions have to meet the conditions specified by kernel functions theory (Schölkopf and Smola, 2001).

4.4.1.3 Regressive Model

The SVM method was defined for regression problems as well. In the input space, a regressive model is defined in the form of a linear approximator (Vapnik, 1995; Schölkopf and Smola, 2001):
$$h(\mathbf{x}) = \mathbf{w}^{\mathrm{T}}\mathbf{x} + w_0, \tag{4.45}$$
where the quality of approximation is measured by means of the ε-insensitive loss function (Vapnik, 1995):

$$L(y, h(\mathbf{x})) = \begin{cases} 0 & \text{for } |y - h(\mathbf{x})| \leq \varepsilon \\ |y - h(\mathbf{x})| - \varepsilon & \text{otherwise.} \end{cases} \tag{4.46}$$

Figure 4.5 presents the idea of SVM formulation for a regressive model along with the consideration of (4.46).

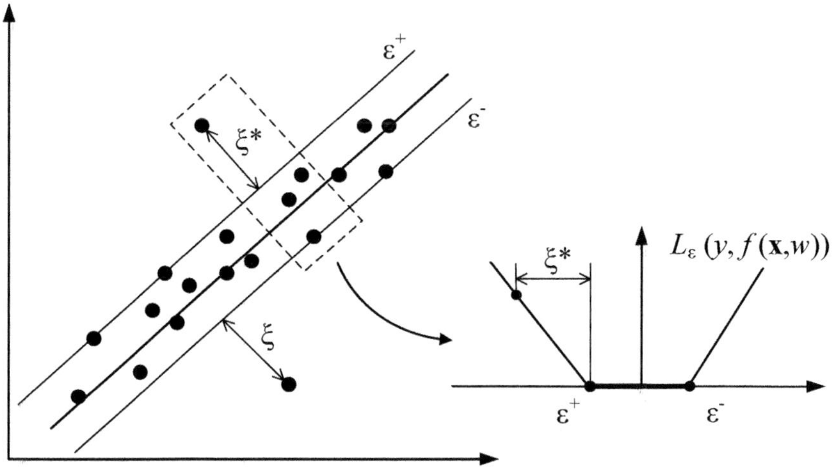

Fig. 4.5 SVM formulation for regression (Schölkopf and Smola, 2001)

The regressive model is accomplished by taking the values of the loss function (4.46) into consideration with simultaneous reduction of the complexity of the model. The reduction of model complexity takes place when the $||\mathbf{w}||^2$ norm is minimized. This problem is solved by means of two (non-negative) ξ_i, ξ_i^*, $i = 1, \ldots, n$ slack variables which measure the deviation of the $\mathbf{x}_i$ vector from the zone determined by the ε-insensitive loss function. The task of the identification of the model (4.45) with the consideration of the loss function (4.46) is examined as the task of the optimization of the following objective function (Cristianini and Shawe-Taylor, 2000; Schölkopf and Smola, 2001):

$$\left(\frac{1}{2} ||\mathbf{w}||^2 + C \sum_{i=1}^{N} (\xi_i + \xi_i^*) \right) \xrightarrow[\mathbf{w}, w_0, \xi, \xi^*, \varepsilon]{} \min, \tag{4.47}$$

subject to

$$h(x_i) - y_i \leq \varepsilon + \xi_i \; i = 1, \ldots, N, \tag{4.48}$$

$$y_i - h(x_i) \leq \varepsilon + \xi_i^*, \tag{4.49}$$

$$\xi_i, \xi_i^* \geq 0. \tag{4.50}$$

Similarly to the case of the linear model (Sec. 4.4.1.1), the task of optimization (4.47) is transformed into a dual optimization problem by means of the method of Lagrange's multipliers. Having taken the space of the base function $\phi_j(x)$ into account, the objective function is presented as

$$L(\alpha,\alpha^*) \xrightarrow[\alpha,\alpha^*]{} \max, \tag{4.51}$$

$$L(\alpha,\alpha^*) = -\varepsilon \sum_{i=1}^{N}(\alpha_i^* + \alpha_i) + \sum_{i=1}^{N}(\alpha_i^* - \alpha_i)y_i$$
$$-\frac{1}{2}\sum_{i,j=1}^{N}(\alpha_i^* + \alpha_i)(\alpha_j^* + \alpha_j)K(\mathbf{x}_i,\mathbf{x}_j), \tag{4.52}$$

subject to

$$\sum_{i=1}^{N}(\alpha_i - \alpha_i^*) = 0 \quad i = 1,\ldots,N, \tag{4.53}$$

$$0 \leq \alpha_i, \alpha_i^* \leq C. \tag{4.54}$$

The problem of optimization (4.52) may be solved, e.g., by means of an appropriately modified Sequential Minimal Optimization (SMO) algorithm (Platt, 1998). Such a modification was proposed, among others, by Schölkopf and Smola (2001) as well as by Flake and Lawrence (2002).

Finally, the regressive model including the function space $\phi_j(x)$ takes the following form:

$$\hat{h}(\mathbf{x}) = \sum_{i=1}^{N}(\alpha_i^* - \alpha_i)K(\mathbf{x}_i,\mathbf{x}) + w_0. \tag{4.55}$$

In Fig. 4.6, a manner of determining the SVM model output is presented.

4.4.2 Methods Involving Case-Based Reasoning

Case-based reasoning is a method which assumes using experience to solve new problems. At a high level of generalization it is possible to compare this method to the way human reason is applied when dealing with a problem or situation. A new problem which is similar to another one solved in the past can also be solved in a similar way. The result should be similar, too. CBR is a method which searches in a database of historical cases for the case most similar (or similar enough) to the current one and applies the solution from the past to this case.

At a high level of generalization, the implementation of the method can be described by four groups of operations (Aamodt and Plaza, 1994), Fig. 4.7:

- **retrieve:** at this stage, searching for the most similar cases from the available database is accomplished. The activity can be divided into feature identification, searching, preselecting and case selection;

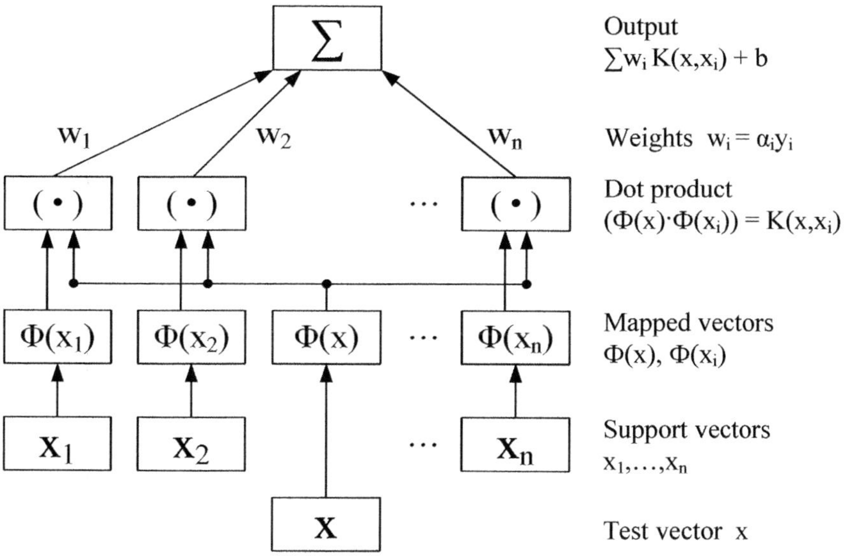

Fig. 4.6 Determining the output of the SVM (Schölkopf and Smola, 2001)

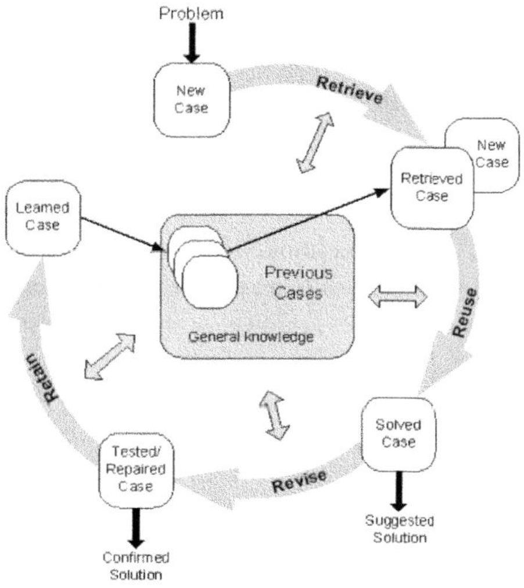

Fig. 4.7 CBR cycle

- **reuse:** during the reuse stage two aspects must be considered, i.e., differences between the current and the discovered case, and which features describing the case found can be applied to the new case. It is possible to use the archival case as a solution or use a method which would lead to the elaboration of a solution;
- **revise:** a solution found in the database of cases at the previous stage and applied to a new case is not always good. Once the case had been used it is possible to compare results obtained after its application. Different approaches can be used as presentation results for the expert to estimate the quality. In the case of differences in results of the stored and the current case, it is possible to modify the case in a database to achieve better fitness to the obtained results;
- **retain:** when the problem is solved, the system can store the new case in the database. If the problem was solved using a historical case, the new case can be stored fully in the database or can only modify the historical one. If the problem was solved in a different way (e.g., by asking the expert), the case will be stored as a brand new case. The explanation of the case can be stored, too.

To search for similar cases in the database, a special mechanism of comparing cases stored in the database is necessary. The result of the comparison should be a value of similarity. To compare the cases, different similarity measures can be used, but many times it is necessary to elaborate a special dedicated similarity measure.

4.4.2.1 Approximation Models of Periodic Processes

Data collected while monitoring cyclical, complex industrial processes may carry information concerning the way of controlling these processes. These data are collected in archival databases and acquired either from different sensors or directly from the operator of the given device.

The control of a complex process can be carried out in an open loop by the operators of the process. The quality of managing the process depends largely on their knowledge and experience. The formalized procedure of conducting such a process is not always sufficiently defined. The operators cannot communicate their knowledge in a way that would enable the identification of a procedure to carry out optimal control of the object/process. However, one can assume that each operator would have his or her own method of managing the process that likely differs from those applied by other operators.

Additional tasks performed by the operators include compliance with certain ranges of the parameters. Exceeding the parameters of both the value and duration of such a state could seriously affect the state of a certain industrial object and/or may have a negative effect on other objects. During the production cycle it is required to maintain synchronization between the objects of the process due to the mass flow between them.

As a result of a review of the state-of-the-art concerning methods of modeling slowly varying processes, it was found to be useful to base modeling processes on simple methods of data processing. New methods of modeling processes were proposed which are represented by a multi-dimensional time series. These methods are based on measures of distance and character sequences to determine the degree of similarity between different realizations of the process. The obtained models of individual cycles of the process are global in nature. They can be used in reasoning based on examples.

The methods outlined below are described in detail by Tomasik (2006). This chapter presents selected parts of the methodology without detailed descriptions.

Method of Time Normalization of Continuous Courses of Process Parameters

Time normalization relates to solving the problem of the need to compare the cycles or phases of different time duration.

In the case of cycles consisting of repeated phases, it was suggested that time normalization would be related to partial phases by designating the adopted parameter of scale for each of the phases on the basis of the following dependency:

$$ScaleT^n(j,k) = \frac{T^n(j)}{T^n(k)}, \tag{4.56}$$

where k, j, n is the index of the compared cycle reference and cycle phase, $ScaleT^n(k, j)$ is the time-scale between various cycles with indices (k, j) for the phase of the cycle with index n, and $T^n(j), T^n(k)$ are the durations of the compared cycle with index (j) and the reference cycle or pattern cycle with index (k) for a given cycle phase with index n. If we consider the whole cycle, then the value of n shall equal 0.

Using a scale defined in such a manner results in the following:

- for cycles or phases longer than the pattern, a scale smaller than 1 was obtained (reducing the sampling frequency of attribute values);
- for cycles or phases shorter than the pattern, a scale larger than 1 was obtained (increasing the sampling frequency of attribute values).

The effect of an increase or a decrease in the sampling frequency of the examined cycles by introducing a fixed scale relates to data approximation of the examined cycles (to be compared and the reference one), resulting in a fixed number of points approximating the cycle (or its specified phase) in the group of tested cycles.

In order to approximately represent cycles for processes attributes, the Bezier curves approximation (free curves) was used.

Method of Determining Similarities Based on the Absolute Distance Measured by the Euclidean Metric

This method is used to determine the values of similarities between the compared cycles using the measure based on the Euclidean distance determined by the formula

$$D^n_{2i}(k,j) = [\sum_{p=1}^{P} (Y(k)^n_{i,p} - Y(j)^n_{i,p})^2]^{\frac{1}{2}}, \tag{4.57}$$

where i is the index of the process attribute, p is the index of the point approximating the cycle ($p = 1 \ldots P$), P is the number of points approximating the cycles for which the Euclidean distance is calculated, $D_{2i}^n(k,j)$ is the Euclidean distance calculated for a given attribute with index i for a given cycle phase with index n; the distance is calculated between the graphs of cycles with indices (k,j), and $Y(k)_i^n; Y(j)_i^n$ are values of the attribute with index i in individual approximation points for the cycles with indices (k,j) and the cycle phase with index n. Absolute distances in this case are calculated as differences between the values of these points (samples) in the adopted reference system.

Based on the Euclidean distance calculated this way, values of similarities between cycles with indices (k,j) for a given attribute with index i and the phase of the cycle with index n on the basis of the similarity measure defined below are estimated (4.58):

$$\begin{aligned} \Pi_i^n(k,j) &= 1 - \frac{D_{2i}^n(k,j)}{\max(D_{2i}^n(k,j))} && \text{for } \max(D_{2i}^n(k,j)) \neq 0, \\ \Pi_i^n(k,j) &= 1 && \text{for } \max(D_{2i}^n(k,j)) = 0, \end{aligned} \tag{4.58}$$

where Π_i^n is the similarity between various individual cycles with indices (k,j) within the individual cycle phase with index n for a given attribute with index i and $\max(D_{2i}^n(k,j))$ is the furthest Euclidean distance for a given attribute of index i, and the cycle phase with index n for cycles with indices (k,j).

The introduced measure for determining similarities yields a linear change of the similarity value versus distance. Such a simple measure of similarity has been adopted to facilitate subsequent modification of similarities.

For the values of similarities for individual phases of the compared cycles determined this way, the values of total similarities are defined based on the following relation:

$$\begin{aligned} \Pi C_i(k,j) &= \frac{1}{N} \sum_{n=1}^{N} w_n \cdot \Pi_i^n(k,j), \\ \sum_{n=1}^{N} w_n &= N, \end{aligned} \tag{4.59}$$

where $\Pi C_i(k,j)$ is the overall similarity between the cycles with indices (k,j) for a given attribute with index i, N is the number of phases of the tested cycle, and w_n represents weights dependent upon the cycle phase, which may be adopted on the basis of an expert's opinion, or, if unknown, take the value $w_n = 1$ (where $n = 1,\ldots,N$); the weights shall determine the impact of the similarity of phases of the cycle on the overall similarity of the cycles.

Since for such a method of determining values of similarities between the cycles with indices (k,j) for a phase with index n the following may occur:

$$\Pi C_i(k,j) \neq \Pi C_i(j,k), \tag{4.60}$$

to fulfill the symmetry axiom for the similarity value $\Pi C_i(k,j) = \Pi C_i(j,k)$, the concept of average similarity $\Pi MC_i(k,j)$ was introduced and expressed by the relation

$$\Pi MC_i(k,j) = \Pi MC_i(j,k) = \frac{\Pi C_i(k,j) + \Pi C_i(j,k)}{2}, \tag{4.61}$$

where $\Pi MC_i(k,j), \Pi MC_i(j,k)$ are average integral similarities between cycles with indices (k,j) for a given attribute with the index i.

The value of the total similarity between the cycles with indices (k,j) for all the chosen parameters is determined based upon the relation

$$\Pi_{ALL}(k,j) = \frac{1}{I} \sum_{i=1}^{I} w_i \cdot \Pi MC_i(k,j),$$
$$\sum_{i=1}^{I} w_i = I, \tag{4.62}$$

where $\Pi_{ALL}(k,j)$ is the overall similarity between cycles with indices (k,j) for all the selected attributes $(i = 1,\dots,I)$, I is the number of process attributes selected for the study, and w_i are weights for each process attributes $(i = 1,\dots,I)$.

Method of Sequences of Signs

The distance method, based on the Euclidean measure in a selected similarity value, does not consider the influence of the shape of the course of the compared cycles but only the differences between the corresponding values of the data describing the cycle.

To express the similarity of shapes of the curves describing the analyzed course, a method has been developed involving the discretization of the courses using straight line segments, whose number is equal to $(P-1)$, where P is the number of approximation points. Any two consecutive points of approximation designate one segment of the description. The greater the number of points of approximation, the more segments and the more accurate approximation of describing the course of the cycle.

A glossary has been introduced describing the membership of the slope angle of the segment to a discretized set of intervals (NS) (an odd number of intervals), described by a predefined alphabet (Cholewa, 2004). The length of the sequence is dependent upon the number of points of approximation P and equals $(P-1)$. Characters that are used in the alphabet are dependent upon the number of intervals NS. For $NS = 3$, characters are from the set (A,B,C), and for $NS = 7$ from the interval $A - G, (A,B,\dots,F,G)$. Comparing corresponding characters in the sequences is based on the ASCII codes of these characters, while similarity values are assigned according to (4.63). For the presented dependency, the similarity value of characters of the accepted glossary has been determined in relation to the character A. An additional assumption is to compare character sequences of the same length.

$$
\begin{array}{c|c|c}
\textit{Signs} & & \textit{Similarity} \\
\hline
A & \rightarrow & (1) \\
B & \rightarrow & 1-(\frac{1}{NS-1}) \\
C & \rightarrow & 1-[2\cdot(\frac{1}{NS-1})] \\
\vdots & & \vdots \\
SIGN(NS-1) & \rightarrow & 1-[(NS-2)\cdot(\frac{1}{NS-1})] \\
SIGN(NS) & \rightarrow & (0)
\end{array}
, \tag{4.63}
$$

where $SIGN(NS)$ is the final character dependent on the number of intervals (e.g., $NS = 3 \rightarrow SIGN(NS) = C$).

Method of Determining Similarities Based on the Relative Distance Measured by the Euclidean Metric

Described below are two ways applied in the discussed method that use the values of relative distances to determine the similarity of the compared cycle values. The first way discussed used relative distances between the points of approximation of two compared cycles for a selected process attribute. The second way of proceeding used the distance values of the relative slope angles of the two segments connecting the next two approximation points of the two compared cycles for a given process attribute.

To determine similarities based on the relative distances between successive points in compared cycles, the relation has the form (4.64)

$$
\begin{aligned}
\Pi D_i^{KS}(k,j) &= 1 - \frac{D_i^{KS}(k,j)}{\max(D_i^{KS}(k,j))} && \text{for } \max(D_i^{KS}(k,j)) \neq 0, \\
\Pi D_i^{KS}(k,j) &= 1 && \text{for } \max(D_i^{KS}(k,j)) = 0.
\end{aligned}
\tag{4.64}
$$

Depending on the different relative location of individual points of approximation, two options for calculating the distance between these points for the curves K and S were introduced:

$$
\begin{aligned}
D_i^K &= Y_{i,p}^K - Y_{i,(p+1)}^K, \\
D_i^S &= Y_{i,p}^S - Y_{i,(p+1)}^S,
\end{aligned}
\tag{4.65}
$$

where D_i^K, D_i^S are values of differences of absolute distances between points with indices p and $(p+1)$ describing the attribute course with index i for the curves K and S, and $Y_{i,p}^K, Y_{i,(p+1)}^K, Y_{i,p}^S, Y_{i,(p+1)}^S$ is the attribute value with index i in approximation points with indices p and $(p+1)$ for the curves K and S describing the compared cycles.

Dependency 1. When $D_K < 0$ and $D_S > 0$ and $D_K > 0$ and $D_S < 0$:

$$D_i^{KS} = |D_i^K| + |D_i^S|; \tag{4.66}$$

Dependency 2. When $D_K < 0$ and $D_S < 0$ and $D_K > 0$ and $D_S > 0$:

$$D_i^{KS} = |D_i^K| - |D_i^S|, \tag{4.67}$$

where D_i^{KS} is the relative distance between the compared segments of the curves K and S for the attribute with the index i.

For the second method based on the values of the relative distances of various slope angles of the segments joining the successive points of approximation of the cycle for the selected process attribute, the similarity is calculated using the following formula:

$$\begin{aligned} \Pi\alpha_i^{KS}(k,j) &= 1 - \frac{\alpha_i^{KS}(k,j)}{\max(\alpha_i^{KS}(k,j))} && \text{for } \max(\alpha_i^{KS}(k,j)) \neq 0, \\ \Pi\alpha_i^{KS}(k,j) &= 1 && \text{for } \max(\alpha_i^{KS}(k,j)) = 0. \end{aligned} \tag{4.68}$$

The relative angular distance between two corresponding segments approximating the compared courses of a given process attribute is defined from the relationship

$$\alpha_i^{KS} = |\alpha_i^K - \alpha_i^S|. \tag{4.69}$$

Verification of the Developed Methods

The developed methods for modeling slowly changing processes have been verified on data collected from an actual object. The available archival database contained registered values of about 200 attributes describing the process run on a furnace de-copperizing slag. This object is one of many in a complex process line. The process conducted on it is cyclic and has a standard cycle time equal to 480 mins. Cycles consist of three phases (loading, reduction, release).

From the database, 26 attributes were selected describing control activities and process states (electrical parameters, electrode position, pressure in the oven), six attributes describing the input (chemical composition and quantity of the entering slag) and six attributes describing the output (chemical composition and quantity of the slag, quantity of the de-copperized slag and the total energy consumed during the cycle). For the verification, 130 complete cycles were selected from the database.

Table 4.1 describes selected results of searching for cycles similar to the reference cycle with ID 1534 for similarities identified by the above-described methods. To optimize the selected values of similarities, a sigmoidal modifying function

Table 4.1 Selected set of cycles with various similarities to the cycle 1534 (weighted importance of the parameters with the time parameter included) for averaged values of similarities for tested cycles with the help of the methods to be verified

	Similarity			
ID Cycle	Input	Slow-changing parameter	Output	**All**
4531	0,97	0,84	0,99	**0,93**
1660	0,92	0,86	0,86	**0,88**
...	...	...	...	...
1474	0,67	0,85	0,67	**0,71**
...	...	...	...	...
4615	0,57	0,73	0,45	**0,59**
1456	0,55	0,71	0,49	**0,58**

was used and weights for selected parameters were determined using the genetic algorithm. After averaging the values of the similarities identified by the described methods, a search of similar realizations was performed. Detailed descriptions and results of studies conducted for diversified methods can be found in the work of Tomasik (2006).

In order to verify the correct operation of the developed methods, the results obtained from averaged values of similarities identified by the developed methods were evaluated. The density of empirical distribution of similarity values was modified by the sigmoidal modifying function in order to achieve a significant number of cycles similar to those specified in the various areas of similarities.

Table 4.2 summarizes the results of the study for different threshold values of the similarities of particular classes of parameters and different count thresholds for these parameters. The evaluation measure of the developed elements of the CBR system was the average value of the defined coefficient EST (4.70) from each sample:

$$EST = (\frac{N_{SUCC}}{N_{TEST}})\,100\%. \tag{4.70}$$

Based on the results, one can state that the adopted criterion is met for the prepared set of examples of cycles. The obtained high values of the estimator allow one to state that the constructed elements of the inference system based on examples, i.e., the way of presenting examples and the measure of similarity of examples, enable the selection of similar cycles of the process, and that this selection is efficient enough.

Table 4.2 Evaluation results of the developed CBR system components

Input		Control		Output	Number of tests	Estimator
$\Pi_{IN} \geq$	$N_{OUT} \geq$	$\Pi_{CONTR} \geq$	$N_{CONTR} \geq$	$\Pi_{OUT} \geq$	N_{TEST}	$EST\,[\%]$
0,9	20	0,8	10	0,7	21	83
0,9	20	0,8	5	0,7	66	85
0,9	20	0,85	10	0,75	5	85
0,9	20	0,8	10	0,75	21	78
0,9	20	0,8	10	0,8	21	71

Application

The developed methods allow constructing a knowledge base as a base for slowly changing cyclical patterns of industrial processes. This database can be used to assist the managing of an industrial process. A suitable system can generate information for process operators about the state of the current phase of the process in relation to archival data represented by properly searched models in the database of patterns. This information may relate to the predicted quantitative values for output attributes for a given process control. Moreover, it is also possible to inform the operators about other possible ways of control which can bring a better final effect in relation to values of output attributes.

4.4.2.2 Models Based on Fuzzy Description of Periodic Processes

In this chapter a method of data acquisition is presented which might be used for many industrial objects realizing complex periodical process with the influence of random factors (Szulim, 2004). The method can be used to build expert systems using archival measurement data. The knowledge base will use CBR for searching for similar cases. The elaborated method of data acquisition defines a systematic approach embracing

- suitable data representation for fast comparing and searching for similar realizations of processes. It has fuzzy character;
- building a similarity measure of the cases which is process oriented and able to compare fuzzy process representations.

Process Data Representation

The elaborated method can be used for industrial objects realizing dynamical processes. The data describing the process can be divided into the following parts:

- *input data of static character describing the process at its start:* these data can be represented by process values like the amount of material and its characteristics together with the data describing the state of the object at a given moment;

- *data describing the control of the process with dynamic character:* these data are often represented as a time series consisting of values of a selected physical quantity describing process control;
- *output data of static character describing the process in the final state:* there can be quantities describing the final product or its quality description.

Cases Representation

The representation of cases in the database is strongly determined by features of the object (or the process) and the type of archival measurement data. The criteria which might be used to define the way of representation are fast searching and updating, effective usage of computer resources, an easy way of presenting the cases for the personnel, and the elimination of random distortions in the data.

The amount of data often imposes using special data preprocessing in order to achieve reduction in the amount of information. This is particulary important especially when comparing time series. Reducing the amount of information should, among others, allow fast storing and searching of cases as well as comparing cases with a different number of measurement points. One possible way of reducing information in the time series is to use approximation, which can be accomplished in many ways. It might be piecewise linear approximation (Keogh et al., 2003), which was used in the presented method of building the knowledge base. In such a case, database structures should be prepared to store data in an approximated way.

Cases can be stored in database tables. Because of complex structures describing one case, it might be necessary to elaborate special tables and the relation between them. Data describing the process may have different character. They might be records about the beginning and the end of the process like the amount of production material with its quality description, etc. Data of this type, because of their static character, can be stored in the main table as records describing one case. For every case it is necessary to store the time series of control activities together with its approximated representation. Every case needs storing information about the control course as a list of records.

Searching for Similar Cases

Searching for similar cases is a key element of the case-based reasoning method. The performance of the expert system strongly depends on that of the search for similar cases.

The search for similar cases can be boosted by means of indexing. The possibility of using this mechanism depends on the way of representing the cases in the database. The index can be built using a fixed number of parameters which describe every case. Using indexing can be difficult when working with the time series. In such a case it is possible to search for cases in a sequential way according to the algorithm presented in Fig. 4.8. It is necessary to elaborate and use a special similarity measure of cases. Every historical case is compared using a similarity measure with the given new case. As a result of the work of the algorithm, a set of cases similar

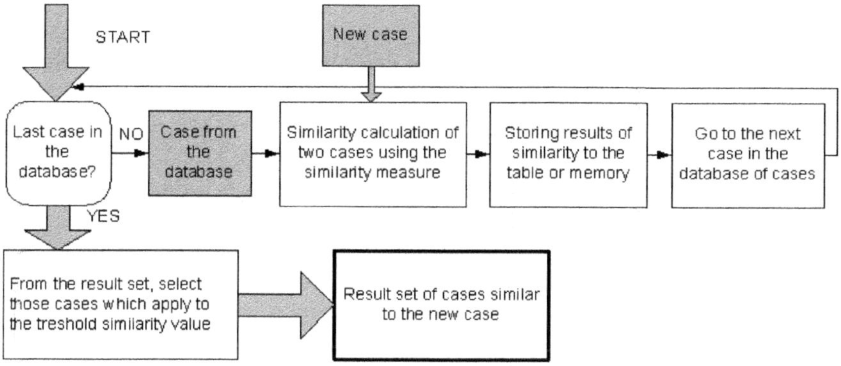

Fig. 4.8 Algorithm of searching for similar cases

enough might be returned. From the set of similar cases, the most similar one might be selected.

Searching for similar cases can be realized in a few stages (Fig. 4.8). During the first stage, searching for cases similar to the new one according to the initial state (input) will be accomplished. The program realizes this operation in a loop. Every case in the database is compared using the similarity measure and certain attributes of the initial state with the new case. The results will be stored in a special table or in the memory. Then they can be sorted according to the similarity value with the threshold value to find cases similar enough.

In the next step, from the set of similar cases according to the input data, searching for similar cases according to process control will be accomplished. The same algorithm is used but a different similarity measure is applied to compare the cases as a series of lines approximating every case. The result can also be sorted and compared with the given threshold value.

In the same way, from the set of similar cases according to the input and control states, a set of similar cases according to the output state can be derived for the given threshold value.

Similarity Measure

The knowledge base as a database of cases requires defining the similarity measure of cases. Processes realized by complex objects of dynamic character require definitions of the similarity measure for data of static character (representing the initial state) and the similarity measure for control represented in a dynamic way. Usually there is no universal similarity measure possible to be used in every situation for static and dynamic case representation. Special similarity measures were elaborated for a given method of representation of cases. These similarity measures have fuzzy character.

Similarity Measure of Static Data

Data known at the beginning of the cycle and at the end of the cycle have static character. These data can be a set of values of attributes describing the process in the space of values of states. For a given set of objects (cases) $\{\mathbf{O}^p\}, p = 1..n$ and their symbolic labels $C\{\mathbf{O}^p\}$, numerical features $X_j^p = X_j(\mathbf{O}^p), j = 1..N$ characterizing those objects are defined. Two realizations are described by vectors of feature values r_1 and r_2.

The similarity of two cases is calculated using a fuzzy similarity measure. The similarity measure is process oriented and strongly depends on properties of a set of process realizations.

Hence, the similarity measure to be defined shall satisfy the following common-sense criteria:

- for realizations that significantly differ between each other their similarity ought to be small;
- two realizations that seem to be similar enough for a human observer shall have a high similarity value.

The similarity measure is defined in the following general form:

$$\Pi(r_1, r_2) = \Pi(r_1, r_2, \underline{p}), \tag{4.71}$$

where $\underline{p} = [p_1, p_2, \ldots, p_m]$ is a vector of parameters that determine specific properties of the similarity measure under consideration. Hence, these properties may have been optimized using, e.g., an evolutionary algorithm.

The similarity of the two realizations r_1, r_2 is calculated as

$$\Pi(r_1, r_2) = \pi_{A_1} \pi_{A_2} \ldots \pi_{A_N}, \tag{4.72}$$

where N is the number of features, and π_{A_j}, $j = 1, 2, \ldots$ is a partial similarity of the same feature for two compared realizations.

For every feature, respective fuzzy sets must be defined. The membership function can be of trapezoid shape. To calculate partial similarity for a given feature A_j, vectors of membership functions for fuzzy sets defined for a given variable are used. If for the attribute A_j there is defined a fuzzy set family $\{\tilde{A}_{j1}, \ldots, \tilde{A}_{j,n_j}\}$, where n_j is defined as number of fuzzy sets for the attribute A_j, then values of the membership function form a vector $[r_{j1}, \ldots, r_{j,n_j}]$.

Partial similarities π_{A_x} for every attribute feature value are calculated according to the formula

$$\pi_{A_j} = \frac{d(r_{1j}, r_{2j})^2}{d(r_{1j}, r_{1j}) d(r_{2j}, r_{2j})}, \tag{4.73}$$

where d is weighted distance between vectors of values of functions of membership to fuzzy sets of features of a pair of realizations:

$$d(x, y) = \mathbf{x}\mathbf{W}\mathbf{y}^T, \quad \text{where } \mathbf{y}^T \text{ is a transposed matrix.} \tag{4.74}$$

W is a given matrix of similarities between fuzzy sets. Values of similarities of two fuzzy sets are in the range 0..1 (where 0 means values of low similarity and 1 stands for fully similar fuzzy sets). Values of similarities on the main diagonal equal to 1 show the similarity of the same fuzzy sets. The **W** matrix is of symmetric character. Its size depends on the number of fuzzy sets used. For every feature there is a definition of its own matrix of similarities between fuzzy sets. An example matrix for an attribute with three fuzzy sets defined is

$$\mathbf{W}_{\mathbf{Cu}} = \begin{pmatrix} 1.0\,0.8\,0.3 \\ 0.8\,1.0\,0.8 \\ 0.3\,0.8\,1.0 \end{pmatrix}. \tag{4.75}$$

The similarity measure returns a value from the range 0..1 as a result. An example measure satisfies

$$0 \leq \Pi(r_1, r_2) \leq 1. \tag{4.76}$$

This relation can be easily proven since

$$0 \leq \Pi_{A_j}(r_1, r_2) \leq 1, \text{ because } \quad d(r_1, r_2)^2 \leq d(r_1, r_1) d(r_2, r_2). \tag{4.77}$$

The similarity measure is not insensitive to differences in ranges of feature values. Values can be either integer or float. Normalizing the values before the calculation of similarity is unnecessary. It is possible to define any number of fuzzy sets for every feature describing a case. Thanks to using fuzzy sets and their similarity matrices, it is possible to tune the similarity measure in order to achieve desired properties of similarities.

Dynamic Data Similarity Measure

The similarity measure of dynamic data is used to estimate the similarity of two control courses. The control courses can have different number of measurements, there might be distortions, values might be off-set along both the axes. The similarity measure should deal with such cases and calculate similarity for them, too.

Reduction of information. The time series can contain many values, they can be contaminated with noise, there may be sudden changes of values. For these data, special partitions of the time interval into subintervals are identified taking into consideration remarkable trends of values of the attribute (e.g., the attribute value is stable, raises or drops), Fig. 4.9. The method consists in approximating the course by means of a series of line segments (*piecewise linear approximation*) according to the momentary changes of the control course of the given process. There are many approximation methods elaborated (Keogh et al., 2003; Batyrshin and Wagenknecht, 2002). For the needs of the similarity measure, the bottom-up algorithm was implemented (Batyrshin and Wagenknecht, 2002).

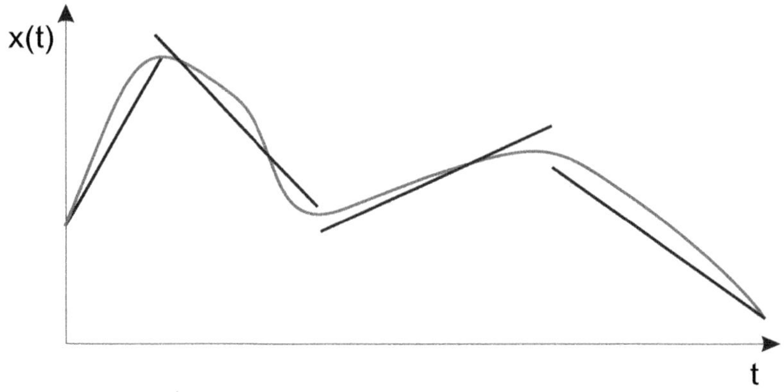

Fig. 4.9 Piecewise linear approximation example

It consists in creating a linear segment approximating consecutive measuring values until the preset threshold MS value of the error of approximation (parameter of the method) is exceeded. If the next value causes the approximation error to exceed the threshold, it needs not be added to the segment under construction. On the contrary, this value has to start the next segment.

Fuzzy description. Every segment approximating the course is represented by a dynamic fuzzy statement called an event which is represented by the following attributes:

$$< Type, Value, Duration > \tag{4.78}$$

that correspond to the event type, the beginning value of the event, and the duration of the event, respectively.

Values of every attribute are represented in a fuzzy way using vectors of values of membership to fuzzy sets.

Fuzzy sets definition. Similarly to the static similarity measure, for every attribute used to describe events approximated by the straight segments there have to be previously defined fuzzy sets and similarity matrices must own matrices.

Attribute *Type*. The fuzzy sets definition for the attribute *Type* must take into account the character of the event. There can occur both positive and negative values corresponding to the rise and fall of the attribute values along the given segment, respectively. Values can be calculated from measurement data based on the formula

$$Type = \frac{l_k - l_p}{d}, \tag{4.79}$$

where l_k is a value at the end of the segment representing the control event, l_p is a value at the beginning of the segment representing the control event, and d is the duration of the event.

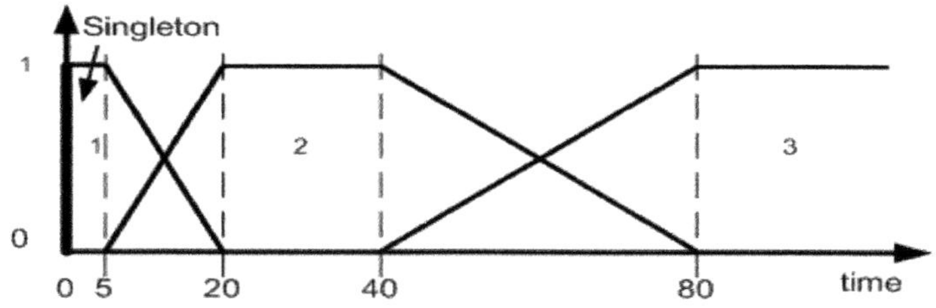

Fig. 4.10 Example fuzzy sets for the *Duration* attribute

Attribute *Duration*. In Fig. 4.10, example fuzzy sets for the attribute *Duration* are presented. In this case, an additional singleton was introduced with duration equal to 0 which is used in a procedure to calculate similarity between process realizations. It is necessary to calculate the similarity of NIL events (described later).

Attribute *Value*. Fuzzy sets for this attribute are defined based on available measurement data. Values of the membership functions are defined at the beginning of every event.

Similarity of two realizations

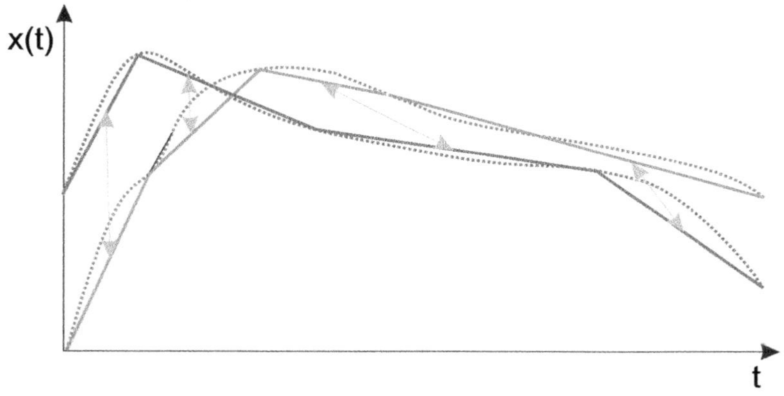

Fig. 4.11 Courses similarity calculation

The total similarity of two process realizations s_1, s_2 represented by a list of events corresponding to segments obtained by means of piecewise approximation is calculated as the weighted average of similarities of complying pairs of events (Fig. 4.11):

$$\Pi(s_1,s_2) = \frac{\sum_{i=1}^{n} \pi(e_{1i},e_{2i})w_i}{\sum_{i=1}^{n}(w_i)}. \tag{4.80}$$

The similarity of a pair of events $\pi(e_{1i},e_{2i})$ is determined as the product of partial similarities of individual attributes that describe the given pair of events (segments) approximating the respective courses:

$$\pi(e_{1i},e_{2i}) = \pi_t \pi_v \pi_d, \tag{4.81}$$

where the similarities π_t, π_v, π_d correspond to the similarity of pairs of attributes: *Type*, *Value*, and *Duration*, respectively.

Partial similarities π_α, where $\alpha\ \varepsilon\ \{\pi_t, \pi_v, \pi_d\}$ for the given attribute, are calculated using the formula

$$\pi_\alpha = \frac{d_\alpha(e_{1i},e_{2i})^2}{d_\alpha(e_{1i},e_{1i})d_\alpha(e_{2i},e_{2i})}, \tag{4.82}$$

where $d_\alpha()$ is the distance between vectors of values of membership functions:

$$d_\alpha(x,y) = \mathbf{x}\mathbf{W}_\alpha \mathbf{y}^T. \tag{4.83}$$

The $\mathbf{W}_\alpha$ matrix defines similarities between individual fuzzy sets. The matrix can be similar to that used in the static data similarity measure.

It is worth stressing that events identified from the data are assigned segments of differing lengths. It is expected that the segments of longer duration should have greater impact on the similarity of two realizations. This manifests itself in the value of the weight w_i in (4.80) that is a function of the attribute *Duration* defined as

$$w_i = MAX(ed_{1i}, ed_{2i}), \tag{4.84}$$

where values *ed* are calculated as

$$ed = \mathbf{d}(\mathbf{wd})^T, \tag{4.85}$$

and **d** is the vector of values of membership functions of the attribute *Duration*, and **wd** is a vector of weights for this attribute that may be defined as follows:

$$\mathbf{wd} = \left[0.01\ 2{,}5\ 30\ 100\right]. \tag{4.86}$$

Similarity for Different Numbers of Events

The equation (4.80), defining the total similarity of realizations, requires that both realizations to be compared be represented by the same number of events. However, in many cases it may be necessary to compare two realizations of differing numbers of events. In this case, the realization represented by the smaller number of segments is supplemented with a special dummy event type NIL (Not In List). NIL has the value of *Duration* equal to 0, while the remaining attributes take their values from the neighbors. For two realizations represented by $n_1, n_2, n_1 > n_2$ events, there are

$$N = \binom{n_1}{n_2}$$

possible insertions of NIL events into the shorter realization.

Let us consider, e.g., two realizations, where the first one includes three events and the other one two. All the possible insertions of the dummy event NIL are shown in Fig. 4.12.

The algorithm for calculating similarity between two realizations with differing numbers of events analyzes all the possible insertions of dummy events into the sequence of events of the shorter realizations. The similarity value is calculated for each combination of insertions, and the combination which brings the greatest similarity is selected. However, due to very small weight of duration of the dummy event (4.86), the insertion of the event adds a very small amount to the total similarity, thus only slightly distorting the actual similarity.

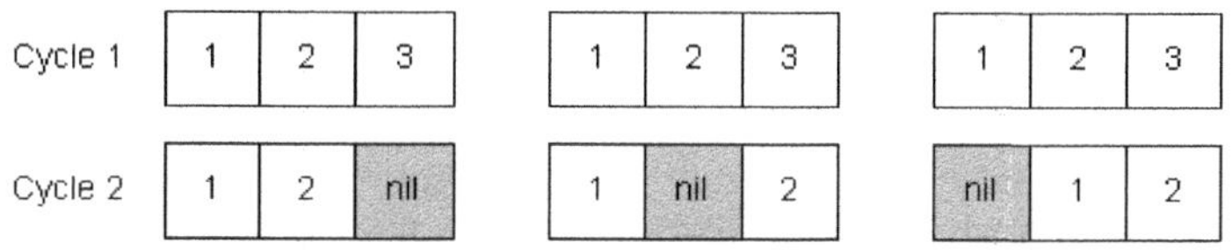

Fig. 4.12 Different possibilities to supplement the shorter realization with the dummy event NIL in order to achieve the same number of events

Verification of the Method

To verify the system built so far, a very intuitive attempt was adopted that consisted in the verification of the following hypothesis:

If among the set of realizations R represented by means of the fuzzy description one selects a subset $R_1 \subset R$ of realizations that have initial states sufficiently similar to each other, and in this subset R_1 again selects a subset $R_2 \subseteq R_1$ of a sufficiently similar course of control (dynamic part of the data), then the final states of realizations will also be sufficiently similar to each other.

The verification was conducted using data coming from a real industrial object, i.e., an electrical furnace working in a copper factory (Szulim, 2004).

Features of the Similarity Measure for Comparing Dynamic Control Data

The presented similarity measure is able to calculate the similarity of two courses in the range [0..1] (similarly to the static data similarity measure). Thanks to using course approximation and the NIL insert mechanism, it is possible to compare courses with a different number of events and the reduction of the influence of shift in the time axis. Owing to fuzzy comparison of values describing particular events, it is possible to minimize chaotic and random changes of course values. The algorithm realizing approximation and similarity calculation is not complicated. Therefore, it is possible to achieve good performance when searching for similar cases, which is a significant advantage of the method. It is possible to adjust features of the similarity measure by tuning parameters of the method like fuzzy sets borders and similarity matrices.

4.5 Conclusion

In this chapter, theoretical fundamentals of the developed methodology of knowledge discovery in databases was presented. The discussed methods concern the selection of attributes, discovering qualitative dependencies (which can be applied for selecting attributes and projecting records contained in source databases), and discovering quantitative dependencies. Descriptions of these methods helped the authors to prepare and implement algorithms, and then allowed implementations of these algorithms in the form of respective modules of the *DiaSter* system.

The modules of *DiaSter*, together with the procedures of the basic layer of the system, allow carrying out the complete process of knowledge discovery, starting from data selection, through data preprocessing, transformation, and exploration. By their implementation, it is possible to extensively automate numerical computations, which is typical for the methodology of automated discovery of knowledge (Żytkow and Zembowicz, 1993). The assumed open structure of the *DiaSter* system facilitates further development of the modules of knowledge acquisition from databases, which is the goal of our further work.

Chapter 5
Diagnostic Methods

Wojciech Cholewa, Józef Korbicz, Jan Maciej Kościelny,
Krzysztof Patan, Tomasz Rogala, Michał Syfert, and Marcin Witczak

5.1 Introduction

The chapter focuses on selected methods of diagnostics which have been implemented in the *DiaSter* system. Taking into account the specificity of industrial processes diagnostics, the robust fault detection problem is presented in the first part of the chapter. Considering the robust neural model, the passive approach based on model error modeling is described. On the other hand, the active robust approach using the dynamic neural model is based on the bounded error approach. Knowing the model structure and possessing knowledge about parameters uncertainty, the adaptive threshold can be defined and the robust fault detection scheme is designed. Next, based on the diagnostic signal generated by detection algorithms, the fault isolation problem with the use of fuzzy logic is presented. Here reasoning algorithms for single and multiple faults described by the binary diagnostic matrix and rules are considered. Then, taking into account the complex structure of industrial plants, reasoning algorithms are extended on the hierarchical structure of the diagnostic system. Such a diagnostic system reflects the structure of the process and/or control system. The rest of the chapter outlines the so-called belief-network-based diagnostic model. This approach is an alternative to the well-known model-based one and is

Wojciech Cholewa · Tomasz Rogala
Department of Fundamentals of Machinery Design, Silesian University of Technology,
ul. Konarskiego 18A, 44–100 Gliwice, Poland
e-mail: {wojciech.cholewa,tomasz.rogala}@polsl.pl

Józef Korbicz · Krzysztof Patan · Marcin Witczak
Institute of Control and Computation Engineering, University of Zielona Góra,
ul. Podgórna 50, 65–246 Zielona Góra, Poland
e-mail: {j.korbicz,k.patan,m.witczak}@issi.uz.zgora.pl

Jan Maciej Kościelny · Michał Syfert
Institute of Automatic Control and Robotics, Warsaw University of Technology,
ul. Św. Andrzeja Boboli 8, 02–525 Warsaw, Poland
e-mail: {jmk,msyfert}@mchtr.pw.edu.pl

called symptom-based diagnostics. Such a model allows simultaneous representation of diagnostics knowledge acquired from passive as well as active experiments, diagnostic relations, and generic knowledge described by physics laws.

5.2 Specificity of the Diagnostics of Industrial Processes

Diagnosing (recognition of states) is treated as a process of fault detection and isolation in a system due to the collection, conversion, analysis and evaluation of diagnostic signals. One can list three different stages of diagnosing (Isermann and Ballé, 1997; Kościelny, 2001; Korbicz et al., 2004):

- *fault detection:* fault existence in the system is noticed and the moment of detection is defined;
- *fault isolation:* the type of the fault, its location and time of appearance are defined. It is carried out after fault detection;
- *fault identification:* the fault size and the character of its change in time are defined. It is carried out after fault isolation.

At the detection stage, values of diagnostic signals are calculated based on the control and measured signals with the use of models of the system. The diagnostic signal carries information on the state of the system. The fault symptom is the existence of such a value of the diagnostic signal that testifies that a fault appeared in the controlled part of the system.

At the fault isolation stage, diagnoses are formulated based on current values of the diagnostic signals as well as the knowledge of relations existing between the faults and the symptoms. The diagnoses show existing faults. Based on the diagnoses, the system can also aid the operators, presenting them with instructions on how to deal with abnormal or emergency states.

Fault identification consists in more precise definition of the fault, e.g., its size and location of a leak from the installation. In practice, the identification stage occurs only rarely in the case of the diagnostics of abrupt faults, and sometimes it is united with the fault isolation stage. However, fault identification has fundamental meaning in the case of incipient (slowly developing) faults. The fact that degrading processes exist is usually well known, but the aim of diagnosing is the monitoring of the degradation degree.

In the *DiaSter* system, the detection and isolation of abrupt faults, as well as the detection and keeping track of the development (identification) of incipient faults, are carried out.

The task of diagnostic systems for industrial processes is the detection and isolation of faults, and aiding the operators in abnormal and emergency states. The implementation of the above tasks is very difficult due to the complexity of diagnosed installations, which have hundreds or even thousands of devices, operating most often in difficult and changing conditions, which increase the diversity of possible faults.

Diagnosis must be carried out in real time with the use of exclusively working data from normal process operation since it is impossible to disturb the process with test stimulations. The process structure often changes, which is related to switch-ons and -offs of the technological apparatus, switch-offs of measuring devices to be serviced, etc. This structure variation is a very difficult factor in the design of the diagnostic system. Moreover, many installations have unique character. Measurement uncertainties are also typical for industrial installations.

Another difficulty is the lack of measurement data for emergency states. In the databases of automatic control systems (DCSs, SCADA systems), large sets of measurement data are available, but archived changes concern normal system operation only together with merely few registered abnormal and emergency states. However, it is required that the first-time faults be isolated or, at least, properly detected. This fact limits the possibility of the application of many known diagnostic methods.

In the diagnostics of complex technological installations, the key problem is the selection of proper methods for fault detection and isolation that would be suitable for industrial applications. The diagnostic methods used in the *DiaSter* system are described in this chapter.

5.3 Fault Detection Methods

Fault detection is the process consisting in the generation of diagnostic signals S based on the generation of known input U and output Y variables of the system in order to detect faults. The task of detection algorithms is therefore symptom extraction. The diagnostic signals should have information on faults. During the detection, the space of process variables $X = \{Y, U\}$ is mapped into the space of diagnostic signals S. Symptom detection can be signaled in the form of an alarm, and it should initiate the fault isolation procedure. In process diagnostics, fault detection is carried out automatically by the diagnosing computer.

There are many fault detection methods known. The most comprehensive description can be found in the books by Basseville and Benveniste (1986), Chen and Patton (1999), Gertler (1998), Isermann (2006), Kościelny (2001), Korbicz et al. (2004), Patton and Frank (2000), Witczak (2007), the survey papers of Calado et al., (2001), Frank, (1987; 1990), Korbicz, (2006) as well as the conference proceedings of SAFEPROCESS and the Workshop on *Principles of Diagnosis*, and others. Methods based on system models have fundamental significance in process diagnostics. They are characterized by high sensitivity, which enables early detection of small size faults.

The most complete models of selected parts of the process can be obtained directly from physical equations, e.g., balance ones. Such a model reflects system properties in the whole range of operation. However, the calculation of analytical models is very difficult or even impossible for many systems. This fact limits the application of this method to systems that are described by relatively simple equations.

Due to the difficulties and limitations of analytical models application, the greatest significance in industrial process diagnostics is assigned to neural and fuzzy models as well as their combinations. Such models are tuned based on measurement data registered during process operation. They map well system operation in the range of signal variations they were trained on the basis of. In the case of fuzzy models, their application is limited to systems having a low number of inputs. This is related to rapid growth of the number of rules together with the growth of the number of inputs and the number of fuzzy sets for particular inputs of the model. However, in industrial processes diagnostics, partial models are applied that have a low number of inputs so the limitation is not so vital.

Heuristic detection algorithms are based on the control of simple relations between process variables. They use hardware redundancy of measurements, the control of feedback signals, the control of relations between variable values, the control of the consistency of signal changes directions, etc. (Kościelny, 2001; Korbicz et al., 2004). Knowledge about such relations is familiar to technologists, automatic control engineers and process operators. Therefore, such methods are especially well suited to be applied at the first stage of diagnostic system construction. They can be quickly and cheaply implemented, while the application of methods that make use of system models requires us to have more time for system identification as well as higher financial expenditure.

A vital problem that must be taken into account during diagnostic system design is that knowledge about diagnosed processes is usually not uniform. For some parts of the process, analytical models can be known, but for some others, it is possible to obtain neural or fuzzy models based on measurement data. On the other hand, for the parts of the system that do not have many measurement data available, only simple heuristic relations or limit control may be used for fault detection. Due to that, diagnostic systems should have the ability of the application and integration of various fault detection methods.

In the *DiaSter* system, some modules exist that make it possible to apply detection algorithms based on different types of models, as well as the procedures of heuristic tests that can be configured with the use of expert knowledge of the process. The following kinds of models can be applied: those that describe physical phenomena existing in the process (e.g., balance models), MLP type neural models, TSK type fuzzy models, linear and multinomial models. Currently, new modeling methods are being developed (see Chapter 3) that will be applied to residual generation. Generated diagnostic signals that make detection algorithms outputs are uniformly interpreted by the fault isolation module.

The *PExSim* module in the *DiaSter* system is applied to the creation of models based on physical equations. The user has several blocks at his/her disposal that implement basic mathematical and logic operations, input and output operations, signal flow control, integration and differentiation operators, filtration, and many others. The conversion structure that implements the system model (Fig. 5.1) is composed of these blocks. Due to the open architecture, easy expansion and creation of new specialized blocks having the form of plug-ins are possible.

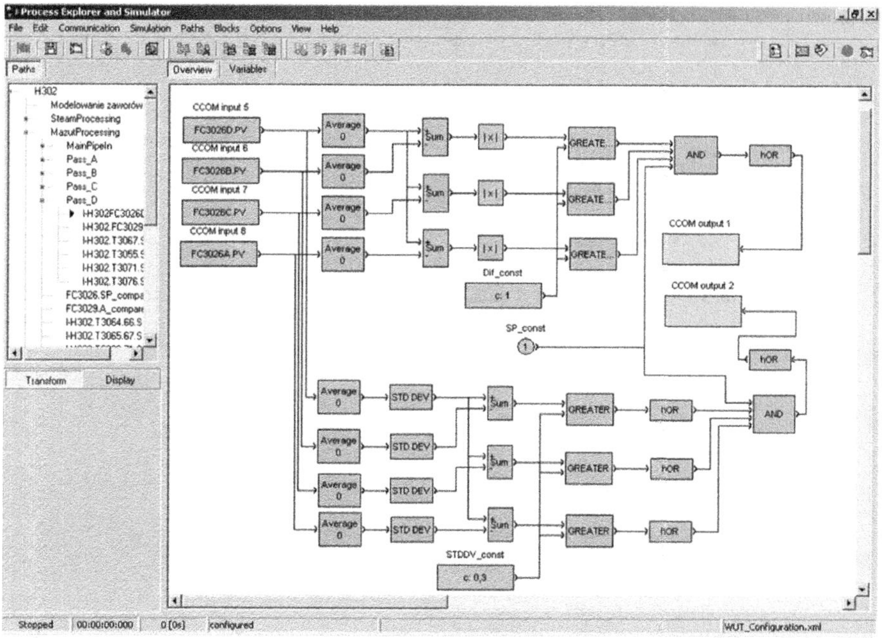

Fig. 5.1 Example realization of the model-based detection algorithm

In the *DiaSter* system, the *MITforRD* module is applied to model identification with the use of measurement data collected and archived in automatic control systems such as DSCs or SCADA systems. The module implements off-line three basic tasks: initial data processing, identification and verification of models.

The *MITforRD* module calculates on-line process variables being the outputs of the models based on current measurement values. Thus, information redundancy is implemented. Reconstructed values are used to generate residuals (Fig. 5.2). In the case of a fault, values of residuals sensitive to this fault stray from zero. The evaluation of the residual value allows the detection of fault symptoms.

Other kinds of tests are implemented in the *PExSim* module with the use of available functional blocks. These are tests that use the evaluation of statistic parameters and heuristic tests based on the control of simple quantity or quality relations between process variables. These test algorithms were implemented as typical functional blocks. Binary diagnostic signals (0 denotes the lack of a symptom, 1 denotes symptom existence) are outputs of these algorithms.

Each detection algorithm, apart from the residual generation procedure, contains the part in which the decision is made on fault symptom detection. The simplest decision algorithm is threshold evaluation of residual values or process variables parameters. In order to increase fault detection robustness to the effect of pulse electromagnetic disturbances, the decision should be made based not on the momentary

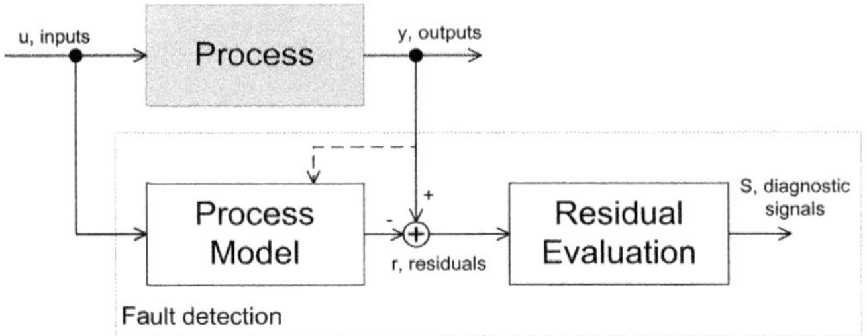

Fig. 5.2 Diagram of fault detection with the use of the process model

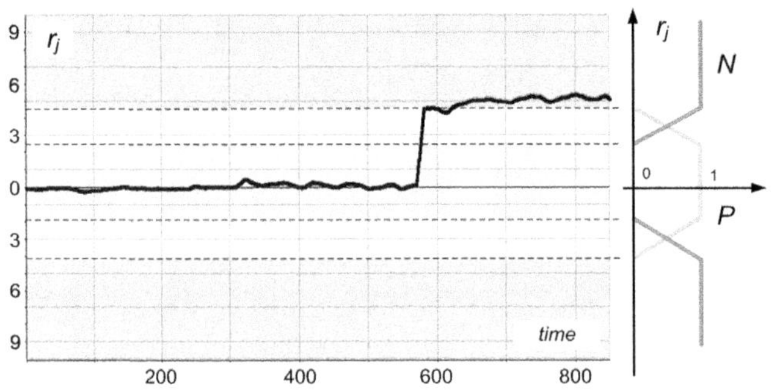

Fig. 5.3 Fuzzy, triple-valued residual evaluation

residual value but on its average value in the moving window including N last values of the residual.

In the *DiaSter* system, fuzzy evaluation of residual values is applied. To each one of the residuals r_j, one may attribute the linguistic variable that describes diagnostic signal values. The space of the linguistic variable V_j is the set of all linguistic values applied to the evaluation of this variable. Fuzzy sets spread on the residual axis correspond to particular values of the linguistic variable. In the case of two-value evaluation of the residual absolute value, the set V_j contains two values $V_j = \{0, 1\}$. "0" denotes the lack of a symptom and "1" denotes the existence of a symptom. With three-value evaluation, the residual sign is also taken into account, which may increase fault distinguishability (Kościelny, 2001; Kościelny et al., 1999; 2006). Figure 5.3 presents fuzzy three-value evaluation of residuals $V_j = \{0, -1, +1\}$. It is assumed that it fulfills the condition $\mu_{j(0)} + \mu_{j(-1)} + \mu_{j(+1)} = 1$.

Fuzzy diagnostic signals are results of fuzzy logic application to residual evaluation. The fuzzy signal value is defined by the coefficients of the membership of the calculated residual value to particular fuzzy sets:

$$s_j = \{\langle \mu_{ji}, \nu_{ji} \rangle : \nu_{ji} \in V_j\}, \tag{5.1}$$

where μ_{ji} is the coefficient of the membership of the j-th residual to the fuzzy set ν_{ji}.

Fuzzy logic application allows us to take into account uncertainties related to modeling errors, disturbances, measurement noise and difficulties with threshold values definition. In the simplest case, fuzzy two-value evaluation of the residual absolute value is applied.

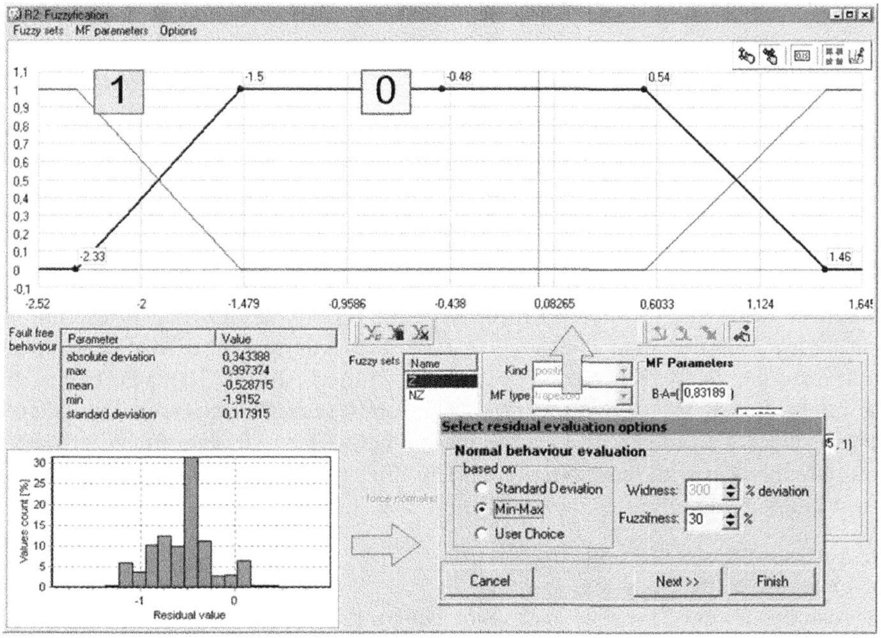

Fig. 5.4 Graphical user interface for setting the parameters of residual fuzzy evaluation in the *DiaSter* system

The system gives us the possibility of automatic definition of parameters of fuzzy sets applied to decision making in relation to the existence of a fault. The parameters are calculated based on the analysis of statistic parameters of the residual value change in the normal state of process operation, which is illustrated in Fig. 5.4. There also exists the possibility to define these parameter values by the system of engineer.

An example of fault detection with the use of the neural model of a control valve is presented in Fig. 5.5.

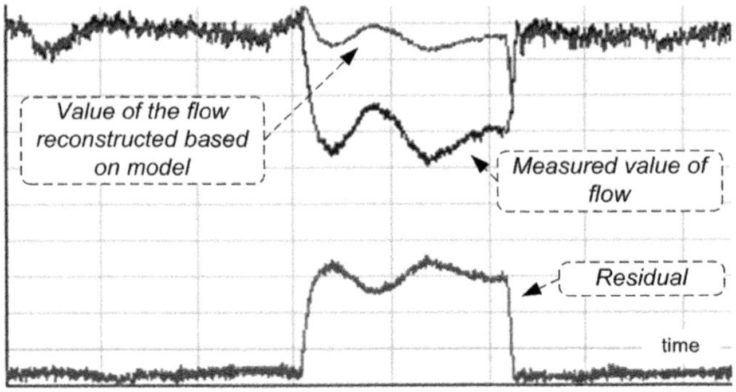

Fig. 5.5 Example of fault detection for the actuator unit consisting of a servomotor and a control valve: changes of signals of the measured and the modeled flow as well as the residual

5.4 Robust Fault Diagnosis

In recent years, great emphasis has been put on providing uncertainty descriptions for models used for control purposes or fault diagnosis design (Blanke et al., 2003; Korbicz et al., 2004; Korbicz, 2006; Mrugalski et al., 2008). These problems can be referred to as robust identification. The robust identification procedure should deliver not only a model of a given process, but also a reliable estimate of uncertainty associated with the model. There are three main factors which contribute to uncertainty in models fitted to data (Quinn et al., 2005):

- noise corrupting the data,
- changing plant dynamics,
- selecting a model form which cannot capture the true process dynamics.

Two main philosophies exist in the literature (Chen and Patton, 1999; Puig et al., 2006; Patan et al., 2008):

- *bounded error approaches* or *set-membership identification*: this group of approaches relies on the assumption that the identification error is unknown but bounded. In this case, identification provides hard error bounds, which guarantee upper bounds on model uncertainty (Gunnarsson, 1993). In this framework, robustness is hardly integrated with the identification process;
- *statistical error bounds*: in these approaches, statistical methods called soft error bounds are used to quantify model uncertainty. In this framework, identification is carried out without examining robustness and then one considers it as an additional step. This usually leads to least-squares estimation and prediction error methods (Reinelt et al., 2002).

In the framework of Fault Detection and Isolation (FDI), robustness plays an important role (Korbicz and Witczak, 2008). Model-based fault diagnosis is built on a number of idealized assumptions. One of them is that the model of the system is a faithful replica of the plant dynamics. Another one is that disturbances and noise acting upon the system are known. This is, of course, not possible in engineering practice. The robustness problem in fault diagnosis can be defined as the maximization of the detectability and isolability of faults and simultaneously the minimization of uncontrolled effects such as disturbances, noise, changes in inputs and/or the state, etc. (Chen and Patton, 1999). In the fault diagnosis area, robustness can be achieved in two ways (Chen and Patton, 1999; Ding, 2008):

- *active approaches*, based on generating residuals insensitive to model uncertainty and simultaneously sensitive to faults;
- *passive approaches*, enhancing the robustness of the fault diagnosis system to the decision making block.

Active approaches to fault diagnosis are frequently realized using, e.g., unknown input observers, robust parity equations or H_∞. However, in the case of models with uncertainty located in the parameters, perfect decoupling of residuals from uncertainties is limited by the number of available measurements (Gertler, 1998). An alternative solution is to use passive approaches, which propagate uncertainty into residuals. Robustness is then achieved through the use of adaptive thresholds. The passive approach has an advantage over the active one because it can achieve the robustness of the diagnosis procedure in spite of uncertain parameters of the model and without any approximation based on simplifications of the underlying parameter representation. The shortcoming of passive approaches is that faults producing a residual deviation smaller than model uncertainty can be missed.

5.4.1 Robust Neural Model: The Passive Approach

The robust identification procedure should deliver not only a model of a given process, but also a reliable estimate of uncertainty associated with the model. Two main ideas exist to deal with such uncertainty. The first group of approaches, the so-called set membership identification (Milanese, 2004; Duzinkiewicz, 2006) or bounded error approaches (Walter and Pronzato, 1997), relies on the assumption that the identification error is unknown but bounded. In this framework, robustness is hardly integrated with the identification process.

In this section, a somewhat different approach is described. The idea behind Model Error Modeling (MEM) is to identify the process without examining robustness first, and then consider it as an additional step. This usually leads to least squares estimation and prediction error methods. MEM employs prediction error methods to identify a model from input-output data (Reinelt et al., 2002). After that, one can estimate the uncertainty of the model by analyzing residuals evaluated from the inputs.

Uncertainty is a measure of unmodeled dynamics, noise and disturbances. The identification of residuals provides the so-called *model error model*. In the original algorithm, a nominal model, along with uncertainty, is constructed in the frequency domain, adding frequency by frequency the model error to the nominal model (Reinelt et al., 2002). Below, an algorithm to form uncertainty bands in the time domain is proposed, intended for use in the fault diagnosis framework (Patan and Korbicz, 2007; Patan, 2008b). The designing procedure is described by Algorithm 1. The model error modeling scheme can be carried out by using neural networks of the dynamic type. Both the fundamental model of the process and the error model can be modeled utilizing neural networks of the dynamic type. Assuming that the fundamental model of the process has already been constructed (according to Algorithm 1, Step 3), the next step is to design the error model. The training process of the error model is illustrated in Fig. 5.6. In this case, a neural network is used to model an "error" system with the input u and the output r. After training, the response of this model is used to form uncertainty bands as presented in Fig. 5.7, where the center of the uncertainty region is obtained as a sum of the output of the system model and the output of the error model. Then, the upper band can be calculated as

$$T_u = y_m + y_e + t_\beta v, \tag{5.2}$$

and the lower band in the following way:

Algorithm 1 Robust model design procedure

Step 1 Construct the fundamental model of a system using measurements $\{u_i, y_i\}_{i=1}^N$, where u represents input, y represents output, and N is the number of samples;

Step 2 Using a model of the process, compute the residual $r = y - y_m$, where y and y_m are desired and model outputs, respectively;

Step 3 Collect the data $\{u_i, r_i\}_{i=1}^N$ and identify an error model using these data. This model constitutes an estimate of the error due to undermodeling, and it is called the error model;

Step 4 Derive the center of the uncertainty region as $y_m + y_e$;

Step 5 If the model error model is not falsified by the data, one can use statistical properties to calculate a confidence region. It forms uncertainty bands around the response of the error model.

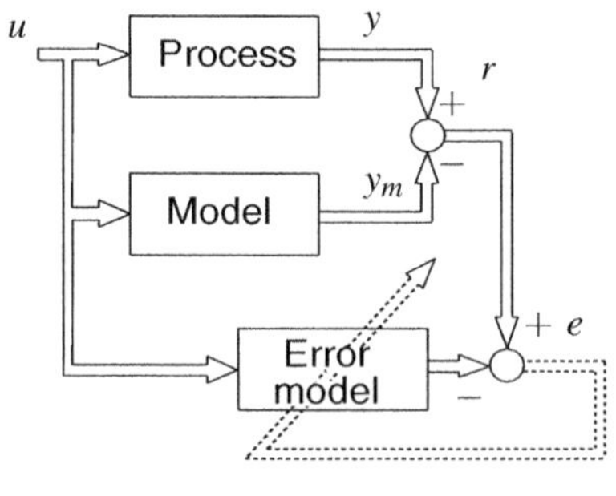

Fig. 5.6 Error model training

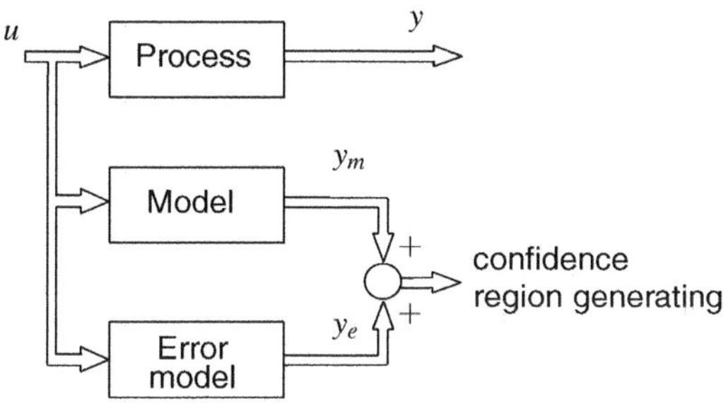

Fig. 5.7 Confidence region construction

$$T_l = y_m + y_e - t_\beta v, \tag{5.3}$$

where y_e is the output of the error model on the input u, t_β is the $\mathcal{N}(0,1)$ tabulated value assigned to the confidence level, e.g., $\beta = 0.05$ or $\beta = 0.01$, v is the standard deviation of y_e. It should be kept in mind that y_e represents not only the residual but also structured uncertainty, disturbances, etc. Therefore, the uncertainty bands (5.2) and (5.3) should work well only assuming that the signal y_e has normal distribution. The center of the uncertainty region is the signal $y_m + y_e \approx y$. Now, observing the system output y, one may make a decision whether a fault occurred or not. If y is inside the uncertainty region, the system is healthy. The idea of model error

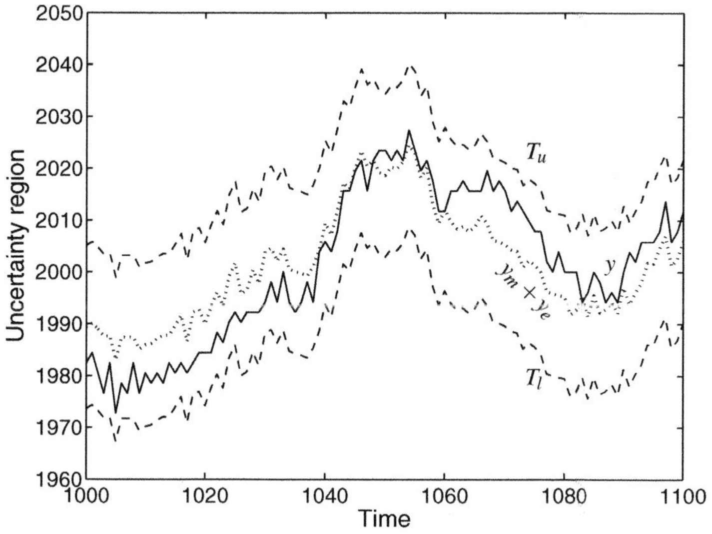

Fig. 5.8 Idea of model error modeling: system output (solid), center of the uncertainty region (dotted), confidence bands (dashed)

modeling in the time domain is presented in Fig. 5.8. The output of the system y is marked with the solid line. In turn, the sum of the outputs of the model y_m and the error model y_e is marked with the dotted line. This signal constitutes the center of the uncertainty region. Using a certain significance level, confidence bands (T_u and T_l marked with the dashed lines) are generated around the center. Thus, the uncertainty region has been determined. As long as the process output lies within the uncertainty region, a fault is not signaled.

The key question is to find a proper structure of the error model. As was discussed by Reinelt et al. (2002), one can start with an a priori chosen flexible structure, e.g., the 10-th order FIR filter. If this error model is not falsified by the data, it has to be kept. Otherwise, model complexity should be increased until it is unfalsified by the data. To construct an error model, a dynamic neural network can be used as well. In Section 5.4.4, a robust model design example using dynamic neural networks is discussed.

5.4.2 Fuzzy Adaptive Threshold: The Passive Approach

Adaptive thresholding can be successfully realized using the fuzzy logic approach. Threshold changes can be described by fuzzy rules and fuzzy variables (Sauter et al., 1993; Schneider, 1993). The threshold is adapted based on the changes of the values of u and y_p. The idea is presented in Fig. 5.9. The inputs u and the outputs y_p are expressed in the form of fuzzy sets by proper membership functions, and then the adaptation of the threshold is performed with the help of fuzzy sets. The resulting relationship for the fuzzy threshold adaptation is given by

$$T(u,y_p) = T_0 + \Delta T(u,y_p), \tag{5.4}$$

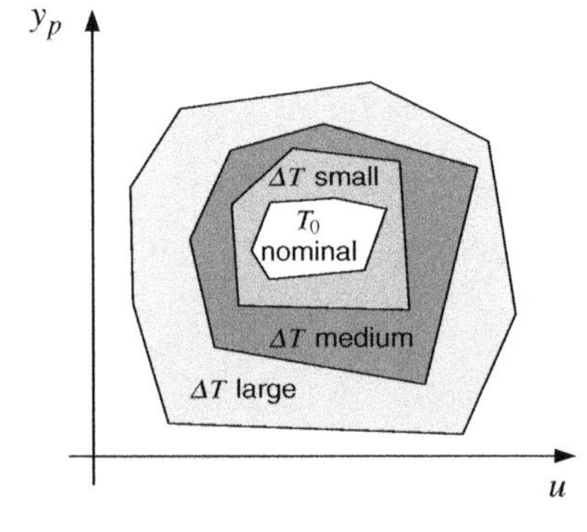

Fig. 5.9 Illustration of fuzzy threshold adaptation

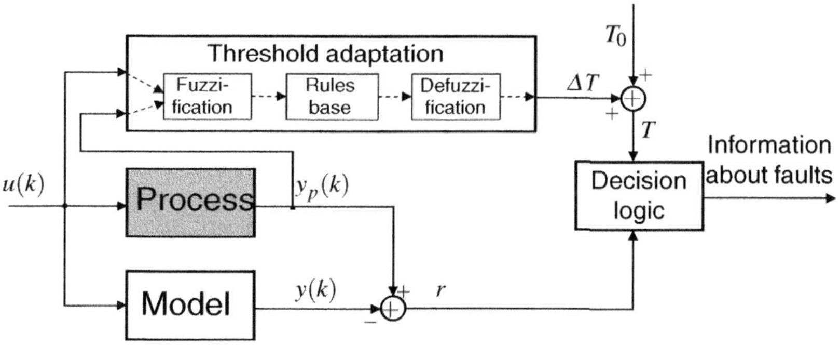

Fig. 5.10 Scheme of a fault detection system with threshold adaptation

where T_0 denotes the constant (nominal) threshold and $\Delta T(u, y_p)$ denotes the effect of modeling errors, due to deviations of the process from its operating point. The value of the nominal threshold T_0 can be set as follows:

$$T_0 = m_0 + v_0, \tag{5.5}$$

where m_0 is the mean value of the residual under nominal operating conditions and v_0 denotes the standard deviation of the residual under nominal operating conditions. Other methods useful for selecting the nominal threshold T_0 can be found in the works of Basseville and Nikiforov (1993) and Patan (2008b).

A general scheme of a fault detection system using the adaptation of the threshold, in the framework of model-based fault diagnosis, is shown in Fig. 5.10. The main idea is to use fuzzy conditioned statements operating on fuzzy sets which represent the inputs u and the outputs y_p. The residual r, calculated as a difference between the process output y_p and the model output y, is compared with the adaptive threshold T in the decision logic block. If the value of the residual r is greater than the threshold, then a fault is signaled. The adaptation of thresholds can be also interpreted as that of membership functions of residuals (Frank and Köppen-Seliger, 1997).

The idea of the fuzzy threshold is presented in Fig. 5.11 The first maximum of a residual represents a disturbance, while the second one a fault. In the classical manner, illustrated in Fig. 5.11(a), the first maximum does not exceed the threshold T, but in the case of a small disturbance a false alarm would appear. Figure 5.11(b) presents fuzzy threshold selection when the threshold is split up into an interval of a finite width, the so-called fuzzy domain, as presented in Fig. 5.11(c). Now, a small change of the value of the first or the second maximum around T causes small changes in false alarm tendency and, consequently, a small change in the decision making process. By the composition of the fuzzy sets *{healthy}* and *{faulty}*, the threshold can be fuzzified as depicted in Fig. 5.11(c). If required, a threshold can be represented by more fuzzy sets, e.g., *{small}*, *{medium}*, *{large}*. In general, the fuzzyfication of the threshold can be interpreted as that of the residual (Frank and Köppen-Seliger, 1997). The application of fuzzy adaptive thresholding is illustrated in Section 5.4.4.

5.4.3 Robust Dynamic Model: The Active Approach

The usual statistical parameter estimation framework assumes that data are corrupted by errors which can be modeled as realizations of independent random variables, with known or parameterized distribution. A more realistic approach is to assume that errors lie between priorily given bounds. This is the case, for example, for data collected with an analog-to-digital converter or for measurements performed with a sensor of a given type. Such reasoning leads directly to the Bounded-Error Approach (BEA) (Milanese et al., 1996; Walter and Pronzato, 1997).

Let us consider the following system:

$$y(k) = r^T(k)p + \varepsilon(k), \tag{5.6}$$

where $y(k) \in \mathbb{R}$ is the system output, $r(k) \in \mathbb{R}^{n_p}$ is the regressor vector, p is the parameter vector, while $\varepsilon(k)$ is the noise/disturbance vector bounded according to

$$\varepsilon^m(k) \leq \varepsilon(k) \leq \varepsilon^M(k), \tag{5.7}$$

while the bounds $\varepsilon(k)_k^N$ and $\varepsilon(k)_k^M$ ($\varepsilon(k)_k^N \neq \varepsilon(k)_k^M$) are known a priori. These bounds can be determined in either an intuitive or an empirical way. The idea underlying the bounded-error approach is to obtain a feasible parameter set (Milanese et al., 1996). This set can be defined as

$$\begin{aligned}\mathbb{P} = \{p \in \mathbb{R}^{n_p} \mid y(k) - \varepsilon^M(k) \leq r^T(k)p \\ \leq y(k) - \varepsilon^m(k)\ , k = 1, \ldots, n_{\mathscr{U}}\}\end{aligned} \tag{5.8}$$

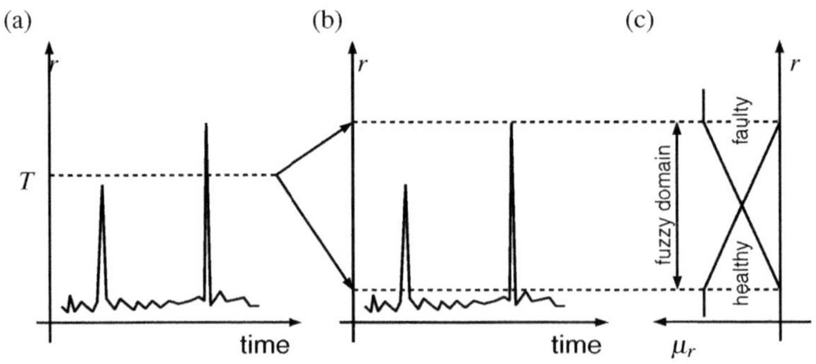

Fig. 5.11 Idea of the fuzzy threshold

and perceived as a region of the parameter space that is determined by n_t pairs of hyperplanes:

$$\mathbb{P} = \bigcap_{k}^{n_{\mathscr{Y}}} \mathbb{S}(k), \tag{5.9}$$

where each pair defines the parameter strip

$$\mathbb{S}(k) = \left\{ p \in \mathbb{R}^{n_p} \mid y(k) - \varepsilon^M(k) \leq r^T(k)p \leq y(k) - \varepsilon^m(k) \right\}. \tag{5.10}$$

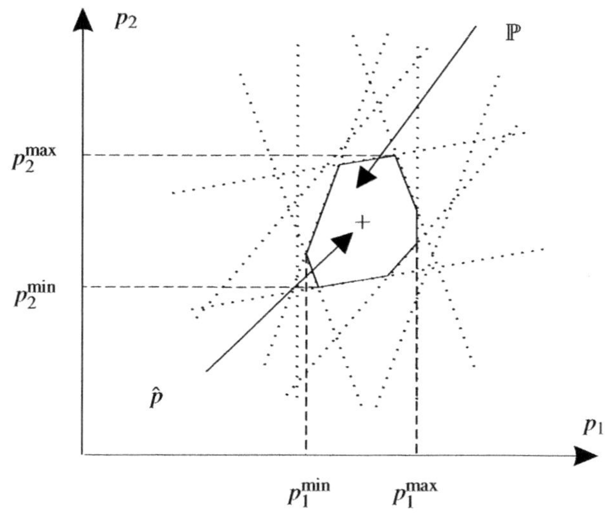

Fig. 5.12 Example feasible parameter set

Any parameter vector contained in $\mathbb{P}$ is a valid estimate of p. In practice, the center (in some geometrical sense) of $\mathbb{P}$ (cf. Fig. 5.12 for $n_p = 2$) is chosen as the parameter estimate $\hat{p}$, e.g.,

$$p_i^{\min} = \arg\min_{p \in \mathbb{P}} p_i, \tag{5.11}$$

$$p_i^{\max} = \arg\max_{p \in \mathbb{P}} p_i, \tag{5.12}$$

$$\hat{p}_i = \frac{p_i^{\min} + p_i^{\max}}{2}, \quad i = 1, \ldots, n_p. \tag{5.13}$$

The problems (5.11) and (5.12) can be solved with the well-known linear programming techniques (Milanese et al., 1996), but when n_t and/or n_p are large, the computational cost may be significant. This constitutes the main drawback of the approach. One way out of this problem is to apply a technique where constraints are executed separately one after another, although this approach does not constitute a perfect remedy for the computational problem being considered. This means that the described BEA can be employed for tasks of a relatively small dimension, as is the case for GMDH neurons (Puig et al., 2007; Korbicz and Mrugalski, 2008). In spite of the above-mentioned computational problems, the technique described by Korbicz and Mrugalski (2008) was implemented and used in *DiaSter*. The main difficulty associated with the BEA concerns a priori knowledge regarding the error bounds $\varepsilon(k)_k^N$ and $\varepsilon(k)_k^M$. However, these bounds can also be estimated (Milanese et al., 1996) by assuming that $\varepsilon(k)_k^N = \varepsilon(k)^N$, $\varepsilon(k)_k^M = \varepsilon(k)^M$, and then suitably extending the unknown parameter vector p by $\varepsilon(k)^N$ and $\varepsilon(k)^M$.

As has already been mentioned, the neurons in the l-th ($l > 1$) layer are fed with the outputs of the neurons from the $(l-1)$-th layer. Thus, the parameters of the neurons in the layers have to be obtained with an approach that solves the problem of an uncertain regressor (Milanese et al., 1996).

In order to modify the above-presented approach, let us denote an unknown "true" value of the regressor $r_n(k)$ by a difference between a known (measured) value of the regressor $r(k)$ and the error in the regressor $e(k)$:

$$r_n(k) = r(k) - e(k), \tag{5.14}$$

where it is assumed that the error $e(k)$ is bounded as follows:

$$e_i^m(k) \leq e_i(k) \leq e_i^M(k), \quad i = 1, \ldots, n_p. \tag{5.15}$$

Substituting (5.14) into (5.15) yields

$$\varepsilon^m(k) - e^T(k)p \leq y(k) - r(k)^T p \leq \varepsilon^M(k) - e^T(k)p. \tag{5.16}$$

Unfortunately, for the purpose of parameter estimation, it is not enough to introduce (5.14) into (5.15). Indeed, the bounds of (5.16) depend also on the sign of each p_i. It is possible to directly obtain these signs only for models whose parameters have physical meaning. For models such as GMDH neural networks this is rather impossible. In the work of Milanese et al. (1996, Chapters 17 and 18), the authors proposed some heuristic techniques, but these drastically complicate the problem (5.16) and do not seem to guarantee that these signs will be obtained properly.

Bearing in mind the fact that the neuron contains only a few parameters, it is possible to replace them with (Witczak et al., 2006)

$$p_i = p_i' - p_i'', \quad p_i', p_i'' \geq 0. \tag{5.17}$$

Although the above solution is very simple, it doubles the number of parameters, i.e., instead of estimating n_p parameters it is necessary to do so for $2n_p$ parameters. In spite of that, this technique is very popular and widely used in the literature (Milanese et al., 1996). Due to the above solution, (5.16) can be modified as follows:

$$\varepsilon^m(k) - \left(e^M(k)\right)^T p' + (e^m(k))^T p'' \\ \leq y(k) - r^T(k)(p' - p'') \leq \\ \varepsilon^M(k) - (e^m(k))^T p' + \left(e^M(k)\right)^T p''. \tag{5.18}$$

Thus, it is possible to show that the model output uncertainty is of the form

$$\tilde{y}^m(k) \leq r_n(k)p \leq \tilde{y}^M(k). \tag{5.19}$$

In order to adapt the above approach to parameter estimation of a non-linear neuron model with an activation function $\xi(\cdot)$, it is necessary to transform

$$\varepsilon^m(k) \leq y(k) - \xi\left((r(k))^T p\right) \leq \varepsilon^M(k) \tag{5.20}$$

with $\xi^{-1}(\cdot)$ into

$$\xi^{-1}\left(y(k) - \varepsilon^M(k)\right) \leq (r(k))^T p \leq \xi^{-1}\left(y(k) - \varepsilon^m(k)\right). \tag{5.21}$$

Knowing the model structure and possessing knowledge about its uncertainty, it is possible to design a robust fault detection scheme with an adaptive threshold:

$$\tilde{y}^m(k) + \varepsilon^m(k) \leq y(k) \leq \tilde{y}^M(k) + \varepsilon^M(k). \tag{5.22}$$

Thus, robust fault detection boils down to checking if the system output satisfies (5.22).

5.4.4 Robust Model Design Examples

5.4.4.1 Neural Model Error Modeling

Let us consider the problem of Fluid Catalytic Cracking (FCC) process modeling (Patan and Korbicz, 2007). Here we focus our attention on determining the uncertainty of the already constructed model. To derive model uncertainty, the MEM technique is applied (see 5.4.1).

The error model was designed using Neural Networks AutoRegresive with eXogenous input (NNARX) (Norgard et al., 2000; Patan, 2005). Many neural architectures have been examined by the trial and error method. The best performing two-layer network consists of four hidden neurons with hyperbolic tangent activation functions and one linear output element. The number of the input delays n_a and the output delays n_b is equal to 5 and 15, respectively. The conclusion is that, to capture, residual dynamics, a high order model is required. The output of the error model (dashed line), along with the residual (solid line), is shown in Fig. 5.13. To determine confidence bands, the 95% significance level was assumed ($\beta = 0.05$). According to (5.3) and (5.2), two adaptive thresholds were generated. The uncertainty region (dashed lines), along with the output of the healthy system (solid line),

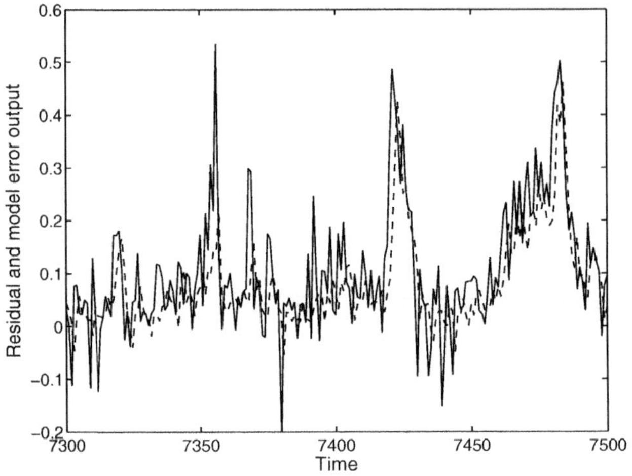

Fig. 5.13 Residual (solid) and the error model output (dashed) under nominal operating conditions

is shown in Fig. 5.14. This case illustrates the work of the system in normal operating conditions. When there are rapid changes of the output signal with a large amplitude, the uncertainty region is relatively narrow. This situation is depicted in Fig. 5.14 at the 40-th time step, when the output signal exceeds the uncertainty region. However, the false detection rate in this case is equal to 4.72%, which is relatively small. For comparison, when using simple thresholding, the false detection

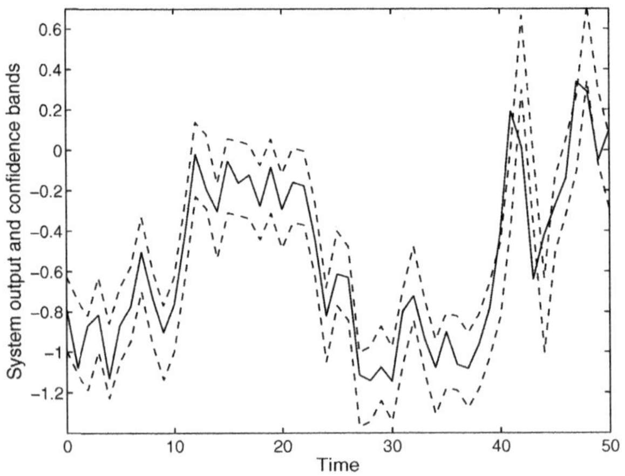

Fig. 5.14 Confidence bands (dashed) and the system output under nominal operating conditions (solid)

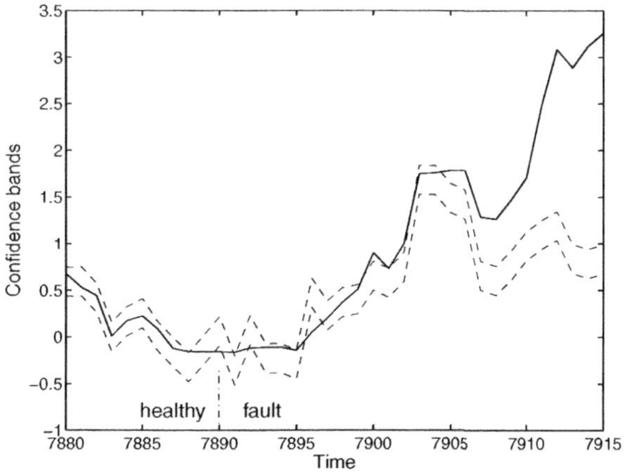

Fig. 5.15 Fault detection results

rate is equal to 7.34%. This is a result more than 1.5 times worse in relation to the adaptive technique based on MEM.

The results of fault detection are presented in Fig. 5.15. The fault start up time is equal to 7890. It is clearly shown that, after 12 time steps, the output of the system permanently exceeds the uncertainty region, which means that a fault is signaled. This experiment shows that the proposed method gives promising results. An open issue here is to find a proper error model. This problem seems to be much more difficult to solve than finding a fundamental model of the system.

5.4.4.2 Fuzzy Adaptive Threshold

Analyzing the residual signal in the fault-free case, one can see that in some time intervals there are large deviations of the residuals from zero. Unfortunately, these deviations, caused by disturbances or modeling errors, can generate false alarms during residual evaluation. In order to avoid false alarms, it is necessary to analyze how changes of inputs and outputs of the process influence deviations of the residual from zero. Such knowledge can be elaborated in the form of the adaptive threshold by means of fuzzy rules (see 5.4.2) .

Let us consider the temperature model of the evaporation station of the Lublin Sugar Factory in Poland (Patan, 2000; Patan and Parisini, 2005). Two sample fuzzy rules, which take into account the modeling mismatch of the vapor model, are given below:

R_1: **If** $\{u$ is *zero*$\}$ **and** $\{y_p$ is *zero*$\}$ **then** $\{\Delta J$ is *large*$\}$;

R_2: **If** $\{u$ is *small positive*$\}$ **and** $\{y_p$ is *zero*$\}$ **then** $\{\Delta J$ is *medium*$\}$.

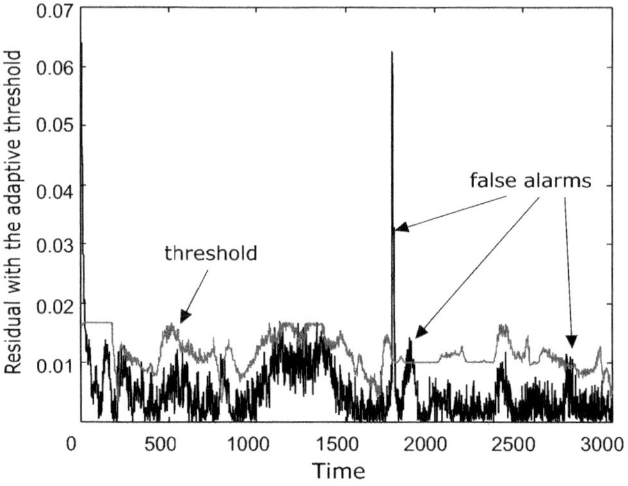

Fig. 5.16 Normal operating conditions: residual, adaptive threshold

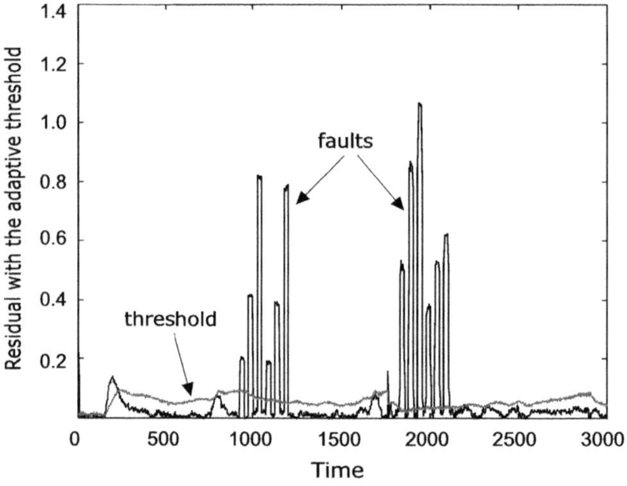

Fig. 5.17 Residual for different faulty situations

The linguistic variables *zero*, *large*, *small positive* and *medium* are defined by the relevant membership functions. To realize such a kind of threshold, the Fuzzy Logic Toolbox for *Matlab 5.3* was used. The number of linguistic variables, as well as the shape of membership functions, is chosen experimentally. The defuzzification process is carried out using the center of area method. The adaptive threshold is presented in Fig. 5.16. It adjusts to the changes of the residual. In this case, the false detection rate is equal to 4.42%. For comparison, using classical simple thresholding, the false detection rate is equal to 11.3% (Patan, 2000; 2008b). The faults were simulated by increasing or decreasing the values of particular signals by 5, 10 and 20% at specified time intervals. Figure 5.17 presents the absolute value of the residual signal for the vapor model. It is observed that the faults are detected immediately and surely. Taking into account the sensitivity of the proposed fault detection system, it can be stated that even sensor failures smaller than 5% can be detected easily. Moreover, using the adaptive threshold technique, the fault detection system can avoid a certain number of false alarms.

Taking into account these experimental results, one can conclude that the proposed robust fault detection system is very sensitive to the occurrence of faults. Using the adaptive threshold technique, it is possible to considerably reduce the number of false alarms caused by modeling errors. However, problems of the selection of fuzzy model components such as the number of linguistic variables, the shape of membership functions or the generation of rules are still open.

5.4.4.3 Design of a Robust GMDH Model

The main objective of this section is to present a comprehensive model design study regarding a valve actuator being a part of the evaporation station of the Lublin Sugar Factory. A detailed description of the valve as well as possible fault scenarios can be found in the work of Bartyś et al. (2006). The valve actuator is shown in Fig. 5.18, while the list of the measured variables is given in Tab. 5.1.

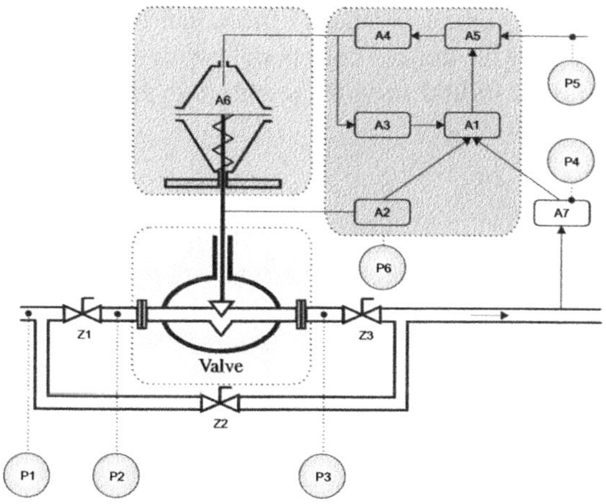

Fig. 5.18 Actuator scheme

The objective of the subsequent part of this section is to design a GMDH model of the form

$$F = r_F(X, P_1, P_2, T_1). \tag{5.23}$$

Table 5.1 List of process variables

Name	Description
F	Juice flow at the outlet of the valve
X	Servomotor rod displacement
C_V	Control value
T_1	Juice temperature at the outlet of the valve
P_1	Juice pressure at the inlet of the valve
P_2	Juice pressure at the outlet of the valve

First, the data collected at the valve actuator were suitably preprocessed. Subsequently, to avoid the saturation of the activation function, this range was further decreased to $[-0.8, 0.8]$. In order to perform data transformation, linear scaling was used. The choice of the neuron structure and the selection method for the neurons in the GMDH network are other important problems of the proposed technique. For that purpose, dynamic neurons were employed. The dynamics in this neuron are realized by the introduction of a linear dynamic system—an infinite impulse response filter. As has previously been mentioned, the quality index of a neuron for the validation data set was defined as

$$Q_{\mathscr{T}} = \frac{1}{n_{\mathscr{T}}} \sum_{k=1}^{n_{\mathscr{T}}} \left| \left(y^M(k) + \varepsilon^M(k) \right) - \left(y^m(k) + \varepsilon^m(k) \right) \right| . \tag{5.24}$$

Based on the achieved numerical values of the quality index $Q_{\mathscr{T}}$, suitable partial models were selected with the constant population method. As a result, the final structure of the GMDH neural network is shown in Fig. 5.19. Figure 5.20 shows

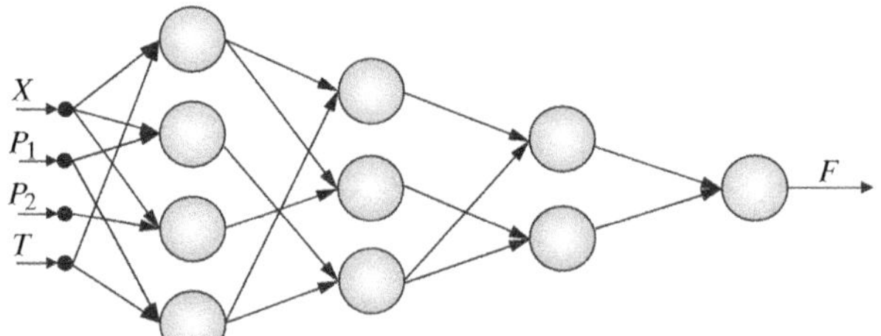

Fig. 5.19 Structure of the GMDH neural network of $F = r_F(\cdot)$

the performance of the achieved GMDH model. As can be observed, the response of the model is not a perfect replica of that of the system. Indeed, this is a result of model uncertainty. On the other hand, it can be observed that the confidence interval covers the system response and hence it can be used as an adaptive threshold for fault detection.

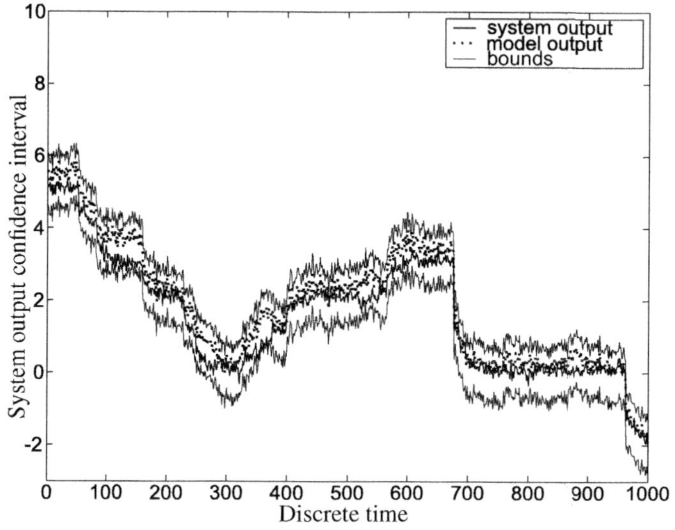

Fig. 5.20 Confidence intervals of $F = r_F(\cdot)$ for validation data

5.4.5 Implementation of Neural Models in the DiaSter System

One of components of the *DiaSter* system is a block named *MITforRD*, which is used for modeling/identification purposes. This block makes it possible to design models of a technical process without knowledge about its mathematical description but using measurably available data. The modeling is realized in an off-line mode. Each identification, modeling or parameter estimation algorithm is implemented in the form of a dynamically loaded plug-in. The plug-in mechanism of the `MITforRD` block is shown in Fig. 5.21. The remainder of this section is devoted to the implementation of the neural network modeling plug-ins.

5.4.5.1 LRGFnet Plug-In

In the framework of the plug-in, the locally recurrent network, together with the training procedure, is implemented. Network parameters are derived using the adaptive random search algorithm (Walter and Pronzato, 1997; Patan and Parisini, 2002). It belongs to the class of global optimization methods. The training algorithm guarantees the stability of the neural model because it uses special stabilizing techniques such as Gradient Projection (GP) and Minimum Distance Projection (MDP) (Patan, 2007b; 2008a). The uncertainty of the neural model is derived using the model error modeling technique by means of neural networks with tapped delay lines. Such a kind of dynamic neural network makes it possible to set the dynamics of the model independently of its approximation abilities. Using both neural networks, the fundamental and error models, a center of the uncertainty region is calculated. The uncertainty bands are calculated assuming a certain confidence level.

Fault detection is realized by checking whether the output of the system lies inside the uncertainty region. If not, the plug-in signals a fault.

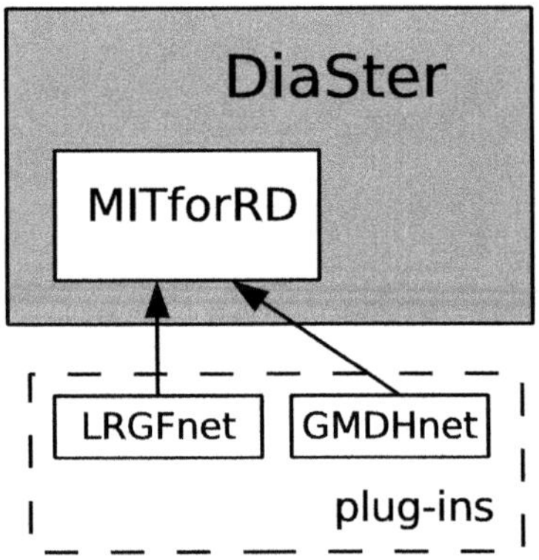

Fig. 5.21 Illustration of the plug-in mechanism in *MITforRD*

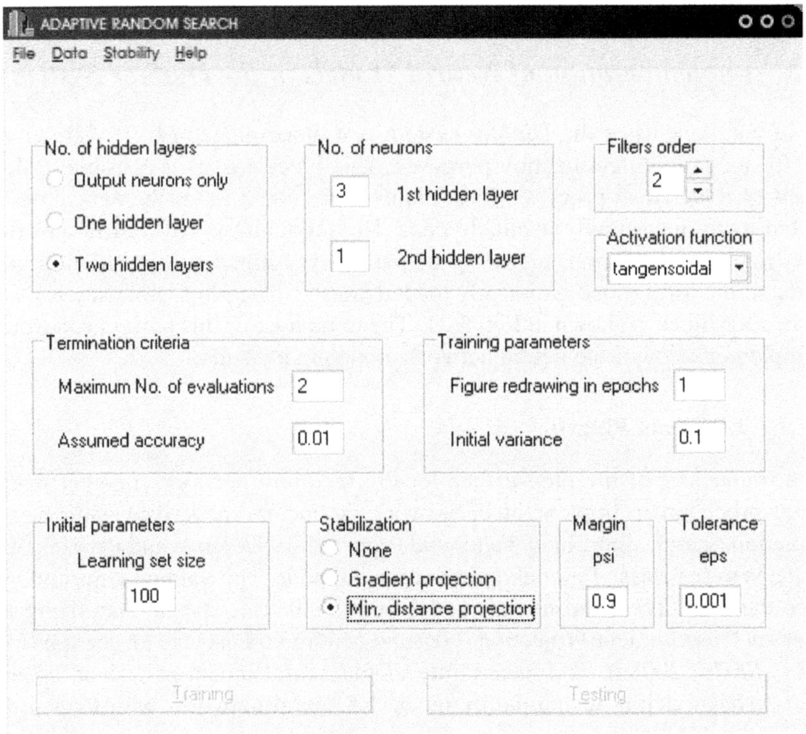

Fig. 5.22 Graphical user interface of neural model training parameters

In order to investigate how the network configuration or training parameters influence modeling results, the plug-in delivers a graphical user interface, which is shown in Fig. 5.22. This interface makes it possible to set the following parameters:

- the number of hidden layers,
- the number of neurons in layers,
- the network order,
- the activation function,
- stopping criteria of the training: the maximal number of training epochs or reaching the assumed modeling accuracy,
- initial variance for the ARS algorithm,
- the selection of the stabilization algorithm,
- parameters of the selected stabilization method.

The plug-in is implemented in the *CodeGear* ™ *C++ Builder*® environment. The C++ programming language makes it possible to implement the plug-in to be portable. This will be very important when the *DiaSter* system is realized for other operating system environments, e.g., *UNIX/Linux*.

The plug-in is fully compatible with the *MITforRD* component, the part of the *DiaSter* system responsible for designing a model in the off-line mode. The model designed in *MITforRD* can be then used in the on-line mode in the simulation component—*PExSim*.

The main part of the plug-in is the class defining the locally recurrent network, including interfaces for parameter setting, network training and network testing. The class possesses also methods for reading parameters of already stored neural networks as well as writing parameters of newly trained neural networks. Additionally, the plug-in implements all required matrix operations.

5.4.5.2 GMDHnet Plugin

The synthesis process of the GMDH neural network can tackle the challenging problem related to the selection of an optimal structure of a neural network. Apart from its unquestionable appeal, also in this case, there are some free parameters that have to be selected by the designer. The module for designing non-linear discrete-time dynamic systems requires parameter setting related to

- the neuron structure,
- the training algorithm.

The GMDH neural network is composed of uniform neurons formed in hidden layers. The neurons in hidden layers are two-input filters with an infinite impulse response (Lyons, 1997). These neurons are configured according to Fig. 5.23 (Section *Neuron configuration*), where n_a is the rank of the output delay, n_b is the rank of the input delay, $f(\cdot)$ is the activation function. The initial parameters of the neuron are randomly selected with $\mathcal{N}(\mathbf{0}, \sigma I_{2n_b+n_a})$, where the standard deviation σ can be set in the *Initial parameters* section. The network synthesis process is as follows. First,

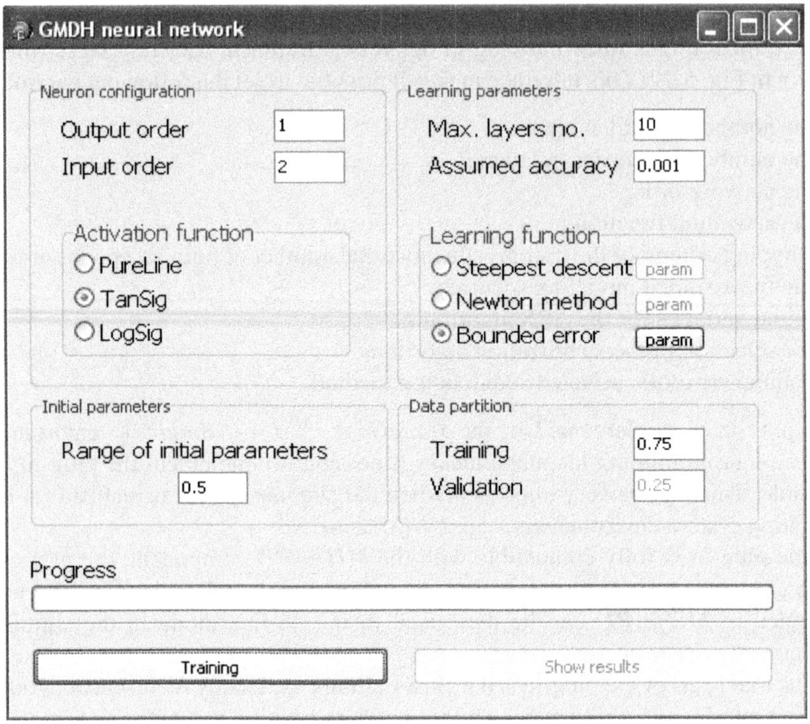

Fig. 5.23 User interface for designing non-linear dynamic models with the GMDH algorithm

all possible couples of input signals $\mathbf{u} \in \mathbb{R}^{n_u}$ are formed (u_i, u_j),$i \neq j$. Each couple is fed to the first layer of neurons. Thus, the first layer is composed of $n_u(n_u - 1)/2$ neurons. The data are divided into two sets, namely, the training and validation data sets, respectively (cf. *Data partition* section). Subsequently, the neurons of the first layer are trained with the selected learning method (cf. *Learning function* section). The next step is to test the neurons with the validation data set and the least square criterion. Next, the best performing $n_u(n_u - 1)/2$ neurons are selected, whose outputs form the inputs to the subsequent layer. The procedure of extending the network structure with new layers is continued until a maximum number of layers is achieved (cf. *Learning parameters* section). Apart from the preferred bounded-error algorithm, there are also two additional learning methods, i.e., the steepest descent and Newton algorithms. In both cases, the gradient and the Hessian are computed in an analytical way, while the line search is realized with the parabolic approximation algorithm. Finally, it should be pointed out that all design parameters have some default values, which facilitate the synthesis of a neural model.

5.5 Process Fault Isolation with the Use of Fuzzy Logic

5.5.1 Forms of Diagnostic Relation Notation

Fault isolation is based on the analysis of diagnostic signals generated by detection algorithms. The diagnosis pointing out the faults is a result of isolation. Knowledge about the relation between diagnostic signal values and faults is necessary for fault isolation. It can be obtained by three methods:

- based on the structure of mathematical models used for fault detection that are designed taking into account the influence of faults on process outputs;
- as a result of learning;
- based on expert knowledge.

The model of the linear process in the form of state equations, which considers fault influences, takes the form

$$\dot{\mathbf{x}}(t) = \mathbf{A}\mathbf{x}(t) + \mathbf{B}\mathbf{u}(t) + \mathbf{E}\mathbf{f}(t), \tag{5.25}$$

$$\mathbf{y}(t) = \mathbf{C}\mathbf{x}(t) + \mathbf{D}\mathbf{u}(t) + \mathbf{F}\mathbf{f}(t), \tag{5.26}$$

where $\mathbf{x}$ is the vector of process state variables, $\mathbf{y}$ stands for the outputs vector, $\mathbf{u}$ denotes the inputs vector, and $\mathbf{f}$ is the faults vector.

The following transmittance model complies with the above one:

$$\mathbf{y}(s) = \mathbf{G}(s)\mathbf{u}(s) + \mathbf{H}(s)\mathbf{f}(s). \tag{5.27}$$

Process modeling that takes into account fault influence is very difficult and expensive or, in many cases, even impossible. Such an approach can be used for relatively simple systems. It is not suitable for the diagnostics of complex technological installations and systems with non-repeatable properties. This is the main reason why theoretically well grounded methods of fault diagnosis with the use of a bank of Luenberger's state observers, Kalman filters or unknown input observers (Chen and Patton, 1999; Frank, 1987; 1990; Korbicz et al., 2004; Patton and Frank, 2000; Witczak, 2007) do not apply, in practice, to such kinds of processes. The directional residual method (Chen and Patton, 1999; Gertler, 1998; Korbicz et al., 2004) has also limited applicability due to the lack of knowledge of unknown matrices $\mathbf{E}$ and $\mathbf{F}$ in (5.25) and (5.26) or transmittances $\mathbf{H}(s)$ in (5.27).

The method of learning (Frank and Marcu, 2000; Isermann, 2006; Korbicz et al., 2004; Patton et al., 1999) is a very attractive way of acquiring knowledge about the relations between residual values and faults. However, measured data characterizing all process states that should be recognized, including the normal process state as well as states with faults, are necessary for learning. Obtaining such data is usually impossible. Introducing faults in real systems is unacceptable and often unrealizable. Thus, the usage of the method is limited by the knowledge of the process model that enables fault simulation. Such models can be built for relatively simple processes.

Collecting measuring data for all process states is impossible in the case of industrial processes. The number of possible faults is very large, while particular abnormal and emergency states occur very seldom. Moreover, technological installations in the chemical, power, food and other industries are most often unique or realized in short series, whereas the diagnostic system should detect and recognize dangerous failures that have never occurred. Thus, the methods that need the determination of the symptoms–faults relation during the learning phase have limited applicability in the diagnostics of industrial processes. However, they can be very useful in the diagnostics of products that are manufactured in series, like engines, pumps, valves, etc.

In the diagnostics of complex technological installations, methods of designing the faults–symptoms relation that utilize expert knowledge play the most important role. Deep knowledge about process operation allows defining this relation in a relatively simple way. Additionally, the diagnostic system designer can utilize the knowledge of process engineers, operators and maintenance stuff.

Expert knowledge about the faults–symptoms relation can be transferred and registered in different forms. In the case of binary residual evaluation, it can take the form of (Isermann, 2006; Kościelny, 2001; Korbicz et al., 2004): logic function, diagnostic trees, a binary diagnostic matrix and rules.

The binary diagnostic matrix is most often used. An example of such a matrix is presented in Fig. 5.24.

S/F	f_1	f_2	f_3	f_4	f_5	f_6
s_1	1	0	1	0	0	1
s_2	0	1	0	1	1	0
s_3	1	1	1	0	1	1
s_4	0	1	1	0	1	1
s_5	1	0	1	1	0	1

Fig. 5.24 Example of a binary diagnostic matrix

This matrix is defined on the Cartesian product of the set of faults

$$F = \{f_k : k = 1,2,\ldots,K\} \tag{5.28}$$

and the set of diagnostic signals

$$S = \{s_j : j = 1,2,\ldots,J\}. \tag{5.29}$$

The matrix element in the j-th row and the i-th column has the value $v_j(f_k) = 1$ if diagnostic signal s_j detects fault f_k, and the value $v_j(f_k) = 0$ otherwise. In other words, the occurrence of fault f_k results in the occurrence of diagnostic signal $s_j = 1$, which is called a symptom.

The diagnostic relation R^{FS} described by the binary diagnostic matrix can be defined by attributing to each diagnostic signal the subset of faults $F(s_j)$ that are detectable by this signal:

$$F(s_j) \equiv F(s_j = 1) = \{f_k \in F : v_j(f_k) = 1\}. \tag{5.30}$$

It can be also defined by attributing to each fault $f_k \in F$ the subset of diagnostic signals $S(f_k)$ that detect a particular fault:

$$S(f_k) = \{s_j \in S : \langle s_j, f_k \rangle \in R^{FS}\}. \tag{5.31}$$

Thus, $S(f_k)$ determines the set of k-th fault symptoms.

The fault signature is defined as the vector of diagnostic signal values corresponding to that fault:

$$\mathbf{V}(f_k) = \begin{bmatrix} v_1(f_k) \\ v_2(f_k) \\ \dots \\ v_J(f_k) \end{bmatrix}. \tag{5.32}$$

The faults are unisolable if their signatures are identical.

Each column of the binary diagnostic matrix determines the rule of the type (5.33), while the matrix rows correspond to the rules of the shape (5.34):

$$\textit{If } (s_1 = 0) \wedge \ldots \wedge (s_j = 1) \wedge \ldots \wedge (s_J = 1), \textit{ then } f_k, \tag{5.33}$$

$$\textit{If } s_j = 1, \textit{ then } f_a \vee \ldots \vee f_k \vee f_k. \tag{5.34}$$

The approximate information system (Kościelny, 2001; Kościelny et al., 1999; 2006; Korbicz et al., 2004) for fault isolation, called a Fault Isolation System (FIS), is a generalization of the binary diagnostic matrix. It is an adaptation of the information system proposed by Pawlak (1983). An example of the FIS is presented in Fig. 5.25.

The extensions of the FIS with respect to the binary diagnostic matrix are as follows:

- the individual set of values V_j can exist for each diagnostic signal s_j;
- the set V_j can contain more than two elements;
- one value or subset of values of the diagnostic signal can be attributed to each pair $\langle s_j, f_k \rangle$.

Multiple-valued diagnostic signals appear as a result of residual values quantization. They can also result from limit checking of the process variable when several threshold values are applied.

S/F	f_1	f_2	f_3	f_4	f_5	f_6	f_7	V_j
s_1	1	0	1	0	0	1	0	{0,1}
s_2	0	1^-	0	1^+	1^-	0	1^-	$\{0,1^+,1^-\}$
s_3	1^-	1^+	$1^+,1^-$	0	$1^-,1^+$	1^-	0	$\{0,1^+,1^-\}$
s_4	0	1	1	0	1	1	1	{0,1}
s_5	1^-	0	1^+	1^+	0	$1^+,1^-$	$1^+,1^-$	$\{0,1^+,1^-\}$

Fig. 5.25 Example of an FIS

The FIS is defined as the following quadruple:

$$FIS = \langle F, S, V_s, q \rangle, \tag{5.35}$$

where $F = \{f_k : k = 1,2,\ldots,K\}$ is the set of faults, $S = \{s_j : j = 1,2,\ldots,J\}$ is the set of diagnostic signals, and V_S is the set of all diagnostic signal values,

$$V_s = \bigcup_{s_j \in S} V_j, \tag{5.36}$$

while q denotes the function defined over the Cartesian product $F \times S$.

The function

$$q : F \times S \to \Phi(V_s) \tag{5.37}$$

attributes to each pair fault–diagnostic signal $\langle f_k, s_j \rangle$ the subset of diagnostic signal values occurring in the case of particular fault existence:

$$V_{jk} = \{v_{jki} \in V_j\}. \tag{5.38}$$

Thus, the FIS is a table that determines pattern diagnostic signal values for particular faults. The FIS simplifies to the binary diagnostic matrix if the set of all diagnostic signal values is identical and equals $V_S = \{0,1\}$.

The signature of the k-th fault corresponds to the FIS column. It is a generalization of the signature (5.32) defined as

$$\mathbf{V}(f_k) = \begin{bmatrix} V_{1k} \\ V_{2k} \\ \ldots \\ V_{Jk} \end{bmatrix}. \tag{5.39}$$

The following rule is equivalent to the above signature:

$$\begin{aligned} r_k : \mathit{If}\, (s_1 \in V_{1k}) \wedge \ldots \wedge (s_j \in V_{jk}) \wedge \ldots \wedge \\ \wedge (s_J \in V_{Jk})\ \mathit{then}\ f_k. \end{aligned} \tag{5.40}$$

The complex signature (5.40) can be represented by the particular number of simple signatures that consist of single diagnostic signal values. The simple signatures are constructed as the combination of different values for particular diagnostic signals. The following rule corresponds to a simple signature:

$$r_{kn} : \mathit{If}\,(s_1 = v_1 \in V_{1k}) \wedge \ldots \wedge (s_j = v_j \in V_{jk}) \wedge \ldots \wedge \\ \wedge (s_J = v_J \in V_{Jk})\ \mathit{then}\ f_k. \tag{5.41}$$

Analogously to the case of the binary diagnostic matrix, the rules corresponding to FIS rows can be defined. The number of rules for a single FIS row equals the number of diagnostic signal values $v \in V_j$ different than zero (where "0" denotes the value corresponding to a fault-free state):

$$r_{wi} : \mathit{If}\, s_j = v_{ji} \neq 0\ \mathit{then}\ f_a \vee \ldots \vee f_k \vee f_n. \tag{5.42}$$

The robustness of the diagnostic system against changes of the diagnosed system structure (including changes of the available measurements) plays a decisive role during the selection of the notation of the relation between faults and diagnostic signals. Logic functions, diagnostic trees, binary diagnostic matrices, information systems designed during the stage of diagnostic system configuration are neither flexible nor robust against process structure changes. In the rules of the type (5.33), (5.40), (5.41), the set of consequences also varies during changes of the set of available measured signals. Moreover, in the case of large scale systems where the number of realized tests is very high, the rules corresponding to the columns of the binary diagnostic matrix or information system are inconvenient due to a very large number of premises.

The rules in the form (5.34) and (5.42) are a robust (with respect to possible changes of the process structure) method of diagnostic relation notation—the subsets of faults are attributed to the symptoms which they cause. This dependency is invariable. Due to changes of the process structure or an earlier generated diagnosis, particular rules can be temporarily excluded from the set of active ones but their from is constant. Besides, such rules have compact shape due to the small number of possible faults pointed out in the conclusion, especially in the case of partial model utilization.

Such notation of the faults–diagnostic signal values relation is used in the *DiaSter* system. This type of rules enables simple utilization of two-valued as well as three-valued residual evaluation. The rules (5.34) and (5.42) can be rewritten in the following simplified form:

$$\mathit{If}\,(s_j = v_{ji} \neq 0)\ \mathit{then}\ f \in F(s_j = v_{ji}), \tag{5.43}$$

while $F(s_j = v_{ji}) = \{f_k \in F : s_j = v_{ji} \in V_{jk}\}$.

Automatic reconstruction of rules that correspond to the fault signatures (5.33), (5.40), (5.41) is possible based on the rules (5.42), (5.43). Such rules can be contradictory, i.e., with the same premises and different consequences. They correspond to unisolable faults.

5.5.2 Reasoning Algorithm for Single and Multiple Faults

Diagnostics deals with the recognition of states of technical systems. Changes of the state are caused by faults. State recognition consists in the indication of existing faults. Fault isolation is carried out based on diagnostic signals generated by detection algorithms. A diagnosis showing existing faults is the result of isolation. It is not always possible to obtain sure and unequivocal information about existing faults. This is caused by incomplete and uncertain knowledge on the system, limited fault distinguishability, diagnostic signals uncertainties, etc.

In the diagnostics of industrial processes, many problems and limitations exist that must be taken into account and solved if the diagnostic system is to successfully distinguish the appearing faults. The most important ones (Kościelny et al., 2006) are stated below.

Ensuring that diagnosing is correct with structure changes. During system operation, the industrial process structure may change. Some technological apparatus may be switched off for a while. Measurement devices may also be disconnected for the time needed for them to be aligned or maintained. Moreover, faults of these devices may appear. As a result, adequate adjusting of the set of useful detection algorithms is required. Changes of the set of implemented algorithms result also from the organization of the diagnosing algorithm itself. Results of detection algorithms that control faults recognized earlier become temporarily useless. They cannot be used until complete efficiency of a given part of the system is restored. Therefore, in the diagnostic system it is not possible to define constant rules of diagnostic inference. They are often modified during system operation, so it is necessary to run all of the operations of diagnostic inference in real time, taking into consideration existing changes of the sets of process variables, diagnostic signals, faults and relations between these sets (Kościelny, 1995; 2001; Kościelny et al., 2006). Diagnostic systems for industrial processes must take this problem into account and solve it efficiently.

Taking into account time delays of fault symptom appearances. The system of diagnosing is a dynamic one so some time passes between fault occurrence and the appearance of a measurable symptom of this event. The time depends, among other factors, on dynamic properties of the tested part of the system. The same fault is detected after various time delays by different diagnostic signals. If the fault isolation algorithm does not have built-in mechanisms that make the inference process robust to symptom delays, it can generate false diagnoses. Various methods of solving this problem were described in the works of Kościelny and Syfert, (2007) as well as Kościelny et al., (2007; 2008b).

Ensuring that multiple faults are correctly isolated. In the case of complex industrial installations, multiple faults may pose a serious problem. Methods of multiple faults isolation were described by Blanke and Staroswiecki (2006), de Kleer and Williams (1987), Kościelny (1995, 2001), Kościelny and Bartyś (2003), Ligęza and

Kościelny (2007) as well as Watanabe and Hou (1992). Multiple faults can exist as a sequence of successive faults or simultaneously. Simultaneous faults are most difficult to isolate. It may appear that such a situation is very rare if independent faults are considered. However, this problem exists practically with every start-up of a system that diagnoses a large technical installation. During the start-up, all earlier faults are seen by the system as simultaneous faults. The lack of a mechanism for recognizing such faults may lead to incorrect operation of the diagnostic system. The method of solving this problem in the *DiaSter* system was described by Kościelny et al. (2006).

Necessity to recognize unknown states of the process. These are situations in which the obtained signal values do not agree with fault signatures. Unknown states of the system are a vital source of information since they show the possibility of the existence of faults that were not taken into account at the design stage. The diagnosing algorithm that allows us to recognize such states is presented in the works of Kościelny (2001) as well as Kościelny and Syfert (2006).

Need for the system to be decomposed into subsystems and for diagnosing in a decentralized structure. The complex process is usually divided into technological nodes. However, in the case of nodes that contain many inter-connected elements, further division is advisable to minimize indices of the inter-connection between particular subsystems. A genetic algorithm (Wnuk et al., 2007) has been applied to solve this problem. During the design of diagnostic systems of complex installations, it is not possible to separate completely independent subsystems of diagnosing. This means that symptoms of faults appearing in one of the subsystems can be observed also in other subsystems. Algorithms of diagnosing in a decentralized structure that take this problem into account are presented by Kościelny, (1998; 2001), Korbicz et al., (2004) and Kościelny et al., (2008b).

The diagnostic inference algorithm for the diagnostics of complex technological installations should take into account the above-mentioned problems and solve them in an efficient way.

There exist many different fault isolation methods. In the works of Isermann and Ballé (1997) as well as Leonhardt and Ayoubi (1997), two basic groups were distinguished: classification methods and automatic inference ones. Due to variations of the diagnosed system structure, the latter are more suitable for the diagnostics of industrial processes. Uncertainties of diagnostic signals induce us to apply fuzzy evaluation of residuals and inference with the use of fuzzy logic. Such a solution is used in the *DiaSter* system. A general diagram of inference based on the application of partial models to fault detection and fuzzy inference on faults is presented in Fig. 5.26. What is characteristic is the lack of the sharpening block. The diagnosis shows faults and rules activation degrees that correspond with the faults. The rules are interpreted as indices of conviction that the particular faults really appeared.

The fault isolation algorithm used in the *DiaSter* system is an expansion of the dynamic tables of the state methods DTS, F-DTS, I-DTS described by

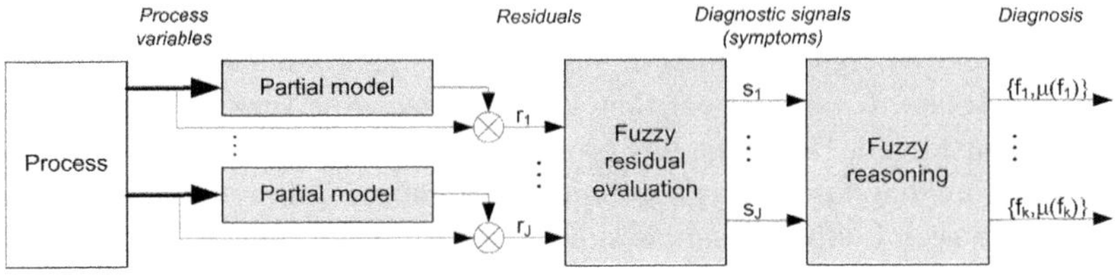

Fig. 5.26 Example FIS system

Kościelny, (1991; 1995; 2001), Korbicz et al., (2004) and Sędziak, (2002). An outline of this algorithm is presented further on. It is implemented in the *iFuzzyFDI* module.

5.5.2.1 Reasoning Assuming Single Faults

In the *DiaSter* system, one of the essential features of the fault detection algorithm is the Method of Dynamic Decomposition (MDD) of the diagnosed process (Kościelny et al., 2006). In this method, the fault is searched in a specially created isolation subprocess. The isolation subprocess is created based on the first observed symptom. It is assumed that the symptom occurred if the coefficient of the membership function of the fuzzy set $v_{xi} \neq 0$ of the diagnostic signal S_x is greater than the threshold A (during n consecutive steps of test realization). The set of possible faults is determined based on the rule (5.43):

$$F^1 = F(s_x = v_{xi}) : \mu_{xi} \geq A. \tag{5.44}$$

This set contains all the faults pointed out in rule conclusion. Then, there are selected such rules from the rule base which point out in their conclusion at least one of the fault from the set F^1. They form the subset RW^1. The subset of rules determined is such a way is used for process state recognition. It unequivocally defines the set of diagnostic signals:

$$S^1 = \{s_j \in S : F^1 \cap F(s_j = v) \neq \oslash\}, \tag{5.45}$$

which will be used for diagnosis elaboration. The diagnostic signal is eliminated from the set S^1 if it is currently unavailable. It usually decreases diagnosis precision (fault isolability) but protects against reasoning errors.

The subsystem used to isolate faults is defined by the subsets F^1, RW^1 and S^1. The isolated subsystem can be represented in the form of an information system. Rules RW^1 of the type (5.43) correspond to table rows (Fig. 5.27). The system defined in this way contains only currently available rules, i.e., corresponding to the actual process structure. Additionally, there are isolated rules of the type (5.40) for the defined subsystem. They are related to table columns. They form the set RK^1. The number of premises in those rules equals the number of diagnostic signals in the set S^1. It is several times smaller than the number of simple premises in the rules of the same form specified for the whole process. The discussed procedure is presented in Example 5.1.

Example 5.1

S/F	f_1	f_2	f_3	f_4	f_5	f_6	f_7	V_j
s_1	1	0	1	0	0	1	0	{0,1}
s_2	0	1^-	0	1^+	1^-	0	1^-	$\{0,1^+,1^-\}$
s_3	1^-	1^+	$1^+,1^-$	0	$1^-,1^+$	1^-	0	$\{0,1^+,1^-\}$
s_4	0	1	1	0	1	1	1	{0,1}
s_5	1^-	0	1^+	1^+	0	$1^+,1^-$	$1^+,1^-$	$\{0,1^+,1^-\}$

Fig. 5.27 Example FIS

The following set of rules corresponds to the FIS presented in Fig. 5.27:

$r_{w1}: \textit{if } s_1 = 1 \textit{ then } f_1 \vee f_3 \vee f_6,$
$r_{w2-}: \textit{if } s_2 = -1 \textit{ then } f_2 \vee f_5 \vee f_7,$
$r_{w2+}: \textit{if } s_2 = +1 \textit{ then } f_4,$
$r_{w3-}: \textit{if } s_3 = -1 \textit{ then } f_1 \vee f_3 \vee f_5 \vee f_6,$
$r_{w3+}: \textit{if } s_3 = +1 \textit{ then } f_2 \vee f_3 \vee f_5,$
$r_{w4}: \textit{if } s_4 = 1 \textit{ then } f_2 \vee f_3 \vee f_5 \vee f_6 \vee f_7,$
$r_{w5-}: \textit{if } s_5 = -1 \textit{ then } f_1 \vee f_6,$
$r_{w5+}: \textit{if } s_5 = +1 \textit{ then } f_3 \vee f_4 \vee f_6.$

As a result of symptom $s_2 = -1$ detection, the set of possible faults $F^1 = \{f_2, f_5, f_7\}$ is created. Then, the set of rules $RW^1 = \{r_{2-}, r_{3-}, r_{3+}, r_4\}$ that point out faults from the set F^1 is determined. Finally, the corresponding subset of diagnostic signals $S^1 = \{s_2, s_3, s_4\}$ is formulated. The application of dynamic process decomposition allowed us to decrease the number of the faults considered from 7 to 3 and the number of rules useful for isolation reasoning from 8 to 4. This illustrates the effectiveness of the method, which increases with the increase of the diagnosed process scale. The FIS shown in Fig. 5.27 was reduced to the form presented in Fig. 5.28.

The conclusions (faults) that do not belong to the set F^1 are eliminated from the rules RW^1:

$r_{w2-}: \textit{if } s_2 = -1 \textit{ then } f_2 \vee f_5 \vee f_7,$
$r_{w3-}: \textit{if } s_3 = -1 \textit{ then } f_1 \vee f_3 \vee f_5 \vee f_6,$
$r_{w3+}: \textit{if } s_3 = +1 \textit{ then } f_2 \vee f_3 \vee f_5,$
$r_{w4}: \textit{if } s_4 = 1 \textit{ then } f_2 \vee f_3 \vee f_5 \vee f_6 \vee f_7.$

Following, the set of rules RK^1 corresponding to FIS columns is created:

$r_{k2}: \textit{if } (s_2 = -1) \wedge (s_3 = +1) \wedge (s_4 = 1) \textit{ then } f_2,$

S/F	f_2	f_5	f_7
s_2	1^-	1^-	1^-
s_3	1^+	$1^-,1^+$	0
s_4	1	1	1

Fig. 5.28 Reduced FIS

$r_{k5}: if\ (s_2 = -1) \wedge [(s_3 = -1) \vee (s_3 = +1)] \wedge (s_4 = 1)\ then\ f_5,$
$r_{k7}: if\ (s_2 = -1) \wedge (s_3 = 0) \wedge (s_4 = 1)\ then\ f_7.$

Two rules with simple premises can be used instead of the rule r_{k5} with complex premises:

$r_{k5a}: if\ (s_2 = -1) \wedge (s_3 = -1) \wedge (s_4 = 1)\ then\ f_5,$
$r_{k5b}: if\ (s_2 = -1) \wedge (s_3 = +1) \wedge (s_4 = 1)\ then\ f_5.$

The rule r_{k5} corresponds to the complex signature (5.40), while the rules r_{k5a}, r_{k5b} correspond to the simple signature (5.41). The rules r_{k2} and r_{k5b} are contradictory because they have the same premises and different consequences. Faults pointed out by these rules are unisolable with respect to the examined symptoms present in rule premises.

Diagnostic reasoning is always conducted with the use of the above-described MDD. It starts the isolation process and finishes itself by determining the sets F^1, RW^1, S^1 and RK^1. Additionally, in further reasoning, the rule for the process fault-free state is used. It has the following form:

$$r_{k0}: If\ (s_1 = 0) \wedge \ldots \wedge (s_j = 0) \wedge \ldots \wedge (s_n = 0)\ then\ OK,\ s_j \in S^1. \tag{5.46}$$

Rules belonging to the set RK^1 have the shape of the conjunction of simple or complex premises. Simple premises have the form $(s = v_i)$, while complex premises are alternatives of simple ones. It is assumed that, in the case of three-valued residual evaluation, only complex premises of the following form occur: $[(s = -1) \vee (s = +1)]$. The premises of the type $[(s = 0) \vee (s = -1)]$ and $[(s = 0) \vee (s = +1)]$ are not used because they testify about the lack of knowledge of test sensitivity for a particular fault. Each rule of the type (5.40) that includes m complex premises in the form $[(s = -1) \vee (s = +1)]$ has corresponding 2^m rules of the type (5.41) with simple premises.

In the *DiaSter* system, the rules with complex premises from the set RK^1 are converted to corresponding subsets of rules with simple premises in the form

$$r_{kn}: If\ (s_1 = v_1) \wedge \ldots \wedge (s_j = v_j) \wedge \ldots \wedge (s_n = v_n)\ then\ f_k, \qquad s_J \in S^1, v_j \in V_{jk}, f_k \in F^1. \tag{5.47}$$

Together with the rule (5.46) they form the set of rules RK^*. Further, rules with the same premises and different conclusions are searched. Faults pointed out in such a subset of rules are grouped into elementary blocks:

$$E_m = \{f_k \in F^1 : \forall_k((s_1 = v_1) \wedge \ldots \wedge (s_j = v_j) \wedge \ldots \wedge \\ \wedge (s_n = v_n))\}. \tag{5.48}$$

Thus, the elementary block contains unisolable faults with respect to diagnostic signal values present in the premises of the rules being grouped. The elementary blocks correspond to the rule in the following form:

$$rz_m : \mathit{If}\,(s_1 = v_1) \wedge \ldots \wedge (s_j = v_j) \wedge \ldots \wedge (s_n = v_n) \\ \mathit{then}\,(f_k \vee f_m \vee \ldots), \qquad s \in S^1, f \in E_m. \tag{5.49}$$

The set of elementary blocks

$$E = \{E_m : m = 0, \ldots, M\} \tag{5.50}$$

contains the block E_0 corresponding to the OK state. Particular elementary blocks do not have to be disjoint in the case of three-valued residual evaluation. In the case of binary evaluation, they are disjoint. This results from the fact that in the case of three-valued evaluation the same faults can be isolable for one combination of diagnostic signal values and unisolable for another one (Kościelny, 2001; Kościelny et al., 2006; Korbicz et al., 2004).

The new set of *RM* rules corresponding to elementary blocks is created when grouping the rules with the same premises. Diagnostic reasoning consists in determining activation levels of particular rules form the set *RM*. The degrees of fulfillment of all premises in a rule must be calculated in order to determine the rule activation level.

The simple premise fulfillment factor $\mu(s_j, E_m)$ depends on the actual value of the membership degree of the fuzzy set corresponding to the pattern diagnostic signal s_j value present in the rule premise. For example, for the premise $(s_j = -1)$ and the membership factor of the "-1" fuzzy set of the j-th diagnostic signal equal to 0.8, the premise fulfillment factor $\mu(s_j, E_m)$ equals 0.8.

The activation level of the rule whose premise is a conjunction of simple premises is calculated according to

$$rz_m : \mu(f_k) = \mu(E_m, s_1) \otimes \ldots \otimes \mu(E_m, s_j) \otimes \ldots \otimes \mu(E_m, s_n), \tag{5.51}$$

where $\otimes$ denotes a generalized operator of fuzzy conjunction.

In the *DiaSter* system, T-norm operators are used as generalized operators of fuzzy conjunction. During system configuration, the type of operator is selected. There are two possibilities: product (PROD) and minimum (MIN) operators. Thus, the activation factor is calculated according to

$$\mu(E_m)_{\mathrm{PROD}} = \mu(E_m, s_1) \cdot \ldots \cdot \mu(E_m, s_j) \cdot \ldots \cdot \mu(E_m, s_n) \tag{5.52}$$

or

$$\mu(E_m)_{\mathtt{MIN}} = \mathtt{MIN}\{\mu(E_m, s_1) \cdot \ldots \cdot \mu(E_m, s_j) \cdot \ldots \cdot \mu(E_m, s_n)\}. \tag{5.53}$$

The rule (5.49) activation level takes the values from the interval $[0,1]$. It is interpreted as the certainty factor of the existence of one of the faults pointed out in the rule conclusion. The rule (5.46) activation level is interpreted as the certainty factor of the lack of faults in the subsystem considered. This information formulates the diagnosis generated at the output of fuzzy fault isolation. It has the shape of a set of pairs (certainty factor of the occurrence of one of the faults belonging to the elementary block, the elementary block):

$$DGN = \{\langle \mu(E_m), E_m \rangle >: \mu(E_m > G), \tag{5.54}$$

where G is some predefined threshold value of the activation level, e.g., $G = 0,1$.

The transformation of the set of rules RK^1 into the set RM makes the new rule base consistent. The application of the PROD operator for rule triggering level calculation has one important advantage. It allows stating the completeness and consistence of the rule base. If the sum of all rule simple premise fulfillment factors equals 1, for any inputs (premises) state, then the rule base is complete and consistent (Piegat, 2001):

$$\sum_{m=0}^{m=M} \mu(E_m)_{\mathtt{PROD}} = 1. \tag{5.55}$$

The rule base used in a particular reasoning process is usually incomplete. It does not include rules for all possible combinations of diagnostic signal values. In practice, not all the combinations are possible. However, the rule base is built under the assumption of single faults. In practice, one cannot be sure that the assumed set of faults includes all possible ones. The rules that are not considered in the rule base also comply with process states with multiple faults and states with omitted faults.

The sum of all rule simple premise fulfillment factors calculated with the use of the PROD operator is a measure of the achieved diagnosis certainty. The diagnosis is more certain as the value of this sum is closer to 1. A low value of the sum can testify to the omission of some faults in the rule base, the occurrence of multiple faults or erasing false diagnostic signal values due to strong disturbances.

The *DiaSter* system calculates an additional output μ_{US}. It is a measure of the uncertainty of the generated diagnosis and, on the other hand, a measure of conviction about the occurrence of an unknown process state. It is calculated in the following way:

$$\mu_{US} = 1 - \sum_{m=0}^{m=M} \mu(E_m)_{\mathtt{PROD}}. \tag{5.56}$$

Reasoning in a separate subsystem is realized by a standalone isolation thread usually conducted under the assumption of a single fault. The set of available diagnostic signals (as well as the set of employed rules) should be decreased by signals which are sensitive to detected faults pointed out in the elaborated diagnosis. Their values are determined by the existence of an isolated fault. They can be again included in the set of available diagnostic signals after automatic detection of the return to the fault-free state of a faulty component (Kościelny, 1991; 1995; 2001).

Several isolation threads can be conducted in parallel. Such a situation takes place when multiple faults occur within very short time periods. The condition for proper fault isolation assuming a single fault with the use of dynamic process decomposition, in such a case, is that the subset of diagnostic signals used in particular threads be disjoint (Kościelny, 2001; Kościelny et al., 2006). If this condition is not satisfied, then the algorithm of fault isolation adjusted to the isolation of multiple faults must be used.

Figure 5.29 shows the fault visualization method used in the *DiaSter* system. The indicators corresponding to particular faults are drawn over the installation mimics. They display the value of certainty factors of fault existence in the range 0–1. The color of a bar graph is connected with the value of the fault certainty factor—faults with high certainty factors are marked with red while those with a smaller factor are displayed in orange, yellow and, finally, in white for values close to zero.

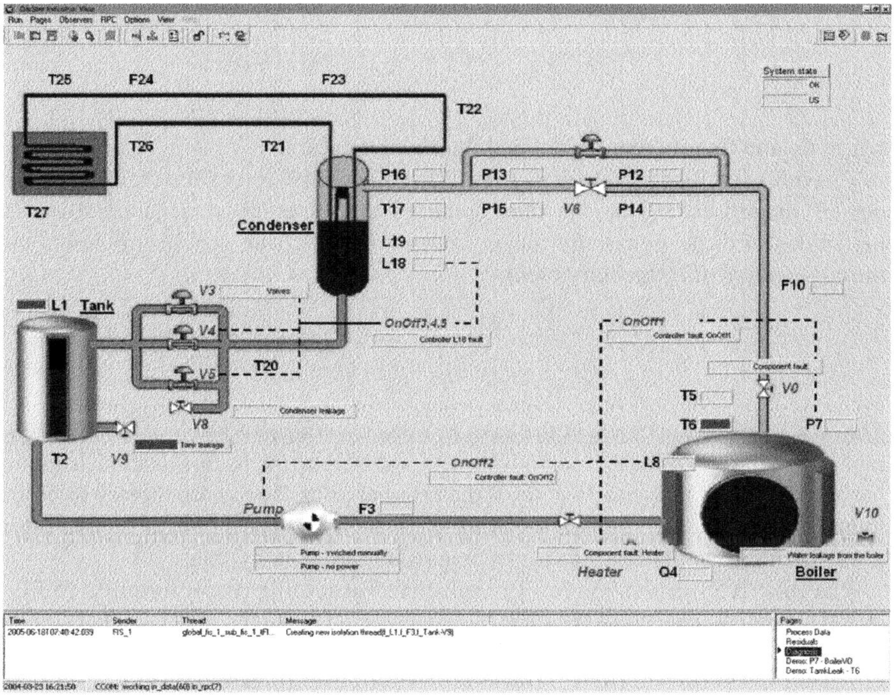

Fig. 5.29 Example of diagnosis visualization in the *DiaSter* system. The bar-graph represents the isolated fault (pipe clogging). The figure displays the comparison of the measured and the modeled value of the flow through the pipe. Additionally, the statement of certainty factors for selected faults is shown on the right.

5.5.2.2 Reasoning under the Multiple Faults Assumption

Reasoning assuming multiple faults, in practice only double ones, is started when the value of an unknown process state factor exceeds some predefined value $\mu_{US} > B$, e.g., $B = 0.5$.

In the general case, the problem of a very high number of possible process states with different combinations of faults appears during multiple fault isolation. The number of possible states equals 2^K, where K is the number of faults. In the *DiaSter* system, this problem is diminished by preliminary defining of the set of possible faults F^*.

For each rule (5.42) and (5.43) it is possible to determine the set of faults $F(s_j)$ pointed out in its conclusion. During reasoning conducted under the assumption about single faults, the values of diagnostic signals from the set S^1 are determined. It is possible to state that the set of possible faults should contain faults for which all the symptoms occurred. Faults whose symptoms are not observed are omitted. In the case of three-valued residual evaluation, the set F^* is defined according to

$$F^* = \bigcup_{j:s_j \in S^1} F(s_j) : \mu_{j-1} > C_1 \cup \\ \bigcup_{j:s_j \in S^1} F(s_j) : \mu_{j+1} > C_1 - \bigcup_{j:s_j \in S^1} F(s_j) : \mu_{j,0} > C_0, \tag{5.57}$$

where C_1 and C_0 are some predefined threshold values.

The rules for double faults are created based on (5.49) for a single fault. The values of diagnostic signals in rule premises are determined as alternatives to the values of these signals for single faults, while the rule conclusion points out the conjunction of elementary blocks:

$$\mathit{If}\,(s_1 = v_{1,k} \vee v_{1,m}) \wedge \ldots \wedge (s_j = v_{j,k} \vee v_{j,m}) \wedge \ldots \wedge \\ \wedge (s_J = v_{J,k} \vee v_{J,m})\ \mathit{then}\ (E_k \wedge E_m). \tag{5.58}$$

The way of rule generation in the case of rules pointing out a subset of unisolable faults is identical, only the conclusion has a different form. In this case it points out the conjunction of conclusions of the original rules. The contradictory rules are joined together by the aggregation of their conclusions. The aggregated conclusion is an alternative to conclusions of the original (contradictory) rules.

The rule (5.58) activation level is calculated according to the formulas (5.51)–(5.53). The diagnosis takes the form

$$DGN = \{\langle \mu(E_m \cap E_n), (E_m \cap E_n) \rangle : \mu(E_m \cap E_n) > G\}, \tag{5.59}$$

where G is some predefined threshold value.

5.5.2.3 Way of Taking into Consideration the Delays of Symptoms Forming

There are four approaches that differently take into account the problem of symptom forming delays in the *DiaSter* system during diagnostic reasoning.

Approach 1. In this approach, in order to avoid false diagnosis, full information about symptoms delays is used (Kościelny et al., 2008a). There are defined minimal θ_{kj}^{1} and maximal θ_{kj}^{2} symptom delays for each pair: a fault and a diagnostic signal. When the time passes, the algorithm takes into account consecutive diagnostic signal values. The observed symptoms, as well as a lack of symptoms, in predefined time intervals allow making the diagnosis more accurate. This approach is very seldom used in the diagnostics of industrial processes. It is almost impossible to achieve such precise knowledge about symptom delays. It practice, this approach can be applied for very simple processes for which mathematical description is known.

Approach 2. In this approach, in order to avoid false diagnoses, conclusions are formulated only after all the symptoms have been determined (Kościelny, 1995; 2001). There is a need to define, at the system configuration stage, the maximal symptom delay θ_j for each diagnostic signal. It is defined as a maximal period from the moment of the occurrence of any of the faults from the set $F(s_j)$ till its symptom appears. After creating the set S^1 (5.45) with the use of the MDD, the time to symptom determination is defined. It equals

$$\theta = \max_{j:s_j \in S^1} \{\theta_j\}. \tag{5.60}$$

Other elements of the reasoning algorithm are consistent with the previously given description.

Approach 3. It is the approach of reasoning called *"reasoning based on symptoms"* (Kościelny and Syfert, 2007). In this case, zero values (representing the fault-free state) of diagnostic signals are not taken into account. Thus, the algorithm does not need information about symptom delays. The reasoning is carried out based on the rules (5.42) and (5.43). The conversion to the form (5.49) is not used. The primary diagnosis $DGN_1 = F^1$ includes all the faults from the set (5.30), which is determined based on the first observed symptom. The occurrence of consecutive symptoms results in diagnosis refining. It points out the faults from the subset determined in the following way:

$$DGN_r = DGN_{r-1} \cap F(s_n). \tag{5.61}$$

Thus, the diagnosis is conducted according to the rules in the form

$$\begin{aligned} &If\,(s_x = -1) \wedge \ldots \wedge (s_n = +1) \ldots then \\ &f \subset F(s_x = -1) \cap \ldots \cap F(s_n = +1). \end{aligned} \tag{5.62}$$

The rule activation level, calculated with the use of fuzzy conjunction, is the same for all the faults pointed out in the diagnosis and equals

$$\forall_{f_k \in DGN_r} : \mu(f_k) = \mu(f_k, s_x) \otimes \ldots \otimes \mu(f_k, s_n). \tag{5.63}$$

The above algorithm protects against false diagnosis generation caused by the delays of symptom forming. However, the achieved fault isolability is smaller. Some diagnoses indicate a higher number of faults than the diagnosis elaborated when "0" diagnostic signal values are considered.

Approach 4. This approach is an extension of Approach 3. The idea of this extension is that additional heuristic knowledge (usually incomplete) about the sequence of symptom forming (Syfert and Kościelny, 2009) is taken under consideration. Declared relations concerning the times of symptom forming of the k-th fault have the form

$$\theta_{k,j} < \theta_{k,m}. \tag{5.64}$$

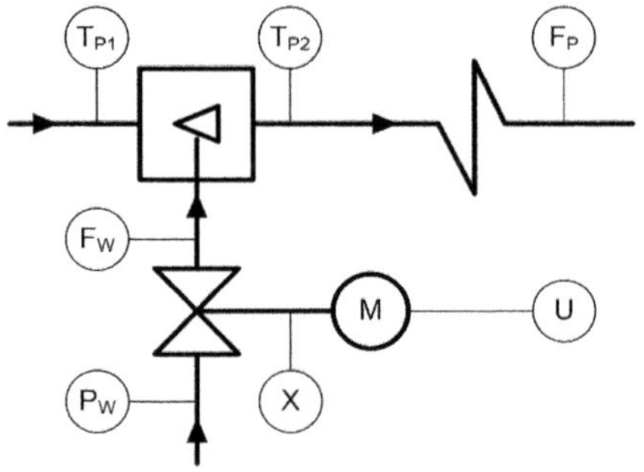

Fig. 5.30 Assembly of a steam attemperator with the cooling water valve. U: control signal, X: valve rod position, P_w: injection water pressure, F_w: water flow stream, T_{p1}, T_{p2}: water temperatures before and after the attemperator, F_p: steam flow stream.

In this case, the rules of reasoning about faults are extended by the relations (5.64) defined in the rule base. The relation between the delays $\theta_{k,j}$ and $\theta_{k,m}$ of two different symptoms forming in the case of the k-th fault can be determined based on knowledge about the process and the structure of detection algorithms. It is illustrated in the example of a part of the steam attemperator of a power block boiler (Fig. 5.30).

Assume the following residuals are used for the diagnostics of this process: $r_1 = X - \hat{X}(U)$, $r_2 = F - \hat{F}(X, P_W)$, $r_3 = F - \hat{F}(U, P_W)$, $r_4 = T_{p2} - \hat{T_{p2}}(T_{p1}, F_p, F_W)$, $r_5 = T_{p2} - \hat{T_{p2}}(T_{p1}, U, P_W)$. Residuals r_2, r_3, r_4, r_5 are sensitive to the control valve fault. This fault (denoted as k) will cause an immediate change of water flow F_w, and shortly after a change of steam temperature T_{p2}. Thus, the following symptom delay relations for that fault are true: $\theta_{k,2} < \theta_{k,4}$, $\theta_{k,2} < \theta_{k,5}$, $\theta_{k,3} < \theta_{k,4}$. Analogously, denoting actuator fault by m, it is possible to state that $\theta_{m,2} < \theta_{m,3} < \theta_{m,5}$.

The algorithm of reasoning founded on the symptom sequence based on the relation (5.64) protects against generating a false diagnosis due to existing symptoms delays. It also allows us to increase fault isolability compared with symptom-based reasoning. The more complete the knowledge about the relations between symptoms delays, the bigger the increase in fault isolability.

5.5.3 Algorithms of Reasoning in a Hierarchical Structure

The structure of industrial plants has often a hierarchical form. The technological process and the installation used for its realization are organized into divisions, sections, control loops and devices. Horizontal and vertical divisions of such structures can be distinguished. The vertical division is used to separate areas inside superior units, while the vertical one mainly refers to the consecutive technological sections. Also, the structure of control systems is often functionally decentralized and geographically distributed.

It is advantageous, in many cases, if the structure of the diagnostic system reflects that of the process and/or control system. This allows elaborating a diagnosis which is much more meaningful. Designing such a diagnostic system is more intuitive. The achieved system structure is easier to design, interpret and reconfigure. Clear reconfiguration procedures are very important in the case of the diagnostics of industrial processes, where a high number of faults and realized diagnostic tests take place.

The possibility to conduct diagnostic algorithm decomposition is also important when multiple-valued, fuzzy diagnostic test results are applied and when multiple faults occurrence is possible. Particular subsystems for separated process sections are defined during reasoning system decomposition. It is usually impossible to completely separate independent (disjoint) subsystems. Fault symptoms visible in one subsystem are often observed in others. There is a necessity to apply specialized reasoning algorithms. Such an algorithm must have the ability to accommodate the diagnosis elaborated in dependent diagnostic subsystems, including certainty degrees of fault free and unknown process states.

On the other hand, the possibility of creating hierarchical reasoning systems does not have to be connected with the hierarchy of the plant structure. The hierarchical lay out of process sections and components can be used to define diagnostic subsystems (with respect to the set of faults and diagnostic tests) designed to analyze particular process sections (Kościelny et al., 2008a; Korbicz et al., 2004), whereas the application of diagnostic reasoning in a hierarchical structure can result from the need for simplifying the reasoning process (Cholewa et al., 2009).

The basic version of diagnostic reasoning in a decentralized structure is presented in the works of Kościelny et al. (2008a) and Korbicz et al. (2004). It utilizes binary logic for diagnostic tests evaluation, the binary diagnostic matrix for the notation of the diagnostic relation and single fault scenarios. Its extension introducing reasoning in a two-level structure and basic elements of fuzzy reasoning is described by Kościelny et al. (2008). However, the extended algorithm does not directly deal with the problem of multiple faults and multiple-valued residual evaluation in hierarchical, decentralized structures of diagnostic reasoning.

Introducing fuzzy, multiple-valued residual evaluation and taking into account multiple faults scenarios make the reasoning algorithm in the hierarchical structure significantly difficult, especially the phase of diagnosis justification between separated subsystems. The algorithm presented in this section (implemented in the *DiaSter* system) is an extension of the above-mentioned algorithms. It introduces a simplified reasoning algorithm that takes into consideration fuzzy, multiple-valued residual evaluation. It also defines the scheme of diagnosis elaboration in the case of detecting the possibility of multiple faults existence, including the way of calculating the degree of certainty of fault-free and unknown process states.

5.5.3.1 Diagnostic Algorithm Decomposition

There are three levels (types) of decomposition used in the isolation module of the *DiaSter* system:

Level 0. Static decomposition into independent reasoning systems (denoted as gFIS). Completely independent subsystems are defined, with respect to the examined set of faults as well as realized set of diagnostic tests. The final diagnosis is a direct composition (sum) of diagnoses generated by particular subsystems. The degrees of fault-free $\mu(OK)$ and unknown μ_{US} process states, calculated according to (5.46) and (5.56), are determined independently for each subsystem. This stage is omitted in this section because its realization is described in Section 5.5.2.1.

Level 1. Defining dependent reasoning subsystems (denoted as sFIS). This stage is also a part of static decomposition but the defined subsystems can have common elements. The division can proceed due to different criteria: the minimization of connections between subsystems, division with respect to logical sections of the plant, etc. There is a need to proceed with accommodation of diagnoses generated by particular subsystems, including the determination of the degrees of fault-free and unknown process states for the whole system (i.e., the subsystem defined at Level 1 of the decomposition). This stage is described in this section.

Level 2. Dynamic decomposition realized according to the MDD described in Section 5.5.2.1.

The reasoning system that utilizes the notation of the faults–symptoms relation in the form of the approximate information system (FIS) expressed as (5.35) and bi- or three-valued fuzzy residual evaluation is considered. The presented mechanism can also be used for complex (more than three-valued) residual evaluation.

The fault isolation system (i.e., gFIS) is decomposed into N subsystems:

$$O = \{o_n : n = 1, 2, \ldots, N\}, \tag{5.65}$$

where each subsystem o_i, defined as the following triplet:

$$o_n = \langle F_n, S_n, R_n^{FS} \rangle, \tag{5.66}$$

has the subsets of faults F_n, diagnostics signals S_n and, finally, the diagnostic subrelation R_n^{FS} defined on the Cartesian product $F_n \times S_n$ attributed. This mapping attributes to each pair $\langle s_j, f_k \rangle$ the set of diagnostic signal values V_{jk}. The subset F_n is a subset of faults monitored in the n-th subsystem by diagnostic signals $s_j \in S_n$ plus a fault representing a fault-free state denoted as "OK":

$$F_n = \{OK\} \cup \left(\bigcup_{s_j \in S_n} F(s_j) \right), \tag{5.67}$$

where $F(s_j)$ denotes the set of faults detected by diagnostic signal $s_j : F(s_j) = \{f_k : \langle s_j, f_k \rangle \in R^{FS}\}$. The degree of unknown process state μ_{US} is calculated for the whole system.

The decomposition into subsystems o_i is realized in such a way that the following conditions are fulfilled:

- subsets of diagnostic signals in all subsystems are disjunctive:

$$\forall_{m \neq n} S_n \cap S_m \neq \varnothing; \tag{5.68}$$

- subsets of isolable faults in all subsystems are not, in general, disjunctive (independently of the OK state):

$$\exists_{m \neq n} F_n \cap F_m \neq \varnothing; \tag{5.69}$$

- each fault is attributed to at least one subsystem:

$$\bigcup_{n=1..N} F_n = F; \tag{5.70}$$

- each pair $\langle s_j, f_k \rangle$ belonging to the diagnostic relation R^{FS} is attributed to at least one subsystem:

$$\forall_j \forall_k \langle s_j, f_k \rangle \in R^{FS} \Rightarrow \exists (s_j \in S_n \wedge f_k \in F_n). \tag{5.71}$$

For the purpose of diagnosis refining, two subsets of faults for each subsystem are defined—the subset F_n^I of faults attributed exclusively to the n-th subsystem and the subset F_n^{II} of faults attributed to at least two subsystems:

$$F_n^I = \{f_k : \forall_{m \neq n} (F_n \cap F_m = \varnothing)\} \tag{5.72}$$

and

$$F_n^{II} = \{f_k : \exists_{m \neq n}(F_n \cap F_m \neq \varnothing)\}, \tag{5.73}$$

while

$$F_n = F_n^I \cup F_n^{II}. \tag{5.74}$$

The basic version of the reasoning algorithm uses diagnostic relation decomposition in a single-level structure. Such a structure is able to conduct correct reasoning. However, the reasoning algorithm has a two-layer structure itself. The lower layer realizes independent reasoning at the level of particular subsystems. The superior layer is responsible for adjusting diagnoses elaborated for particular subsystems. This stage is necessary due to subsystems dependency.

Example 5.2

The example system used in this section contains six faults and six diagnostic signals. The diagnostic matrix for this system and its decomposition onto two subsystems are shown in Fig. 5.31.

The following set of subsystems is defined:
$O = \{o_1, o_2\}$,
with the following subsets of faults:
$F_1 = \{f_1 \dots f_6, OK\}$, $F_1^I = \{f_1, f_2\}$, $F_1^{II} = \{f_3 \dots f_6\}$,
$F_2 = \{f_3, f_4, f_7 \dots f_{10}, OK\}$, $F_2^I = \{f_7 \dots f_{10}\}$, $F_2^{II} = \{f_3, f_4\}$,
and subsets of diagnostic signals:
$S_1 = \{s_1 \dots s_5\}$, $S_2 = \{s_6 \dots s_8, s_{101}, s_{102}\}$,
while bi- and three-valued evaluation is applied:
$\forall_{j:1\dots J, j \neq 7} V_j = \{0, 1\}$, $V_7 = \{1^-, 0, 1^+\}$.

S/F	f_1	f_2	f_3	f_4	f_5	f_6	f_7	f_8	f_9	f_{10}	f_{OK}
s_1		1	1	1							0
s_2	1	1	1	1							0
s_3	1										0
s_4		1			1	1					0
s_5					1	1					0
s_6								1	1	1	0
s_7							1^+		1^+	$1^-,1^+$	0
s_8							1				0
s_{101}				1				1			0
s_{102}			1							1	0

Fig. 5.31 Example system decomposed into two subsystems

5.5.3.2 Reasoning Realized at the Subsystem Level

Reasoning at the subsystem o_i level is realized completely independently for each subsystem in accordance with the diagnostic transformation R_n^{FS} with the use of the algorithm described in Section 5.5.2.1. With respect to the possibility of the existence of unisolable faults (with the same signature), the final diagnosis for the n-th subsystem takes the shape of a set of elementary blocks (set of unisolable faults in a particular subsystem):

$$DGN_n = \{\langle E_m^n, \mu_m^n \rangle : m = 1 \dots M^n, \forall_m (\mu_m^n > T_I)\}, \tag{5.75}$$

$$E_m^n = f_k, \vee \dots f_1 : f_k, \dots f_l \in F_n, \tag{5.76}$$

where T_I is the isolation algorithm threshold, E_m^n is the m-th group of unisolable faults (denoted as $[f_k, \dots f_l]$), μ_m^n denotes the degree of certainty of the m-th fault group ($\forall_{k:f_k \in E_m^n}(\mu^n(f_k) = \mu_m^n)$, M^n is the number of groups of unisolable faults in the n-th subsystem.

The fault is pointed out in the diagnosis (in the elementary block of unisolable faults) if its certainty factor exceeds the defined threshold evaluation value T_I. The value $T_I > 0$ can be used to avoid useless calculations in the case of very small values of fault certainty factors; however, it will influence the accuracy of the elaborated diagnosis.

Unisolable faults can appear in particular subsystem even if they are isolable in respect to the global diagnostic relation R^{FS}. "Local" unisolability is connected with the shape of the relation R_n^{FS}, which does not include all diagnostic signals.

The diagnosis points out one fault or elementary block of faults ($M^n = 1$) when uncertainty is not related to diagnostic test evaluation (degree of membership to fuzzy sets equals '0' or '1'). In a particular case it can be a fault-free or an unknown process state, e.g., $DGN_n = \langle OK, 1 \rangle$.

When multiple fault scenarios are considered, each fault in the elementary block can be a block of multiple faults, called a "multiple fault". Such a group is denoted as $|f_k, f_l|$ or $f_{|k,l|}$ (an example for a "double fault").

Multiple faults are interpreted as simultaneous faults, i.e., related by the logic conjunction "and" ($|f_k, f_l| \equiv f_k \wedge f_l$), while unisolable faults are treated as alternative faults, i.e., connected with the logic conjunction "or" ($[f_k, f_l] \equiv f_k \vee f_l$). Particular sets of unisolable faults appearing when the uncertainty of test result evaluation takes place are also treated as alternative faults ($\{\langle f_k, \mu(f_k) \rangle, \langle f_l, \mu(l) \rangle\} \equiv (f_k \; with \; \mu_k) \vee (f_l \; with \; \mu_l)$). In this case, it is important to pass the information about elementary blocks to the superior layer. This information is important for proper calculation of certainty factors of unknown process states realized in the superior layer.

Example 5.3

Assume the following symptoms:

$s_1 = \{\langle 1,1\rangle\}$, $s_2 = \{\langle 1,1\rangle\}$, $s_3 = \{\langle 1,1\rangle\}$,
$s_{101} = \{\langle 0,0.2\rangle, \langle 1,0.8\rangle\}$.

S/F	f_1	f_2	f_3	f_4	f_5	f_6	f_7	f_8	f_9	f_{10}	f_{OK}
s_1		1	1	1							0
s_2	1	1	1	1		O_1					0
s_3	1										0
s_4		1			1	1					0
s_5					1	1					0
s_6								1	1	1	0
s_7			O_2				1^+		1^+	$1^-,1^+$	0
s_8							1				0
s_{101}				1				1			0
s_{102}			1							1	0

Fig. 5.32 Example of partial diagnosis elaboration in subsystems. Shaded diagnostic signals are those for which symptoms are observed.

The elaborated diagnoses at the subsystem level take the following form:

$DGN_1 = \{\langle E_1^1 = [f_{|1,3|}, f_{|1,4|}, 1\rangle\}$,
$DGN_2 = \{\langle E_1^2 = [f_4], 0.8\rangle, \langle E_2^2 = [OK], 0.2\rangle\}$.

The diagnosis for the first subsystem points out elementary block of two unisolable multiple faults: ($f_1 \wedge f_3$) and ($f_1 \wedge f_4$). The diagnosis for the second subsystem points out the fault f_4 with the degree of certainty equal to 0.8 *or a fault-free state with the degree of certainty equal to* 0.2.

5.5.3.3 Diagnosis Adjustment at a Superior Level

The final diagnosis for the whole system is elaborated at the supervisory level. It is realized by the justification of the partial diagnosis generated for particular subsystems. It is possible to formulate a more accurate final diagnosis if faults belonging to more than one subsystem were pointed out at the basic level.

The resulting certainty factors for elementary blocks of unisolable faults are calculated in the first stage. Then, it is tested if faults from the sets of unisolable faults for a particular subsystem do not become isolable as a result of diagnosis justification with another subsystem. Faults that remain unisolable are still treated as a group in further steps. This is important with respect to the phase of calculating unknown process state coefficients. The calculations are proceeded with according to the following formula:

$$\mu(E_m) = \prod_{n=1...N} \begin{cases} \mu^n(E_m) : E_m \in F_n \\ \mu^n(OK) : E_m \notin F_n \end{cases} , \tag{5.77}$$

where $E_m \in F_n$ means that all the faults from the elementary block E_m belong to the set F_n:

$$E_m \in F_m \Leftrightarrow \forall_{f_k \in E_m} f_k \in F_n. \tag{5.78}$$

The formula (5.77) is written down in the general form. In practice, most of the groups E_m contain only one fault.

The fault-free state coefficient is determined based on fault-free certainty factor values calculated for all subsystems:

$$\mu(OK) = \prod_{n=1...N} \mu^n(OK). \tag{5.79}$$

In the last stage, the coefficient of an unknown process state is determined according to the following dependency:

$$\mu(US) = 1 - (\mu(OK) + \sum_m \mu(E_m)). \tag{5.80}$$

Example 5.4

Assume that for the symptoms
$s_6 = \{\langle 1,1 \rangle\}$, $s_7 = \{\langle 0,0.2 \rangle, \langle 1^+,0.8 \rangle\}$,
$s_{101} = \{\langle 0,0.2 \rangle, \langle 1,0.8 \rangle\}$,
the following partial diagnoses were elaborated at the subsystem level:
$DGN_1 = \langle OK, 1 \rangle$,
$DGN_2 = \{\langle f_8,0.2 \rangle, \langle [f_9, f_{10}],0.8 \rangle\}$,
$DGN_3 = \{\langle [f_4, f_8],0.8 \rangle, \langle OK,0.2 \rangle\}$.
Diagnosis adjustment according to (5.77) was carried out in the following way:
$\mu_4 = \mu_4^1 \cdot OK^2 \cdot \mu_4^3 = 0 \cdot 0 \cdot 0.8 = 0$,
$\mu_8 = OK^1 \cdot \mu_8^2 \cdot \mu_8^3 = 1 \cdot 0.2 \cdot 0.8 = 0.16$,
$\mu_9 = OK^1 \cdot \mu_9^2 \cdot OK^3 = 1 \cdot 0.8 \cdot 0.2 = 0.16$,
$\mu_{10} = OK^1 \cdot \mu_{10}^2 \cdot \mu_{10}^3 = 1 \cdot 0.8 \cdot 0 = 0$,
$OK = OK^1 \cdot OK^2 \cdot OK^3 = 1 \cdot 0 \cdot 0.2 = 0$.

All the elementary blocks of unisolable faults were decomposed into components because the faults pointed out were globally isolable. The final value of the unknown process state degree was calculated as

$$US = 1 - (\sum \mu_m + OK) = 0.68,$$

and the final diagnosis took the following shape:

$$DGN = \{\langle f_8,0.16 \rangle, \langle f_9,0.16 \rangle, \langle US,0.68 \rangle\}.$$

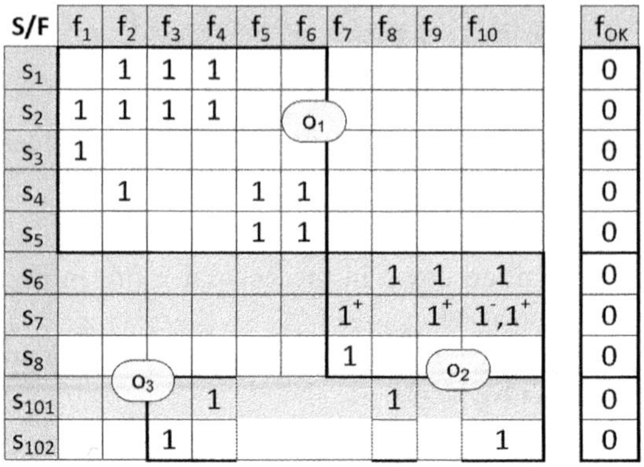

Fig. 5.33 Example of diagnosis justification for three subsystems: case with symptoms uncertainty

Due to uncertain symptoms, the reasoning system pointed out two potential faults but also indicated an unknown process state as another possibility. It is interesting that the faults f_8 and f_9 are isolable. Their simultaneous indication as alternative faults results directly from uncertain evaluation of diagnostic test results and not from the diagnostic relation.

5.5.3.4 Multiple Faults Issue

In the *DiaSter* system, the following assumptions considering multiple fault isolation are used:

- symptoms of particular faults are not compensating each other while they are simultaneously observed;
- the final diagnosis consists of the minimal set of faults that explains the observed symptoms. There can be more than one such a minimal set indicated. In such a case, they form a group of unisolable, multiple faults.

In the case of multiple fault analysis it is worth distinguishing two cases: (a) multiple faults that appeared in the framework of one subsystem, (b) multiple faults that occurred in separate subsystems. In the case (a), the necessity to analyze the possibility of multiple fault existence appears if the diagnosis elaborated at least in one of the subsystems is contradictory. In the case (b), diagnosis inconsistency is revealed at the stage of the final diagnosis elaboration at the superior level. Inconsistent diagnosis takes place when neither faults (or a group of unisolable faults) nor fault-free states are pointed out. In practice, it is assumed that a diagnosis is inconsistent if the value of the certainty coefficient of an unknown process state exceeds some threshold value $T_M(\mu(US) \geq T_M)$. A general description of the algorithm taking into account the multiple faults issue is presented in this section.

In the case (a), reasoning for a subsystem is realized according to the description presented in Section 5.5.2.2. The diagnosis formulated for a particular subsystem with multiple faults takes the most general form $|f_k,\dots f_l|$. Diagnosis justification at a superior level proceeds according to the rules presented in Section 5.5.3.3, but the group of multiple faults is treated as one, cumulative fault.

In the case (b), potential multiple faults are created as combinations of faults pointed out in the primary diagnosis. The diagnosis is formulated after these combinations are tested against observed symptoms and the defined diagnostic relation. When faults of different multiplicity are possible (i.e., their combination explains the observed symptoms), then the final diagnosis is formulated according to the following assumptions:

- the diagnosis contains multiple faults with the smallest multiplicity:

$$DGN_n = \{\langle D^n_m, \mu^n_m\rangle : m = 1\dots M^n, \forall_m(\mu^n_m > T_l)\}, \tag{5.81}$$

$$D^n_m = \{f_k, \wedge \dots f_1\} : f_k, \dots f_l \in F_n, \forall_{s_j \in S_n : s_j = 1} \exists_{f_p \in D^n_m} (f_k \in F(s_j)), \tag{5.82}$$

  where D^n_m is the m-th group of multiple faults, μ^n_m stands for the degree of certainty of the m-th fault group, M^n is the number of groups of multiple faults in the n-th subsystem;
- multiple faults with higher multiplicity are also pointed out if their certainty factor is significantly higher than that of the fault with smaller multiplicity (see Example 5.5).

Example 5.5

Assume that for the symptoms

$s_1 = \{\langle 1,1\rangle\}$, $s_2 = \{\langle 1,1\rangle\}$, $s_3 = \{\langle 1,1\rangle\}$,
$s_{101} = \{\langle 0,0.2\rangle, \langle 1,0.8\rangle\}\}$,

the following partial diagnoses were elaborated at the subsystem level:

$DGN_1 = \langle [f_{|1,3|}, f_{|1,4|}], 1\rangle$,
$DGN_2 = \langle OK, 1\rangle$,
$DGN_3 = \{\langle [f_4, f_8], 0.8\rangle, \langle OK, 0.2\rangle\}$.

The diagnosis for the first subsystem includes multiple faults. Its justification proceeds in the following way:

$f_{|1,3|} = f^1_{|1,3|} \cdot OK^2 \cdot f^3_3 = 1 \cdot 1 \cdot 0 = 0$,
$f_{|1,4|} = f^1_{|1,4|} \cdot OK^2 \cdot f^3_4 = 1 \cdot 1 \cdot 0.8 = 0.8$,
$f_8 = OK^1 \cdot f^2_8 \cdot f^3_8 = 0 \cdot 0 \cdot 0.8 = 0$,
$OK = OK^1 \cdot OK^2 \cdot OK^3 = 0 \cdot 1 \cdot 0.2 = 0$.

Finally, the unknown process state coefficient is calculated as

$US = 1 - (\sum \mu_m + OK) = 0.2$,

and the final diagnosis takes the shape

$DGN = \{\langle f_{|1,4|}, 0.8\rangle, \langle US, 0.2\rangle\}$.

S/F	f_1	f_2	f_3	f_4	f_5	f_6	f_7	f_8	f_9	f_{10}	f_{OK}
s_1		1	1	1							0
s_2	1	1	1	1							0
s_3	1										0
s_4		1			1	1					0
s_5					1	1					0
s_6								1	1	1	0
s_7							1^+		1^+	$1^-,1^+$	0
s_8							1				0
s_{101}				1				1			0
s_{102}			1							1	0

Fig. 5.34 Example of reasoning in the case of multiple faults occurrence in the framework of one subsystem

Example 5.6

Assume that for the symptoms

$s_2 = \{\langle 1,1\rangle\}$, $s_3 = \{\langle 1,0.2\rangle, \langle 1,0.8\rangle\}$,

$s_7 = \{\langle 1,1\rangle\}$, $s_8 = \{\langle 1,0.2\rangle, \langle 1,0.8\rangle\}$,

the following partial diagnoses were elaborated at the subsystem level:

$DGN_1 = \langle f_1, 0.8\rangle$,

$DGN_2 = \langle f_7, 0.8\rangle$,

$DGN_3 = \langle OK, 1\rangle$.

S/F	f_1	f_2	f_3	f_4	f_5	f_6	f_7	f_8	f_9	f_{10}	f_{OK}
s_1		1	1	1							0
s_2	1	1	1	1							0
s_3	1										0
s_4		1			1	1					0
s_5					1	1					0
s_6								1	1	1	0
s_7							1^+		1^+	$1^-,1^+$	0
s_8							1				0
s_{101}				1				1			0
s_{102}			1							1	0

Fig. 5.35 Example of reasoning in the case of multiple faults occurrence in different subsystems

Diagnosis justification proceeds in the following way:

$f_1 = f_1^1 \cdot OK^2 \cdot OK^3 = 0.8 \cdot 0 \cdot 1 = 0,$
$f_7 = OK^1 \cdot f_7^2 \cdot OK^3 = 0 \cdot 0.8 \cdot 0 = 0,$
$OK = OK^1 \cdot OK^2 \cdot OK^3 = 0 \cdot 1 \cdot 0.2 = 0,$
$US = 1 - (\Sigma f_m + OK) = 1.$

The elaborated final diagnosis is contradictory. The system searches for the solution analyzing the possibilities of multiple faults consisting of faults pointed out in the primary diagnosis. Diagnosis justification proceeds in the following way:

$f_{|1,7|} = f_1^1 \cdot f_7^2 \cdot OK^3 = 0.8 \cdot 0.8 \cdot 1 = 0.64,$
$OK = OK^1 \cdot OK^2 \cdot OK^3 = 0 \cdot 1 \cdot 0.2 = 0.$

Finally, the unknown process state coefficient is calculated as

$US = 1 - (\Sigma f_m + OK) = 0.36,$

and the final diagnosis takes the shape

$DGN = \{\langle f_{|1,7|}, 0.64\}.$

In some cases, symptom uncertainty causes difficulties in diagnosis interpretation, especially in determining if a single or a multiple fault occurred (in general, if an m-valued or an $(m-1)$-valued fault occurred). In the *DiaSter* system, in such situations both scenarios are considered during diagnosis justification at a superior level. In this particular case, the diagnosis containing the minimal set of faults that explains observed symptoms is not chosen for further analysis. The decision concerning which elementary block of faults should be taken into account during further calculations is based on parallel analysis in which fuzzy information is omitted. The diagnostic signal values are converted into crisp signals. It is assumed that the symptom is observed (with the certainty factor equal to 1) if the value of its membership function exceeded some threshold value T_D.

Example 5.7

Assume that for the symptoms

$s_1 = \{\langle 1,1 \rangle\}$, $s_2 = \{\langle 1,1 \rangle\}$, $s_3 = \{\langle 0,0.2 \rangle, \langle 1,0.8 \rangle\}$,
$s_{101} = \{\langle 0,0.2 \rangle, \langle 1,0.8 \rangle\}$,

the following partial diagnoses were elaborated at the subsystem level:

$DGN_1 = \{\langle [f_{|1,3|}, f_{|1,4|}], 0.8 \rangle, \langle [f_3, f_4], 0.2 \rangle\},$
$DGN_2 - \langle OK, 1 \rangle,$
$DGN_3 = \{\langle [f_4, f_8], 0.8 \rangle, \langle OK, 0.2 \rangle\}.$

The diagnosis for the first subsystem includes multiple faults as well as single ones. Possible rejection of multiple faults is not correct because the observed symptoms point out, somehow, the possibility of their existence. Such a decision is made because crisp evaluation for the threshold value $T_D = 0.2$ points out multiple faults, not single ones:

$$\begin{array}{lcl} s_1 = \{\langle 1,1 \rangle\} & & \bar{s}_1 = 1 \\ s_2 = \{\langle 1,1 \rangle\} & \xrightarrow{T_D = 0.5} & \bar{s}_2 = 1 \\ s_3 = \{\langle 0,0.2 \rangle, \langle 1,0.8 \rangle\} & & \bar{s}_3 = 1 \\ s_{101} = \{\langle 0,0.2 \rangle, \langle 1,0.8 \rangle\} & & \bar{s}_{101} = 1 \end{array}$$

S/F	f_1	f_2	f_3	f_4	f_5	f_6	f_7	f_8	f_9	f_{10}	f_{OK}
s_1		1	1	1							0
s_2	1	1	1	1		o_1					0
s_3	1										0
s_4		1			1	1					0
s_5					1	1					0
s_6								1	1	1	0
s_7			o_2				1^+		1^+	$1^-,1^+$	0
s_8							1				0
s_{101}				1				1			0
s_{102}			1							1	0

Fig. 5.36 Reasoning in the case of multiple faults with strong symptoms uncertainty

$$D\bar{G}N_1 = [f_{|1,3|}, f_{|1,4|}],$$
$$D\bar{G}N_2 = OK,$$
$$D\bar{G}N_3 = [f_4, f_8].$$

Diagnosis justification proceeds in the following way:

$$f_3 = f_3^1 \cdot OK^2 \cdot f_3^3 = 0.2 \cdot 1 \cdot 0 = 0,$$
$$f_4 = f_4^1 \cdot OK^2 \cdot f_4^3 = 0.2 \cdot 1 \cdot 0.8 = 0.16,$$
$$f_{|1,3|} = f_{|1,3|}^1 \cdot OK^2 \cdot f_3^3 = 0.8 \cdot 1 \cdot 0 = 0,$$
$$f_{|1,4|} = f_{|1,4|}^1 \cdot OK^2 \cdot f_4^3 = 0.8 \cdot 1 \cdot 0.8 = 0.8,$$
$$OK = OK^1 \cdot OK^2 \cdot OK^3 = 0 \cdot 1 \cdot 0.2 = 0.$$

The final value of the unknown process state coefficient is calculated according to

$$US = 1 - (\sum \mu_m + OK) = 0.04,$$

and the final diagnosis takes the shape

$$DGN = \{\langle f_4, 0.16\rangle, \langle f_{|1,4|}, 0.8\rangle, \langle US, 0.04\rangle\}.$$

5.6 Application of Belief Networks in Technical Diagnostics

The current section outlines the so-called *belief-network-based diagnostic model*. It describes a generic structure of the model as well as selected topics regarding model identification and application. The model allows simultaneous representation of knowledge acquired from passive as well as active diagnostic experiments, that collected from diagnostic relations, and generic knowledge made available with physics laws, for example. One major advantage of the above model is its ability to account for partial diagnostic knowledge expressed in the form of approximate, subjective, context-based expert evaluations.

5.6.1 Introduction

Technical object diagnostics can be executed in the form of (Natke and Cempel, 1997)

- model-based diagnostics,
- symptom-based diagnostics.

The following section reveals the prospects of developing a diagnostic model for supporting both model-based diagnostics processes and symptom-based diagnostics ones.

5.6.1.1 Model-Based Diagnostics

The essence of model-based diagnostics is the analysis of residua between object observation results and model results. The model should account for object simulations in a specified technical state of the object. Assuming that the simulation inputs correspond to the observed ones, the simulation results can be compared with those coming from the object. The operation is performed for the purpose of recognizing a particular state class of the object under observation. The reasoning process on the state of the object can be carried out as follows:

- every class of the state of an object corresponds to a different model, and model application results from simultaneous simulations are compared with object observation results. Then, the most congruent simulation results point at object specific states;
- the reasoning process incorporates one model that is a representative of a well-functioning object. Then the residua between the object and model observation results are considered as inputs to symptom-based diagnostics.

To summarize, the quality of the model affects the efficiency of diagnoses. The above methods are widely used in control system diagnostics as well as process diagnostics (Korbicz et al., 2004). In particular, they can be used whenever a capable deterministic model of an object can be developed.

5.6.1.2 Symptom-Based Diagnostics

An alternative to model-based diagnostics is symptom-based diagnostics. This particular branch of diagnostics assumes that object state changes are accompanied by relevant symptoms. Most methods assume the processes under observation are residual ones (Cempel, 1991) whose features are object state information carriers. The relationships occurring between the specific features and the object state are then formed as diagnostic relations. The essence of symptom-based diagnostics is in the use of expert knowledge for indicating specific symptoms giving evidence to a particular state of an object. Methods in this field of diagnostics involve various techniques for comparing observation results with symptom specifications.

The methods can be regarded as classification tasks in which class definitions are symptom specifications pointing at a particular state of an object. Given known symptom specifications, the problems can be solved. Initially, the specifications were expert sourced; however, the complexity of analyzed objects and limitations in formulating symptom specifications by experts prompted the development of research methods for determining symptom specifications. The devised methods involve knowledge acquisition methods (Moczulski, 2002b) as well as symptom identification methods based on passive and active diagnostic experiments. However, the acquired diagnostic relations are often unreliable and inaccurate. Symptom-based diagnostics is often used in scenarios where developing a good quality numerical model is difficult. As a result, no straightforward application of model-based diagnostics is then possible.

5.6.1.3 Diagnostic Model

The purpose of diagnostic experiments is technical state recognition of an object under observation, based on object operation information made available as observation results. The model projecting data on mutual interactions between the environment and the object (or data on process variables of a process performed by the object) is called the *diagnostic model*. The investigated model D (5.83) concerns a model in which input and output values are recorded as the vectors $\mathbf{x}_D$ and $\mathbf{y}_D$, respectively:

$$D : \mathbf{x}_D \rightarrow \mathbf{y}_D. \tag{5.83}$$

The inputs $\mathbf{x}_D$ of the diagnostic model are attribute values of the observed interactions and values of parameters defining the object structure as well as operating conditions. The model inputs include the following items:

- variables (variable values) describing object operating conditions (rotational speed, loads, etc.);
- constant variables (e.g., object design parameters);
- variables whose values present object operation effects. The values can be determined via measurements or simulations (e.g., vibration amplitude).

The diagnostic model outputs $\mathbf{y}_D$ incorporate variables whose values need to be determined as a result of a diagnostic process. Then, the model output is the object estimated state. Therefore, object state information can be represented with an attribute (feature) value set of a state of an object under observation.

5.6.1.4 Diagnostic Model Concept

The diagnostic model D as shown in (5.83) can be identified as a result of the object inverse modeling process (Cholewa and White, 1993). Inverse modeling is used for determining input-to-output transformations, where a particular state of an object is the transformation parameter. The method requires the knowledge of a relevant model of the object to describe precisely the state influence. However, in most scenarios such models are not known or they are difficult to define. In particular, definition problems concern objects in which destructive processes correspond to object component wear processes. One basic method of diagnostic model identification assumes an adequate and representative set of learning data V_L (called the example set) as shown in (5.84) in the form of pairs of the input and output values $< \mathbf{x}_D, \mathbf{y}_D >_k$ of the sought-after model:

$$V_L = \{< \mathbf{x}_D, \mathbf{y}_D >_k\}. \tag{5.84}$$

Among others, having a set of examples allows one to identify basic classes of diagnostic models in the form of

- a neural network to describe the following transformation:

  $$neur : \mathbf{x}_D \rightarrow \hat{\mathbf{y}}_D. \tag{5.85}$$

  The network can be used for determining the estimated value of the model output $\mathbf{y}_D$ based on the known input value $\mathbf{x}_D$;
- joint distribution of the multi-dimensional random variable (X,Y) described by the following distribution:

  $$F(\mathbf{x}_D, \mathbf{y}_D) = P(X < \mathbf{x}_D, Y < \mathbf{y}_D), \tag{5.86}$$

  which can be used for determining the boundary distribution of the random variable Y representing the model output

  $$F(\mathbf{y}_D) = P(Y < \mathbf{y}_D). \tag{5.87}$$

One common issue with the models given by (5.85) and (5.86) is the requirement of a sufficiently numerous (large) learning data set V_L (5.84) for model identification. A drawback of the model (5.85) is its inability to account for generic knowledge as well as domain knowledge expressed in an explicit form. The required knowledge can be incorporated to a model by means of pre-processed learning data (examples) only. Due to the above, straightforward tuning of the model (5.85) is difficult.

The model (5.86) allows partial interpretation of its parameters and their modification in line with the model tuning process. Moreover, any application of the model (5.86) is bound by two essential issues, namely,

- determining probability values requires expensive experiments that may be difficult to execute. Therefore, there is a need to extend the probability definition and its scope of application in subjective terms;

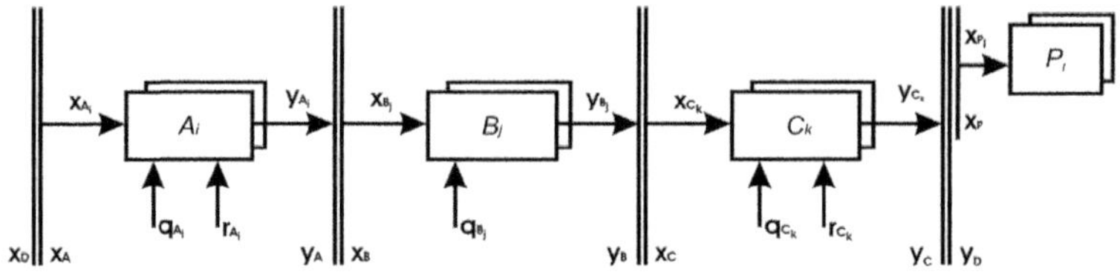

Fig. 5.37 BNBM structure: model inputs and outputs

- the application of the multi-dimensional model (5.86), i.e., recording the multi-dimensional distribution value requires huge data sets even with high-level granularity of data.

5.6.2 Belief-Network-Based Diagnostic Model

One essential advantage of symptom-based diagnostics is the ability to explicitly utilize the possessed knowledge in line with developing the above models and applying them in real problems. For comparison, in model-based diagnostics the knowledge can be applied while tuning simulation models. Both symptom-based diagnostics methods and model-based diagnostics ones utilize the development of efficient computer-aided diagnostic systems. As such, numerous advantages of independently applied methods prompted the development of a generic method combining both model-based diagnostics and symptom-based one. For example, one method under consideration implied serial application of characteristic methods for both classes of diagnostic experiments. The research resulted in the so-called Belief-Network-Based Model (BNBM). Note that the BNBM is a heuristic one. By definition, a heuristic model guarantees correct results in a majority of scenarios; however, it does not warrant good results in all cases. Heuristic procedures are most successful when applied to solve routine tasks (e.g., classic umbrella problem).

5.6.2.1 Structure of the Model

The BNBM is a multi-stage model. Fig. 5.37 presents the general structure of the model, model inputs and outputs $(\mathbf{x}_D, \mathbf{y}_D)$, inputs and outputs of model consecutive stages $(\mathbf{x}_A, \mathbf{y}_A)$, $(\mathbf{x}_B, \mathbf{y}_B)$, $(\mathbf{x}_C, \mathbf{y}_C)$ and the corresponding blocks A_i, B_j, C_k, block inputs and outputs $(\mathbf{x}_{A_i}, \mathbf{y}_{A_i})$, $(\mathbf{x}_{B_j}, \mathbf{y}_{B_j})$, $(\mathbf{x}_{C_k}, \mathbf{y}_{C_k})$, as well as fundamental configuration parameters $\mathbf{q}_{A_i}$, $\mathbf{q}_{B_j}$, $\mathbf{q}_{C_k}$ and tuned configuration parameters $\mathbf{r}_{A_i}$, $\mathbf{r}_{C_k}$. The model was developed in order to allow straightforward application of knowledge related to an object under consideration, or domain and expert-originated with expert knowledge or any knowledge repository.

The diagnostic model (5.83) maps the input data $\mathbf{x}_D$ into the output data $\mathbf{y}_D$. As shown in Fig. 5.37, the BNBM comprises the stages A, B and C, which consist of parallel blocks:

- the first stage incorporates the transformation blocks A_i carrying out data preprocessing tasks;
- the second stage comprises the mapping blocks B_j;
- the third stage includes the belief networks C_k;
- additional external blocks cooperating with the model are the presentation blocks P_l for post-processing and result visualization.

Block outputs are selected outputs of the stage incorporating the given block as follows:

$$\mathbf{x}_{A_i} \subseteq \mathbf{x}_A; \ \ \mathbf{x}_{B_j} \subseteq \mathbf{x}_B; \ \ \mathbf{x}_{C_k} \subseteq \mathbf{x}_C; \ \ \mathbf{x}_{P_l} \subseteq \mathbf{x}_P. \tag{5.88}$$

The sets of consecutive stage outputs are then determined as the respective sums of block outputs sets that are included in that stage:

$$\mathbf{y}_A = \bigcup_i \mathbf{y}_{A_i}; \ \ \mathbf{y}_B = \bigcup_j \mathbf{y}_{B_j}; \ \ \mathbf{y}_C = \bigcup_k \mathbf{y}_{C_k}. \tag{5.89}$$

By turns, the inputs of a particular stage (and the model outputs) are contained in the preceding stage outputs (or the model inputs) as follows:

$$\mathbf{x}_A \subseteq \mathbf{x}_D; \ \ \mathbf{x}_B \subseteq \mathbf{y}_A; \ \ \mathbf{x}_C \subseteq \mathbf{y}_B; \ \ \mathbf{y}_D \subseteq \mathbf{y}_C; \ \ \mathbf{x}_P \subseteq \mathbf{y}_D. \tag{5.90}$$

In order to meet BNBM needs, the set $\mathbf{z}_D$ of the so-called additional variables was introduced for representing attribute (feature) values occurring in real-life operating conditions (yet being non-observable), or virtual interactions taking place between an object and its environment or among particular object components. Note that introducing the additional variables $\mathbf{z}_D$ requires from a researcher extending the learning data set V_L definition (5.84) as follows:

$$V_L = \{< \mathbf{x}_D, \mathbf{y}_D, \mathbf{z}_D >_k\}. \tag{5.91}$$

From the assumption that the additional variables are non-observable it follows that the learning data set V_L (5.91) cannot be acquired as a result of observations. The learning sets of the additional variables $\mathbf{z}_D$ can be determined in a fairly straightforward fashion by means of the object model or, alternatively, by using available inputs and outputs of the object model with other models representing specific knowledge in a given domain. In particular, one important assumption is the ability to interpret the values of additional variables by an expert in a fairly simple manner. The values can be restricted by numerous constraints existing for an object under consideration. Moreover, one major advantage of the BNBM is its ability to incorporate specific knowledge by means of the additional variables $\mathbf{z}_D$. Also, another interesting feature of selected additional variables is their high diagnostic sensitivity.

An example of additional variables $\mathbf{z}_D$ added to a data set for determining a diagnostic model used for estimating actual alignment conditions of a multi-supported rotor from support vibrations can be reaction forces of rotor supports (Kiciński, 2006; Cholewa and Wojtusik, 2005; Cholewa and Urbanek, 2005; Rogala et al., 2009). Such forces may be very useful for the recognition of the actual alignment, but they are not observed (measured) directly. As a result, they are not included in the input value set. However, their values can be determined during the process of learning data acquisition in simulation experiments.

In BNBMs, additional variables are reconstructed from the input data $\mathbf{x}_D$ using first-stage (purpose-built) blocks.

5.6.2.2 Components of Successive Model Stages

The first-stage blocks perform data pre-processing operations. The blocks can carry out numerous standard tasks, namely, data normalization, multi-dimensional scaling, etc. BNBMs feature an extended set of such tasks and the so-called *One-Class Classifiers* (*OCCs*).

For each set of input values, an OCC_i determines the rate of its membership to the i-th class. The membership rate can be interpreted as a characteristic function value of a particular fuzzy set. As a result, a one-class classifier can be interpreted as the definition of a fuzzy set. Therefore, an application of OCCs allows one to transform input data into a space where subsequent coordinates assume values corresponding to the rates of their membership to selected classes. The space is called an *image space*.

During the identification process of a diagnostic model based on learning data, neither explicit generic knowledge nor domain one is accounted for. The type of knowledge can be accounted for while developing a simulator for learning data generation and learning data set completion. In order to enable explicit specification of constraints for variables under consideration in the acquired knowledge context, another stage is introduced into the BNBM. The new stage is used mainly for adjusting additional variable values.

An example of the essence of the matter can be the above-mentioned additional variables in the form of support reaction forces of a multi-supported rotor. In a diagnostic model for current rotor alignment estimation, the reaction forces are estimated from model input data incorporating support and shaft vibration features. The estimated reactions can be constrained due to rotor quasistatic equilibrium conditions. Accounting for such restrictions in such a model increases estimation process accuracy of reaction forces and improves the quality of diagnosis on the rotor alignment state compared to diagnoses based on input data only, i.e., diagnoses estimated without any knowledge of rotor quasistatic equilibrium conditions.

The outputs $\mathbf{y}_{B_j}$ of the second stage blocks take values equal to the adjusted input values $\mathbf{x}_{B_j}$, or they are replicas of the input values $\mathbf{x}_{B_j}$ requiring no adjusting at all.

The BNBM's third stage contains the blocks C_k represented by *belief networks*. The network initial state is referred to default values of the network nodes.

The values of the input nodes C_k are determined based on the network inputs $\mathbf{x}_{C_k}$. The application of the network results in the values $\mathbf{y}_{C_k}$ of the network output nodes for which the data set is the model's third stage output $\mathbf{y}_C$.

Finally, a selection of the third stage outputs is then made available to the external blocks P_l for presenting (visualizing) network operation results to end users.

5.6.3 Input Data Images

The present section outlines operations carried out by selected blocks of the BNBM's first-stage (as shown in Fig. 5.37) for data pre-processing. Here, the model contains the blocks A_i for calculating input data images in the following form:

$$A_i : \mathbf{x}_{A_i} \rightarrow \mathbf{y}_{A_i}. \tag{5.92}$$

The transformation was introduced primarily to reduce redundant attributes (features) of diagnostic signals and processes variables. The blocks A_i have the inputs $\mathbf{x}_{A_i}$ which accept quantitative values (e.g., values of signal attributes). The approach facilitates representing input values in a metric space by a point whose coordinates correspond to the input values. The essence of the proposed approach is not to determine the point location in the input value space (values of signal features, values of process variables) by its coordinates $\mathbf{x}_{A_i}$, but with similarity measures of the point to distinct element classes in that space. Such similarity values can be interpreted as the element coordinates in the new space called the *image space*.

The derived method of representing additional variables influences the characteristics of the transformation A_i. Moreover, it is possible to identify transformations for additional variables stored in a continuous form, e.g., by using quantitative models. Also, the operation can be performed for high-level granulated additional variables by developing qualitative models, fuzzy models, belief models, etc. In the described model, the level of granularity for an additional variable is determined by the number of examined classes of values. Recent research has shown that the grained models can be regarded as some sort of a compromise between the accuracy of the models, stability, and interpretability (Pedrycz, 2001), thus leading to generic characteristics of models, given a careful selection of the data granularity level.

The transformation meets the following requirements imposed by technical diagnostics:

- significant reduction of the number of space dimensions and effective distinguishability of its elements at the same time;
- the similarity of points. Points that are located close to one another in one space preserve that distance in another space upon data transformation;
- distant points in one space remain such in another space upon data transformation operations.

Considering the BNBM's first stage as a set of parallel blocks facilitates the replacement of a global model with easier-to-determine local models. Both spatial-oriented global model decomposition (subsystems) and local-description-based decomposition can be considered here. The introduction of the additional variables $\mathbf{y}_A$ in the form of the so-called instrumental variables in BNBMs facilitates the decomposition of such models (Cholewa and Urbanek, 2005; Cholewa and Wojtusik, 2005; Rogala et al., 2009).

The following sections describe in detail the data transformation method which can be implemented using parallel classifiers. Moreover, a number of classifiers for use with BNBM first-stage blocks, including an interesting group of one-class classifiers, are described in sections that follow below.

5.6.3.1 Classifier-Based Data Transformation

With the number of examined classes as the primary criterion, classifiers can be categorized as multi-class classifiers, binary classifiers and one-class classifiers. The majority of tasks requires classifiers that are capable of recognizing multiple classes. Therefore, one particular group of classifiers are multiple ones. They are particularly well suited for object assignation to appropriate classes. Simultaneous examination of a large number of classes may lead to the development of complex boundaries among particular classes. As a result, the classification process may be unsatisfactory. One solution to eliminate the inefficiency is the replacement of one multi-class classifier with a group of binary (two-class) classifiers or even one-class classifiers. Moreover, it is possible to design a procedure where the classification process for k classes can be performed using

- k binary classifiers for distinguishing objects that do not belong to a particular class,
- $k(k-1)/2$ binary classifiers defined for every pair of classes,
- k one-class classifiers used for determining object class membership rates with a particular type classifier.

The data transformation (5.92) can be considered here in the form of a function mapping every element $\mathbf{x}_{A_i} \in \mathbb{R}^N$ into one element $\mathbf{y}_{A_i} \in \mathbb{R}^M$. An interesting group of data transformation operations (due to numerous feature sets) includes transformations which facilitate the process of transforming multi-dimensional spaces into spaces of lower number of dimensions $N > M$ such as variable projection, grouping, classification, etc. A major advantage of such operators is their ability to define transformation functions during machine learning. An example of such a transformation is classifiers.

In the area of technical diagnostics, classifiers can act as diagnostic models mapping a quantitative variable observation space into a space of value classes defined with qualitative variables. The selection process of the number of class values in an additional variable space depends on the employed global-to-local model decomposition based on their local description. In turn, the above-mentioned images of feature values can be interpreted as vectors of belief degrees of their membership to sets of additional variable value classes or fuzzy sets of value classes. The solution requires the application of approximate one-class classifiers for determining image space coordinates where input variable values are mapped onto belief degrees. The transformation operation with an approximate one-class classifier is of the following form:

$$A_i : \mathbf{x}_{A_i} \to y_{A_i} \in [0;1]. \tag{5.93}$$

Assuming that the multi-class classification problem for M classes of additional variable values can be accomplished with M one-class classifiers $A_1, \cdots, A_M$, the solution is an attribute (feature) image in the form of a vector:

$$\mathbf{y}_A = [y_{A_1}; y_{A_2}; \cdots; y_{A_M}]. \tag{5.94}$$

The transformation is shown in Fig. 5.38. The approach has a number of advantages, namely,

- the ability to obtain a simple and reduced description of belief degrees of membership to value classes of additional variables,
- the ability to apply various classifiers based on simple linear transformations as well as complex non-linear operations.

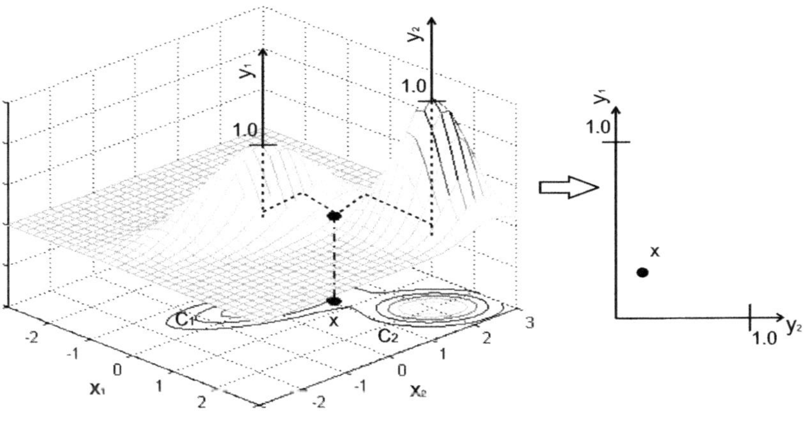

Fig. 5.38 Transformation of a 2D attribute value space into a 2D image space with two one-class classifiers defined by means of potential functions

5.6.3.2 Selected Examples of Classifiers

Note that the application of classifiers as transformation operators (5.93) restricts the set of applicable classifiers that are likely to be within the application scope. In many scenarios, classifiers are defined in such a manner as to enable their straightforward

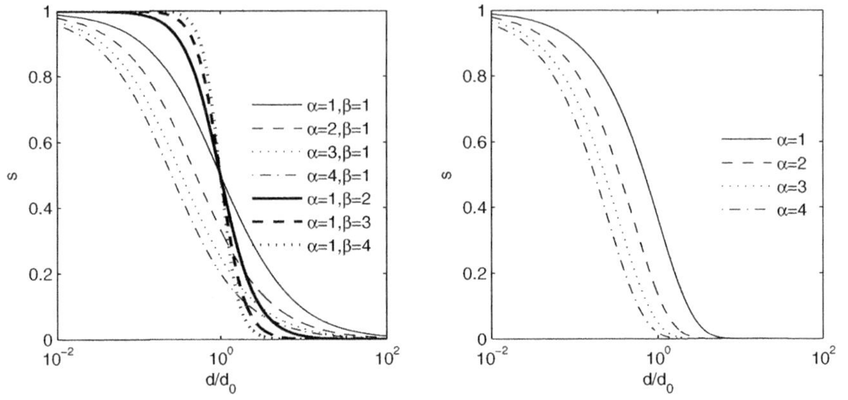

Fig. 5.39 Family of functions $s(d)$ as defined in (5.95) and (5.96)

association with the destination class (i.e., the class that a specific object belongs to), and their major purpose is not performing the task (5.93) for input data image generation.

Examining the transformation (5.93) requires explicit interpretation of identified function value meaning. By definition, it should be the class assignation belief degree. The measure value does not have any formal definition. However, it can be interpreted in the form of a fuzzy set characteristic function value used for describing such classes. One convenient interpretation for engineering applications assumes the belief degree is a subjective probability measure.

Due to the above, the identification task of the transformation (5.93) requires data fitting operations applied to available learning data of the function A_i (5.91) defined with input data. The desired effect can be obtained with a fairly routine procedure for neural network identification. As the function requires multiple adjustment operations, including the use of expert knowledge in a straightforward fashion, the application of selected type classifiers should be accounted for in the process. Also, note that at this point the procedure should consider the objective for which the transformation (5.93) should utilize the function value A_i as estimated with a selected classifier. The value is required only for the transformation operation shown in Fig. 5.38. Within the scope of the block A_i, the value cannot be compared to a threshold value or another (analogous) function value calculated for another class; the main purpose of the block A_i is not class identification. The purpose of the value A_i is to ensure the efficiency of the whole reasoning process performed with a complete BNBM using all three stages of the model. The process is completed on identifying an appropriate class in the blocks C_k—belief networks.

Classifier definitions can be based on either distance measures $d(\mathbf{x}_1,\mathbf{x}_2) \geqslant 0$ or similarity measures $s(\mathbf{x}_1,\mathbf{x}_2) \in [0;1]$ between the elements $\mathbf{x}_1$ and $\mathbf{x}_2$. Assuming classification in the space $\mathbb{R}^N$, various distance measures can be defined and used.

Moreover, various similarity measures can be used. The measures can be defined in a fairly straightforward manner or estimated based on the distance measures $d(\mathbf{x}_1,\mathbf{x}_2)$ (as shown in Fig. 5.39):

$$s(\mathbf{x}_1,\mathbf{x}_2) = s\left(d(\mathbf{x}_1,\mathbf{x}_2)\right) = \frac{1}{1+\alpha\left(\frac{d(\mathbf{x}_1,\mathbf{x}_2)}{d_0}\right)^{\beta}}, \tag{5.95}$$

$$s(\mathbf{x}_1,\mathbf{x}_2) = s\left(d(\mathbf{x}_1,\mathbf{x}_2)\right) = \exp(-\alpha\,\frac{d(\mathbf{x}_1,\mathbf{x}_2)}{d_0}), \tag{5.96}$$

where d_0 is the constant base distance. Selecting a relevant similarity measure as well as a distance one influences classification efficiency. However, identifying optimal metrics can be difficult.

The majority of standard statistical classifiers for estimating the rate of the membership of an object to a class can be adapted for use with the transformation (5.92). However, some major issues are encountered with binary classifiers applied in the transformation (5.93). The classifiers assume positive values for elements assigned to one class and negative values for other class elements. A convenient group of classifiers in such applications covers one-class classifiers that are defined using a set of representatives (or examples) for the i-th class under examination of the following form:

$$\mathbf{R}_i = \{\mathbf{r}_1,\mathbf{r}_2,\cdots,\mathbf{r}_{M_i}\}. \tag{5.97}$$

The function (5.93) can be defined as the similarity function for the representatives $\mathbf{R}_i$ as shown by the equation (5.96):

$$A_i(\mathbf{x}) = \exp\left(-\alpha\,\frac{d(\mathbf{x},\mathbf{R}_i)}{d_0}\right), \tag{5.98}$$

where the distance $d(\mathbf{x},\mathbf{R}_i)$ between the examined element $\mathbf{x}$ and the representative set $\mathbf{R}_i$ (5.97) can be accounted for in the following form:

- the minimum distance:

$$d(\mathbf{x},\mathbf{R}_i) = \min_{m=1}^{M_i} d(\mathbf{x},\mathbf{r}_m), \tag{5.99}$$

- the maximum distance:

$$d(\mathbf{x},\mathbf{R}_i) = \max_{m=1}^{M_i} d(\mathbf{x},\mathbf{r}_m), \tag{5.100}$$

- the average (mean) distance:

$$d(\mathbf{x},\mathbf{R}_i) = \frac{1}{M_i}\sum_{m=1}^{M_i} d(\mathbf{x},\mathbf{r}_m). \tag{5.101}$$

Assuming that there exists a basis for associating the representatives (5.97) with relevant weights as follows:

$$\mathbf{w}_i = \{w_1, w_2, \cdots, w_{M_i}\}, \tag{5.102}$$

the distance $d(\mathbf{x}, \mathbf{R}_i)$ can be estimated as the weighted mean value,

$$d(\mathbf{x}, \mathbf{R}_i) = \frac{1}{\sum_{m=1}^{M_i} w_m} \sum_{m=1}^{M_i} \left(w_m d(\mathbf{x}, \mathbf{r}_m)\right). \tag{5.103}$$

A particular category of weight coefficients $\mathbf{w}_i$ that can be used to change the example influence covers the similarity measures (5.96) estimated for each example on an individual basis as follows:

$$w_m(\mathbf{x}) = \exp\left(-\alpha_m \frac{d(\mathbf{x}, \mathbf{r}_m)}{d_0}\right). \tag{5.104}$$

Another potential modification of the above group of classifiers is restricting the representative set (5.97) to a subset containing only the number M_0 of pattern examples. The examples are the nearest neighbors of the element $\mathbf{x}$.

Learning data applied to one-class classifier definitions do not have to be restricted to a single class. For example, they may incorporate two subsets $\mathbf{R}_1$ and $\mathbf{R}_2$, of which the subset $\mathbf{R}_1$ contains target examples (assigned to a particular class), and the subset $\mathbf{R}_2$ includes outlier values (not assigned to that class). By using the subsets $\mathbf{R}_1$ and $\mathbf{R}_2$, it is possible to interpret the function A_i as some form of relative density of target examples in the nearest neighborhood of the element $\mathbf{x}$ under examination.

5.6.3.3 Classifier Learning and Defining

Classifiers can be interpreted as models subjected to supervised machine learning. Their learning methods change according to the classifier type. For example, discriminative-function-based methods apply to non-parametric classifiers, whereas probability-calculus-based algorithms are used with parametric classifiers. Therefore, the methods for classifier learning are as follows:

- discriminative-function-based learning methods, where no direct implementation of the knowledge of data partitioning in the output value class space is possible;
- generative methods, based on a model allowing new learning example acquisition. The new data form the basis for conditional probability estimation $p(\mathbf{x}|c_n)$ for which a posteriori probability is calculated using Bayesian calculus rules. The method allows implementing the knowledge of data partitioning in the output value class space given prior information on specific classes as well as probability distribution of output value classes;
- discriminative-model-based probabilistic methods, where the conditional probability $p(c_n|\mathbf{x})$ is sought after instead of accurate estimation of discriminative

functions. As a whole, the methods can be used for implementing knowledge of the type of conditional probability density distribution;
- specialized learning methods using a hybrid model. The model combines both the discriminative and the generative approach. Here, the latter type model is not used for estimating the probability $p(\mathbf{x}|c_n)$ but for generating a new learning example. The examples can be then applied for defining the classifier discriminative part. Such solutions have the ability of classifier learning based on incomplete data sets where some elements lack class assignation information (labels).

It is possible to use classifiers that require no supervised learning. In such scenarios, explicit formulations of discriminative functions are necessary, given the possessed knowledge or direct observations of examples in a feature space. A major limitation of such methods is, however, the substantial size of the attribute space in which data partitioning is defined.

5.6.4 Additional Variables and Opportunities for Their Adjustment

Additional variables (additional signal features) in BNBMs can be considered as instrumental variables which

- allow one to introduce additional knowledge, e.g., in the form of a physics-based equation system or a set of constraints whose variables are additional signals features. One necessary pre-condition for carrying out such tasks is possessing relevant and well-understood (physics-based) interpretation of examined attributes of additional signals;
- allow global model decomposition in order to reduce the number of examined input features. The global-to-local model approach delivers substantial benefits in terms of reducing the number of usable features as well as the number of value classes of additional variables. The approach leads to fewer model parameters as well as a model identification process based on a smaller set of learning examples, and it improves classifier efficiency.

One remaining solution is the method of identifying model additional variables. In recent studies related to multi-stage inverse model identification (Cholewa and Wojtusik, 2005; Cholewa and Urbanek, 2005; Rogala et al., 2009), features of additional signals were often physical variables not accounted for in object state definitions. They were concerned with observable (but not observed) real interactions. The signals were acquired using high-fidelity numerical models of dynamic objects.

To emphasize, additional variables can be the result of processing attributes which can be considered BNBM type model inputs. Such operations can be carried out using various techniques to apply relevant transformations in order to ensure higher-to-lower dimension transformation corresponding to classes of values of additional variables. In such scenarios it is necessary to apply any available knowledge on the sought-after additional variables. Any available knowledge on object behavior, as well as domain knowledge, needs to be accounted for in the study. However, a random search for such attributes is a difficult task that does not guarantee any satisfactory results.

5.6.4.1 Adjustment Blocks

The employed and formalized knowledge of additional variables is registered in the BNBM's second stage (see Fig. 5.37). The model blocks follow the transformation rule

$$B_j : \mathbf{x}_{B_j} \rightarrow \mathbf{y}_{B_j}, \tag{5.105}$$

where additional variables are images of the selected additional variables $\mathbf{x}_{B_j}$, whereas the output variables $\mathbf{y}_{B_j}$ are adjusted belief degrees of membership to the additional variable class under examination. The belief degrees are obtained using an appropriate adjustment procedure. The purpose of this stage is to include constraints conditions due to domain type knowledge on a particular object. If such knowledge is missing, then the transformation (5.105) simplifies to

$$B_j : \mathbf{y}_{B_j} = \mathbf{x}_{B_j}. \tag{5.106}$$

If a need for adjustment procedures has arisen, constraint equations are formulated. The equation variables are the additional features. Note that applying the BNBM's first stage results in a vector of belief degrees, not their exact values. This implies there exist two options for registering constraint equations:

- the adjusted values $\mathbf{x}_{B_j}$ are considered as approximate values, and the boundary equations are then approximate equations as well;
- the constraint equations are formulated as exact equations for the unknown exact adjusted values $\mathbf{x}^*_{B_j}$.

The operation requires a pair of transformations of the following form:

$$\mathbf{h}_{B_j} : \mathbf{x}_{B_j} \rightarrow \mathbf{x}^*_{B_j} \tag{5.107}$$

$$\mathbf{h}^{-1}_{B_j} : \mathbf{x}^*_{B_j} \rightarrow \mathbf{x}_{B_j} \tag{5.108}$$

to estimate the exact value of the additional variable $\mathbf{x}^*_{B_j}$, corresponding to the image $\mathbf{x}_{B_j}$, and vice and versa.

5.6.4.2 Application of Adjustment Calculus Methods

Adjustments using the transformations (5.107) and (5.108) are possible provided that the images $\mathbf{x}_{B_j}$ represent sets of approximate values of an additional variable, for example, low load, medium load, high load. Provided that constraint equations are linear, the adjustment process can be carried out using generic methods of adjustment calculus. Let us assume the block B_j (for a vector of N adjusted values $\mathbf{x}_{B_j}$) contains $R < N$ exact constraint equations of the following form:

$$g_r\left(\mathbf{x}^*_{B_j} + \mathbf{v}\right) = 0. \tag{5.109}$$

Therefore, accounting for (5.107),

$$g_r\left(\mathbf{h}_{B_j}(\mathbf{x}_{B_j}) + \mathbf{v}\right) = 0, \tag{5.110}$$

where $\mathbf{v}$ is a sought-after vector of adjustments fulfilling the constraint equations (5.109). Equations can have infinitely many solutions. Therefore, the least squares criterion for identifying the values $\mathbf{v}$ can be used as

$$f(\mathbf{v}) = \mathbf{v}^T\mathbf{v} \Longrightarrow \min. \tag{5.111}$$

In turn, for low-level data granularity it could be necessary to replace the equation (5.111) with a weighted least-squares criterion. For linear constraint equations of the form revealed by (5.110), the error equations become

$$\begin{aligned}
\mathbf{a}_1^T\mathbf{v} + \omega_1 &= a_{1,1}v_1 + a_{1,2}v_2 + \ldots + a_{1,N}v_N + \omega_1 = 0,\\
\mathbf{a}_2^T\mathbf{v} + \omega_2 &= a_{2,1}v_1 + a_{2,n}v_2 + \ldots + a_{2,N}v_N + \omega_2 = 0,\\
&\ldots\\
\mathbf{a}_R^T\mathbf{v} + \omega_R &= a_{R,1}v_1 + a_{R,2}v_2 + \ldots + a_{R,N}v_N + \omega_R = 0,
\end{aligned} \tag{5.112}$$

where

$$\omega_r = g_r\left(\mathbf{h}_{B_j}(\mathbf{x}_{B_j}) + \mathbf{v}\right) - g_r\left(\mathbf{h}_{B_j}(\mathbf{x}_{B_j})\right). \tag{5.113}$$

Then, the system of equations (5.112) can be derived in the following form:

$$\mathbf{A}\mathbf{v} + \omega = 0. \tag{5.114}$$

Given the condition (5.114), the extremum of the function (5.111) can be estimated with the Lagrange multiplier method. Then, the Lagrange function takes the form

$$F_L(\mathbf{v},\lambda) - f(\mathbf{v}) - 2\sum_{r=1}^{R}\lambda_r(\mathbf{a}_r\mathbf{v} + \omega_r) \tag{5.115}$$

Also, the optimum values of the adjustments v_n are estimated from

$$\frac{\partial F_L}{\partial v_n} = 2v_n - 2\sum_{r=1}^{R}\lambda_r a_{r,n} = 0, \tag{5.116}$$

leading to

$$v_n = \sum_{r=1}^{R}\lambda_r a_{r,n} \tag{5.117}$$

and

$$\mathbf{v} = \mathbf{A}^T \lambda, \tag{5.118}$$

where $\mathbf{v}$ is the vector of adjustments for minimizing/maximizing the function (5.115), whereas λ is the vector of Lagrange multipliers given as

$$\lambda = (\mathbf{A}\mathbf{A}^T)^{-1} \omega. \tag{5.119}$$

Taking into consideration the equation (5.118), the transformation (5.105) implemented by the block B_j is as follows:

$$B_j : \mathbf{x}_{B_j} \rightarrow \mathbf{y}_{B_j} = \mathbf{h}_{B_j}^{-1} \left(\mathbf{h}_{B_j}(\mathbf{x}_{B_j}) + \mathbf{v} \right). \tag{5.120}$$

5.6.5 Belief Networks

The present section outlines fundamental concepts of belief networks, their operating principles as well as algorithms for belief network development and application.

5.6.5.1 Graph-Based Models

The application of graphs in the form of a model for knowledge representation is a well known approach (Wright, 1934). Other key concepts involve the so-called Markov networks (Isham, 1981; Lauritzen, 1982), often investigated in the form of contingency tables, and Bayesian networks (Pearl, 1988; Charniak, 1991; Henrion et al., 1991), called belief networks. Over the years, belief networks have gained more and more interest as an efficient and proven tool for approximate reasoning processes.

A Markov type network is an undirected graph, whose every edge is assigned a symmetric probability of the occurrence of states associated with each edge. The network is not capable of knowledge representation of cause-and-effect relationships in a direct fashion even if such relationships occur between the network nodes.

By definition, a belief network is an acyclic directed graph composed of nodes and connecting directed edges. Each node is assigned a complete set of mutually excluded states as well as vectors of node values to determine the state distribution represented by probabilistic values (a belief that a node is in a specified state). Moreover, the nodes are assigned tables of conditional probability values for all elements of the Cartesian product of superior node state sets and a node that the table is assigned to. The conditional probability tables describe the relationships between particular nodes. The relationships do not have to be cause-and-effect ones.

Any reasoning process using belief networks, including complete tables of conditional probabilities, is based on adjustments of probability values assigned to subsequent nodes in order to establish an equilibrium state. In this state, the Bayesian theorem on conditional probability is met. Any attempts to search for global solutions may lead to NP-hard problems. One efficient solution method is based on the identification of conditionally independent nodes, followed by restricting the scope of a given task and a search process including subsequent nodes, their parent nodes as well as child nodes. Some interesting concepts for formulating and solving such tasks are reported in the works of Pearl (1988), Jensen (2002).

5.6.5.2 Random Variable Probability Distribution

Bayesian networks are not models of cause-and-effect relationships. Instead, they are models of joint probability distributions of random variables. Consider the set **X** of discrete random variables X_n as follows:

$$\mathbf{X} = \{X_1, X_2, \cdots, X_n, \cdots, X_N\}, \tag{5.121}$$

where each random variable is defined over the set $\Omega(X_n)$ of atomic (elementary) events $x \in \Omega(X_n)$. The probability distribution of each random variable X_n is defined by the following function:

$$p(x) = Pr[X_n = x] \quad \text{and} \quad \sum_{x \in \Omega(X_n)} p(x) = 1, \tag{5.122}$$

where $Pr[A]$ is the probability of the event A. Joint, discrete probability distribution for a set of random variables **X** (5.121) is defined as follows:

$$p(x_1, \cdots, x_N) = Pr[X_1 = x_1, \cdots, X_N = x_n]. \tag{5.123}$$

Taking into account the conditional probability definition, the equation (5.123) can be rewritten as

$$p(x_1, \cdots, x_N) = p(x_N | x_1, \cdots, x_{N-1}) \cdot p(x_1, \cdots, x_{N-1}) \tag{5.124}$$

and

$$p(x_1, \cdots, x_N) = p(x_N | x_1, \cdots, x_{N-1}) \cdot p(x_{N-1} | x_1, \cdots, x_{N-2}) \cdot p(x_3 | x_1, x_2) \cdot p(x_2 | x_1) \cdot p(x_1). \tag{5.125}$$

The joint distribution of a set of the random variables **X** can then be revealed in the form of a directed graph where each random variable X_n corresponds to one node. The parent nodes are the nodes $X_1, \cdots, X_{n-1}$ corresponding to variables specified in the conditional part of the conditional distribution in (5.125). The conditional distributions are stored here as value tables. The tables are assigned to specific nodes. For example, the parents of the node X_3 are the nodes X_1 and X_2, respectively. In general, this leads to a large number of edges that associate particular nodes with their parents. Another key properties of random variables are statistical

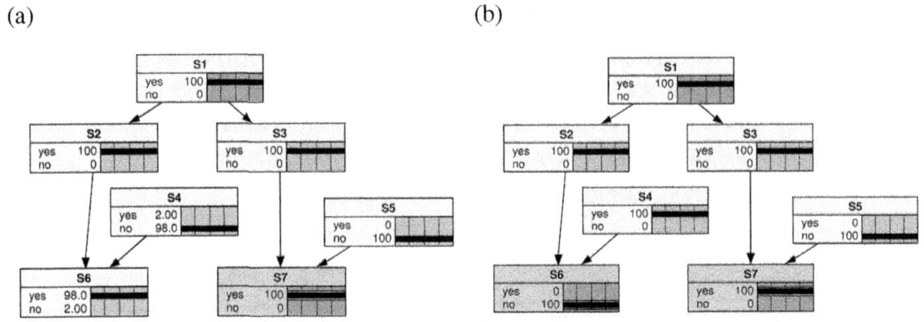

Fig. 5.40 Example of a belief network with various known, fixed-value node sets: **S7**: *Control light is ON* (100% yes) (a), **S6**: *Kettle is warming up water* (100% no) and **S7**: *Control light is ON* (100% yes) (b)

independence and conditional independence. The properties facilitate the omission of selected edges associating independent nodes in a graph.

Note that the above-mentioned directed graph is an image of the distribution (5.125). It defines the direction of the edge between nodes. However, the right-hand-side of the equation (5.125) depends on the sequence of nodes, and it can be expressed in numerous ways. This implies a group of different yet equivalent graphs illustrating the equation (5.125). The observation justifies the lack of any basis for interpreting such graphs as cause-and-effect models.

Taking the opportunity to extend the probability definition, and to use it in the context of the so-called subjective probability, the concept of a belief degree is introduced. It is not based on any formal definition. By modifying the graph description (random variable probability distribution image), it is possible to conclude that the probabilities in the graph can be replaced with equivalent belief degrees. Such extension leads to networks called belief networks.

5.6.5.3 Example of a Basic Belief Network

Figure 5.40 illustrates an example of a belief network. The network is a very basic example of a diagnostic model of an electric kettle. The illustration reveals values of the belief degrees (expressed in percentage values) assigned to the states { *yes*, *no* } of the nodes $\{S1,\cdots,S7\}$ to represent the following variables:

S1 *Kettle is switched on,*

S2 *Heater is powered on,*

S3 *Control light is switched on,*

S4 *Heater is out of order,*

S5 *Control light is out of order,*

S6 *Kettle is warming up water,*

S7 *Control light is on.*

The very first version of the network states as shown in Fig. 5.40(a) has the default value of the variable *S4 Heater is out of order* equal to *yes* = 2%, *no* = 98%, and the variable *S5 Control light is out of order* that is equal to *yes* = 1%, *no* = 99%. The only variable whose value is known is the variable *S7 Control light is on yes* = 100%. The remaining variables assume default values; no other information on the kettle is known. Given the network initial state, it implies the variable *S6 Kettle is warming up water yes* = 98%, *no* = 2%.

The other version of the network states as shown in Fig. 5.40(b) assumes the value of the variable *S6 Kettle is warming up water* that is equal to *no* = 100%. The additional information on the kettle results in the value change of the variable *S4 Heater is out of order* from *yes* = 2% to *no* = 100%.

Note that the network shown in Fig. 5.40 accepts default values of selected variables that are approved as the reasoning process outcome provided the available data and knowledge result in unknown values of statements.

5.6.5.4 Belief Network Application

A belief network is prescribed over a set of nodes, directed edges connecting selected nodes, and tables of conditional probabilities assigned to them. It defines the probability distribution of the following form:

$$p(x_1,\cdots,x_N) = \prod_{n=1}^{N} p\left(x_n|\text{parents}(x_n)\right). \qquad (5.126)$$

The application of the network is related to a reasoning process that is based on determining the boundary distribution of a random variable as shown in (5.126) or variables given known values of other variables. Any straightforward application of the equation (5.126) may lead to an NP-hard task. An efficient method is to exercise operations that may change the graph shape, and to transform it into the form of a tree or a set of trees for easy-to-use calculations. By definition, the basis for such operations is identified independence of the network nodes and conditional independence of the nodes. A key algorithm is the so-called junction tree approach. The majority of detailed algorithm descriptions are available in the accessible literature (Jensen, 2002).

5.6.5.5 Statement Networks

The present section introduces the concept of a *statement*. It permits the examination of *statement networks* (Cholewa, 2007; 2008) defined as complex (extended) belief networks in which node sets are associated with corresponding statement sets. Moreover, the section emphasizes the need for an accurate interpretation of examined facts and declared statements.

5.6.5.6 Facts, Statements and Their Interpretation

By definition, diagnostic systems facilitate technical state recognition processes based on available information on an object under observation. Such information can be stored in the form of various statements describing events, processes and regularities, as well as beliefs and speculations. For the observed system's needs, the stored information can take the form of a logical sentence, i.e., sentences resulting in a logical output, for example, true, false, the degree of possibility, the degree of necessity, the belief of pertinence.

In the case of systems based on classical logical reasoning processes, such sentences are represented by logical variables associated with their corresponding interpretations. The reasoning process in such systems is deductive or reductive, depending on the reason–conclusion relationship. Logical reasoning schemes can be modified and extended to allow incomplete and uncertain reasoning. However, most research studies in this domain are concerned with the formal side of the reasoning process. They assume that variable interpretation does not require any extra operations. In the area of technical diagnostics such an assumption is groundless.

Research on applications of various expert systems shows that the issue of variable (information) interpretation is a key factor for system successful operation. It is fairly easy to forget that expert system developers, knowledge base architects and system end users may utilize a different set of base concepts and conversation rules for the communication flow control between them. Not adhering to one specific standard of communication may lead to unexpected and false results. The information that is transferred to a user as the system operation result may then be unclear and incomprehensible.

While discussing interpretation issues it is critical to emphasize the decision-making system specifics. Clearly, it is the end user who makes the final decision (based on the information contained in the system knowledge database). Therefore, one key question to answer would be as follows: Who is responsible for the system operation end effect—users, knowledge database developers or system architects? In this case, the law clearly states the issue is in the hands of the end user. It is fairly obvious to assume here no end user has the ability to verify the knowledge database content. Therefore, the system's main task is to convince the user about the proposed solution. In such scenarios expert systems utilize the so-called messaging functions for providing insight into the reasoning process.

Also, such systems should feature the so-called *statements* for accelerating the communication process with end users. Such statements can be considered objects in the form of logical sentences. The sentences would contain a description of their meaning (interpretation) in the form of a complex statement or an opinion on a given topic. In the area of computer-aided technical diagnostics, statement application is a convenient feature for solving tasks related to a unique interpretation of values of observed process variables as well as diagnostic signal attributes and attribute changes.

5.6.5.7 Object State Statements

The output $\mathbf{y}_D$ of the diagnostic model given by (5.83) can be considered a set of state attribute (feature) values. However, many real-life applications do not require any accurate knowledge of attribute values—approximate values are sufficient for successful system operation. Instead, a convenient approach for representing approximate values is to indicate a class (of accurate values). The information on single attribute (or a set of attributes) membership can then be stored as the statement examined in the form of the pair

$$s = < c, b >, \tag{5.127}$$

where c denotes the statement content and b is the statement value. In general, the content c of the statement s is a sentence describing a given fact or representing an opinion on it. The value b of the statement s is the acceptance measure of the statement content. It is a measure of genuineness or a measure of conviction on the genuineness of the statement s. It is assumed in statement-based systems that the *statement content* is constant, whereas the *statement values* may be subject to changes.

Let us assume the examined statement values are real values from the range $[0; 1]$. Such statement values would then be approximate statements. Accurate statements values belong to the 2-element set $\{0; 1\}$. The mutually excluding statements (5.127) on various values of an attribute or a set of attributes can be grouped into statement sets of the following form:

$$\mathbf{s} = \{s_n\}. \tag{5.128}$$

The statement sets can be stored as the vectors $[s_n]$. If the statement set (5.128) is a complete set, i.e., it contains statements on all values of an examined attribute or a set of attributes, it is then specified by a multi-variant statement. The multi-variant statement is stored in the form of the pair (5.129) including the vector of component statement contents $\mathbf{c}$ and a vector of the statement values $\mathbf{b}$:

$$\mathbf{s} = < \mathbf{c}, \mathbf{b} > = < [c_n], [b_n] > . \tag{5.129}$$

The above multi-variant definition is an extension of the statement definition (5.127).

Moreover, any statement application process emphasizes the difference between an objective fact and a fact-related sentence. Every so often statements are incorrectly identified with facts. The application of statements has not introduced any essential changes in reasoning process organization or course. However, they are convenient tools for developing a complex help system including commentary, cross-references, and various comments.

5.6.5.8 Output Data Presentation

The BNBM's third stage as shown in Fig. 5.37 contains the belief networks $\mathbf{C}_k$. The networks are associated with their corresponding statement sets (5.127) including, for example, declarations on technical state class assignation for an observed object, thus forming a statement network. According to the membership (5.127), the statement is a content-and-statement-value pair. The content of a statement remains constant (and is not subject to any changes), once established while building a network. The network operation result is values of their output nodes. Given the output values, their (equivalent) output statement values are adopted. The output statements constitute the network output $\mathbf{y}_{C_k}$.

A complete set of output statements, including both contents as well as statement values, is made available to the blocks P_l for visualization purposes.

In the case of small BNBMs (including a relatively small number of the outputs $\mathbf{y}_{C_k}$), the contents and the statement values can be relayed to a user in any acceptable and comprehensive form. Accepted techniques include messaging or report printing, including an information list in the following form:

Unstable oil film occurs in the bearing no. 3: 0.71. (5.130)

A more complex model for determining values of a numerous statement set requires a specialized purpose-built interface for selecting observed statements. For instance, the interface could permit different features:

- presenting statement current values,
- presenting statement values across a prescribed time span,
- hierarchical presentation of statements to enable users to select the inspected content level of detail.

Applying a BNBM does not require any access to configuration data of the subsequent stages of the model. This type of access is required only throughout the system development process. In complex statement sets, system operation can be aided with appropriate explanation systems. The shell expert system *DIADYN* introduces the concept of an explanation system, where a statement content is accompanied by a commentary keyword. Both statement contents and keywords can be related to one another, thus developing a network of linked text messages. Moreover, the structure of the BNBM as shown in Fig. 5.37 contains a separate space for the external presentation blocks P_l for the visualization of the system operation result. Their ability to present historical data as well as to utilize a complex help system requires database application. The presentation blocks cooperating with the databases are often subject to an independent development process (regardless of the primary block processes). One advantage of the "external" location of these blocks is their ability to be developed independently without interfering with the core package.

5.6.6 Model Identification and Tuning

Assuming that learning data are made available, they can be considered data representing an instance of an unknown diagnostic model. The data incorporate the inputs $\mathbf{x}_D$, the outputs $\mathbf{y}_D$ and the additional variables $\mathbf{z}_D$. Due to BMBM complexity, any thorough process of model development should incorporate the following stages:

- Stage I: Developing a complete model,
- Stage II: Model iterative improvement and tuning.

5.6.6.1 Model Initial Version Identification

The process of developing the initial version of a BNBM (Stage I of the development process) is carried out for the stored set of learning data V_L (5.91) in which a subset of training data and a subset of test data are defined. The process is concerned with implementing the following tasks:

- defining a set of model inputs and an initial statement output set due to foreseen model application, as well as statement content definition and model generic structure selection;
- analyzing the examined object model and defining the additional variable set;
- the development of the adjusting blocks B_j (based on the available knowledge);
- independent development (identification) of every classifier A_i based on the training data $\{\mathbf{x}_D, \mathbf{z}_D\}$;
- independent development (identification) of the belief networks C_k based on the training data $\{\mathbf{x}_D, \mathbf{y}_D\}$ modified by the blocks A_i and B_j, respectively. The blocks are identified and developed before the belief networks;
- model testing—the application of subsequent stages of the model for data testing purposes.

5.6.6.2 Model Tuning

The developed model can be subject to a continuous improvement process (Stage II of the model development process). A multi-stage model incorporating parallel and non-linear blocks is difficult to optimize and requires the use of relevant heuristic algorithms. One generic group of methods for solving such problems includes the so-called genetic algorithms (Michalewicz, 1996; Goldberg, 1989). The methods are based on biological genetics concepts and allow optimizing the solution set. The process is based on examining the optimized set in the form of the t-th population $\mathbf{P}(t)$ incorporating a number of individuals. By definition, attributes of individuals are represented by single chromosomes composed of genes in linear/serial order. According to the criterion in the fitness function, the parents $\mathbf{R}(t)$ of the examined population $\mathbf{P}(t-1)$ are subject to a selection process, followed by genetic operations, namely, crossover and mutation. The children $\mathbf{D}(t)$ inherit advantageous

```
procedure Genetic_Algorithm
{  t ← 0 ;
   P(t) ← initialization( ) ;
   while(not condition)
   {  t ← t + 1 ;
      R(t) ← evaluation_and_selection( P(t − 1) ) ;
      D(t) ← reproduction( R(t) ) ;
      N(t) ← evaluation_and_selection( P(t − 1) ) ;
      P(t) ← ( P(t − 1) + D(t) − N(t) ) ;
   }
}
```

Fig. 5.41 Simplified layout of a genetic algorithm

attributes of their parents and complement the output population. At the same time the prescribed number of individuals $\mathbf{N}(t)$ is eliminated from the output population. The operations result in the new generation $\mathbf{P}(t)$. The algorithm iterations are illustrated in Fig. 5.41.

Some well-documented advantages of genetic algorithms include convergence to a global solution and tolerance to local extremes of the test function. The disadvantages are a low or very-low convergence rate, limited accuracy of the indicated solutions and a lack of the ability of individual self-learning.

Assuming that inheritance improves population individuals' attributes, genetic algorithms are capable of efficient optimization. However, research on live organism evolution processes indicates that direct inheritance of selected attributes from parents is not the only mechanism for population development. The so-called soft inheritance (concerned with culture, religion, fashion) is not limited to child–parent type relationships. The study of memetics introduces the term "meme" (Dawkins, 1976) as the basic unit of information that is reproduced on abstract concept exchange. Using the ideas of memetics and the analogy to the terms "gene" and "meme", a new class of genetic algorithms has been coupled with an individual learning procedure. The operation utilizes the selection step in the generation $\mathbf{P}(t)$ of the elitist individuals $\mathbf{E}(t)$, followed by altering the individuals with a local solution search procedure. The algorithms are called memetic algorithms (Moscato, 1999). Their structure, shown in Fig. 5.42, is fairly similar to that of genetic algorithms, illustrated in Fig. 5.41.

Memetic algorithm application in various domains (Krasnogor and Smith, 2005; Wang et al., 2009; Hart, 1994) has shown that their performance is superior to that of standard genetic methods.

Therefore, BNBM optimization should incorporate the model blocks of the stages A, C as well as $A+C$. The block optimization procedure can then be carried out with any memetic algorithm. Memetic algorithms employ a set of base configuration parameters $\mathbf{q}$. The parameters do not change, whereas the tuning configuration parameter $\mathbf{r}$ can be modified during memetic algorithm application. The elements $\mathbf{r}$ can be interpreted as an equivalent of genes and memes. Optimization should

```
procedure Memetic_Algorithm
{  t ← 0 ;
   P(t) ← initialization( ) ;
   while(not condition)
   {  t ← t + 1 ;
      R(t) ← evaluation_and_selection( P(t − 1) ) ;
      D(t) ← reproduction( R(t) ) ;
      N(t) ← evaluation_and_selection( P(t − 1) ) ;
      P(t) ← ( P(t − 1) + D(t) − N(t) ) ;
      E(t) ← evaluation_and_selection( P(t) ) ;
      E(t) ← local_optimisation( E(t) ) ;
   }
}
```

Fig. 5.42 Simplified layout of a memetic algorithm

include the granularity level of the value set of additional variables (Bargiela and Pedrycz, 2003). The graininess change will lead to the first stage block structure change. The problem of granularity level optimization remains to be solved in future research studies.

5.6.7 Implementation in the DiaSter Environment

The BNBM package of the *DiaSter* system includes an empty BNBM instance as well as tools for model management. The package components allow model development, identification, tuning and documentation development in the *MITforRD* platform environment. Moreover, the *DiaSter* system includes an implementation of selected BNBM components. The model's first stage is represented by blocks for developing a system of parallel one-class classifiers of the nearest neighbor type. The classifier learning algorithm accounts for the possibility of defining erroneously classified examples. The second stage features adjustment calculus blocks for adjusting additional variables. Finally, the third stage contains blocks for implementing a belief network. The blocks utilize the junction tree. The system has a maximization expectation algorithm for network training. Moreover, all configuration parameters of the BNBM, as well as subsequent stages parameters and the belief network are defined in the *MITforRD* platform environment, and a *BNBM*-package-based model can be applied in that environment, too. Also, the model can be used in the *PExSim* platform environment upon importing a complete set of configuration data from the *MITforRD* platform.

Chapter 6
Supervisory Control and Optimization

Piotr Tatjewski, Leszek Trybus, Maciej Ławryńczuk,
Piotr Marusak, Zbigniew Świder, and Andrzej Stec

In the first part of this chapter, structures and algorithms of Model-based Predictive Control (MPC) and on-line process set-point optimization are presented, corresponding to implementations in the *DiaSter* system. The principle of MPC is first recalled, as a special case of the general principle of "open loop with feedback optimal control". MPC is now the most important advanced feedback control technique, widely used for supervisory feedback control and also implemented as a direct control algorithm, mainly where classical PID structures cannot deliver required control performance, due to difficult dynamics, strong interactions, active constraints. It is well known that a prerequisite for a successful MPC application is the use of a sufficiently accurate process model. Different models can be used, leading to different realizations of MPC algorithms. Two basic algorithms with linear models are presented: the Dynamic Matrix Control (DMC) algorithm, being one of the first to be implemented in the industry and still very popular there, and the Generalized Predictive Control (GPC) algorithm. The algorithms are given both in analytical versions (control laws, suitable for simple controllers) and in full numerical versions with constrained optimization problems solved at every sampling instant. A selected MPC algorithm with a non-linear process model, based on non-linear prediction and optimization using linearized models, is presented, considered by the authors to be both simple and very efficient in practice. Next, on-line set-point optimization for the MPC controller, integrated with its operation, is described. In the final part of the section, two example applications using the *DiaSter* modules implementing the presented techniques are discussed (obtained in the simulation mode, using the

Piotr Tatjewski · Maciej Ławryńczuk · Piotr Marusak
Institute of Control and Computation Engineering, Warsaw University of Technology, ul. Nowowiejska 15/19, 00–665 Warsaw, Poland
e-mail: {p.tatjewski,m.lawrynczuk,p.marusak}@ia.pw.edu.pl

Leszek Trybus · Zbigniew Świder · Andrzej Stec
Department of Computer and Control Engineering, Rzeszów University of Technology, ul. W. Pola 2, 35–959 Rzeszów, Poland
e-mail: {ltrybus,swiderzb,astec}@prz-rzeszow.pl

PExSim package): DMC control of concentration in a Continuous Stirred-Tank Reactor (CSTR) and two-dimensional GPC control of composition and temperature.

The second part of the chapter describes two basic methods for automatic tuning (self-tuning) of PID loops, i.e., the step response and relay control, followed by an adaptive algorithm. Having the plant step response, 1-st and 2-nd order models with delay are identified using least-squares approximation. Depending which one turns out better, a PI or a PID controller is selected. The settings are chosen by time constant cancellation and from the specification of overshoot or the settling time. Relay control gives a single point on the frequency characteristic of the plant. Having phase and gain margin specifications, PID settings can be calculated. However, for a no-overshoot case, a non-standard structure of the PID controller is needed. Relay control admits asymmetry. The adaptive algorithm employs three template functions expressing overshoot, damping and frequency in terms of two settings of the controller. The target point on such a plane is given by overshoot and damping specifications, or by overshoot and frequency. The actual operating point corresponds to a recent transient. The distance between the two points specifies corrections of the settings made by the adaptive controller. The last part presents input/output structures of PID, self-tuning and adaptive control blocks, together with a set-point generator of internal or external set-points. The latter comes from the higher level of the *DiaSter* software, e.g., from MPC or optimization.

6.1 Predictive Control and Process Set-Point Optimization

There are a few reasons for the tremendous success of the MPC technology in the last decades, both in industrial applications as well as in the reception by the research community, (see, e.g., the works of Morari and Lee (1999), Allgöwer et al. (1999), Eder (1999), Mayne et al. (2000), Maciejowski (2002), Qin and Badgwell (2003), Rossiter (2003), Blevins et al. (2003) or Tatjewski (2007)). MPC algorithms can directly take into account constraints on both process inputs and outputs, which often decide on the quality, effectiveness and safety of production. They generate process inputs taking into account internal interactions within the process, due to the direct use of a model. Thus, they can be applied to processes with difficult dynamics and to multivariable control, even when the numbers of manipulated and controlled variables are uneven. Last but not least, the principle of operation of these algorithms is comprehensible and relatively easy to explain to engineering and operating staff, which is an important aspect when introducing new techniques into industrial practice.

The most important factor regarding a successful application of an MPC algorithm is the quality of the process model. Depending on the kind of the model used, different realizations of MPC algorithms are obtained. The DMC algorithm, one of the first and most successful in the industry, uses step response process models. In the GPC algorithm, process models in the form of discrete transfer functions are utilized. These two algorithms will be described in the following sections.

When feedback control is combined with on-line process optimization, frequent and possibly large changes of set-points for feedback controllers can occur, which can lead to the necessity to apply non-linear feedback control to maintain high performance, at least for strongly non-linear processes. Therefore, non-linear MPC techniques have been developed, (see, e.g., the works of Mayne et al. (2000), Maciejowski (2002), Qin and Badgwell (2003), Rossiter (2003) or Tatjewski (2007)). One of the most practical ones seems to be the MPC-NPL (MPC with Non-linear Prediction and Linearization) algorithm structure, with non-linear prediction and on-line model linearization for dynamic optimization (which is then a quadratic programming problem).

The application of powerful computers and introduction of advanced control algorithms increased the performance of feedback control systems and made it possible to apply on-line economic process optimization. In particular, on-line adjustment of set-point values used in MPC algorithms as target values is of primary importance.

6.1.1 Principle of Model-Based Predictive Control

The general principle of model-based predictive control can be described as follows: At each consecutive sampling instant k (i.e., continuous time kT_p, where T_p denotes the controller sampling period), $k = 0, 1, \ldots$, having

- a dynamic process model, together with an assumed model of disturbances (uncontrolled process inputs) and models of constraints;
- measurements of current and past process outputs, together with past values of control inputs (manipulated variables);
- known or assumed trajectories of set-points for the controlled variables (controlled process outputs) for an assumed horizon of prediction,

the control inputs $\mathbf{u}(k) = \mathbf{u}(k|k), \mathbf{u}(k+1|k), \ldots, \mathbf{u}(k+N_u-1|k)$ are calculated, assuming $\mathbf{u}(k+p|k) = \mathbf{u}(k+N_u-1|k)$ for $p \geq N_u$, where N_u is the *control horizon*. The employed notation "$\mathbf{u}(k+p|k)$" means a prediction of the control inputs for the future sampling instant $k+p$, performed at the current instant k. The control inputs are calculated in such a way as to minimize differences between the predicted controlled outputs $\mathbf{y}(k+p|k)$ and the required reference values—the future set-points $\mathbf{y}^{sp}(k+p|k)$, over the *prediction horizon* N ($p = 1, 2, ..., N$). The minimization of differences is understood in the sense of minimizing a selected criterion of the control quality. Then, only the first element of the calculated sequence of control inputs is applied to the process, i.e., the control input $\mathbf{u}(k) = \mathbf{u}(k|k)$. At the next sampling instant $(k+1)$, there occurs a new measurement of the process outputs and the whole procedure is repeated, with the prediction horizon of the same length N but shifted by one step forward. Thus, the principle of a *receding horizon* is used, called also the repetitive control principle, (see, e.g., the work of Findeisen (1974)).

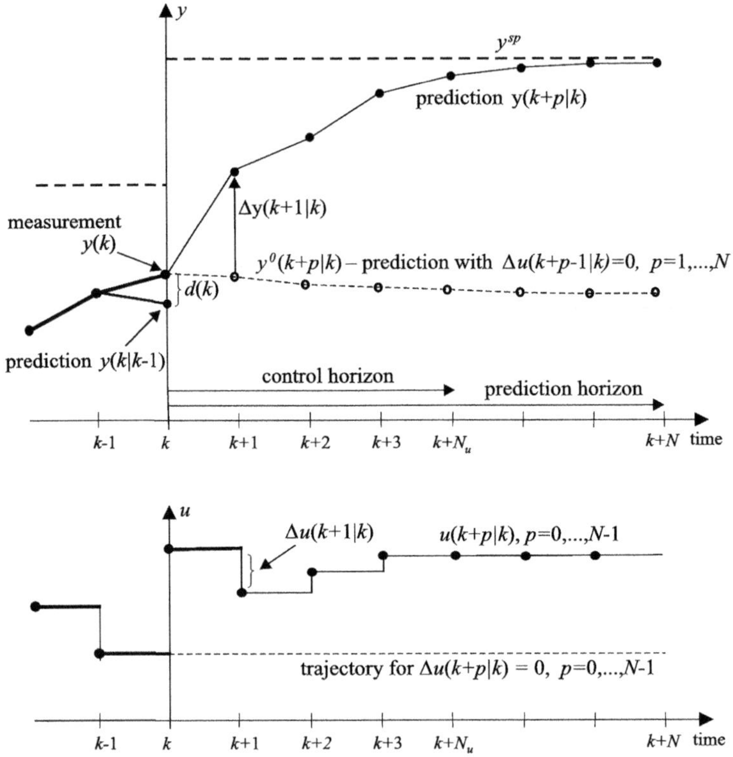

Fig. 6.1 Illustration of the principle of predictive control

The principle of predictive control, for the case of a SISO (single-input single-output) process, is presented in Fig. 6.1, where the horizontal axis represents the discrete time with k being a current sampling instant at which a decision about the current process input signal $u(k) = u(k|k)$ is to be made (with the value constant over the whole sampling time interval $[kT_p, (k+1)T_p)$. The variables which are needed for calculating the input value $u(k)$ are presented as appropriate trajectories of the process control input and the controlled output. The figure presents two trajectories of both the controlled output and the control input as well as a trajectory of the set-point:

- a predicted controlled output trajectory $y^0(k+p|k)$, $p = 1,2,...,N$, corresponding to the situation where the process input is kept constant over the entire prediction horizon, with the value $u(k-1)$ calculated at the preceding sampling instant, i.e., $u(k+p-1) = u(k-1)$ for each $p = 1,2,...,N$. Trajectories corresponding to this case, of both the output and the input, are presented as dashed lines. The trajectory $y^0(k+p|k)$ defined in this way presents the future process outputs as dependent on the *previous* inputs only; we have no influence on this trajectory at a current time k. Therefore, it is often described as a *free component of the predicted output trajectory* (in short, a free output trajectory);

- a predicted controlled output trajectory $y(k+p|k)$, dependent both on *past* and *future* control inputs, i.e., on past inputs up to the last one $u(k-1)$ and future inputs $u(k+p-1|k)$, $p=1,2,...,N-1$, which are calculated at a current time k. It is assumed that the control horizon N_u (precisely, the horizon of process control input variability) can be shorter than the prediction horizon N, $N_u \leq N$. The trajectories listed are presented as continuous curves in Fig. 6.1. Thin fragments of the curves denote predicted parts of the input and output trajectories, whereas thick parts denote the output trajectory which has already been realized (measured) and the input trajectory applied earlier together with the last element being applied to the plant at the current sampling instant;
- a known or foreseen trajectory of the set-points for the controlled output $y^{sp} = y^{sp}(k+p|k)$, $p=1,2,...,N$, presented as a thick dashed line. The set-point for the process output presented in Fig. 6.1 underwent a step change at sampling instant k and then remains constant. Generally, it can be varying over the prediction horizon.

A model of the process used for calculating the future process control inputs is usually only an approximation of reality. Further, there is uncertainty in the uncontrolled inputs, which can be inaccurately measured or not measured at all. Therefore, the output predictions usually differ from the (later) measured values. This fact is depicted in Fig. 6.1 as an unmeasured disturbance $d(k)$, $d(k) = y(k) - y(k|k-1)$, occurring at the process output at sampling instant k, where $y(k|k-1)$ is the process output value predicted for the sampling instant k at the preceding one, $(k-1)$. The dependency of the trajectory of the process output (evaluated at sampling instant k) on the currently measured value $\mathbf{y}(k)$, and not on the value $y(k|k-1)$, means that *discrete output feedback* is applied in the control system.

The determination of the control inputs for the present sampling instant k and for the future instants $k+p$ in the control horizon, $p=1,...,N$, is realized in predictive algorithms on the basis of a process model, by minimizing a selected cost function describing the control quality over the prediction horizon. A prime component of this function is the *cost of predicted control errors*, i.e., deviations of the predicted outputs from the set-points. Moreover, it is also typical to include into the cost function penalties for control input changes. Considering the two mentioned components, the following *cost function* (*objective function*) of the predictive control can be formulated, for the calculation of the optimal process input trajectory over the control horizon:

$$J(k) = \sum_{p=N_1}^{N} (\mathbf{y}^{sp}(k+p|k) - \mathbf{y}(k+p|k))^T \mathbf{Q} (\mathbf{y}^{sp}(k+p|k) - \mathbf{y}(k+p|k)) + \sum_{p=0}^{N_u-1} \triangle\mathbf{u}(k+p|k)^T \mathbf{R} \triangle\mathbf{u}(k+p|k)$$
$$= \sum_{p=N_1}^{N} \|\mathbf{y}^{sp}(k+p|k) - \mathbf{y}(k+p|k)\|^2_{\mathbf{Q}} + \sum_{p=0}^{N_u-1} \|\triangle\mathbf{u}(k+p|k)\|^2_{\mathbf{R}}, \quad (6.1)$$

where the vectors $\mathbf{y}^{sp}(k+p|k)$ and $\mathbf{y}(k+p|k)$ are of dimensionality $n_y = \dim y$ (number of the controlled outputs), while the vector of process input increments $\triangle\mathbf{u}(k+p|k)$ is of dimensionality $n_u = \dim u$.

In (6.1), the predicted control errors $\mathbf{y}^{sp}(k+p|k) - \mathbf{y}(k+p|k)$ are considered, starting from $k+N_1$ until the end of the prediction horizon N, where $1 \leq N_1 \leq N$. A value $N_1 > 1$ is reasonable if there is a delay in the process causing a lack of reaction of the outputs at first $N_1 - 1$ sampling instants, $k+1, ..., k+N_1-1$, to the change of the control input at instant k. However, often $N_1 = 1$ is assumed even for a general case with delays, to simplify the notation. Certainly, it is not an error, but if implemented, it causes additional, useless computations. The length of the control horizon N_u must satisfy the constraints $0 < N_u \leq N$. It is usually assumed that $N_u < N$, as this results in decreased dimensionality of the controller dynamic optimization problem and thus leads to reduced computational load.

The matrix $\mathbf{Q}$ is a diagonal matrix of weights enabling the scaling of the influence of different components of the vector of predicted control errors in the cost function, it may be also time dependent, in general,)see, e.g., the work of Tatjewski (2007)). The role of the matrix $\mathbf{R}$ is, in turn, not only to introduce a scaling of individual components of the vector of control input moves, but first of all to introduce a scaling of the whole second sum in (6.1) against the first one representing the predicted control errors. In the simplest case without scaling between individual components of the vector of the input moves, we obtain $\mathbf{R} = \lambda\mathbf{I}$, where $\mathbf{I}$ is a unity matrix of dimension $n_u \times n_u$. In the case without scaling also in the first sum ($\mathbf{Q} = \mathbf{I}$), the cost function (6.1) takes the following simpler and often met form (see, e.g., the work of Rossiter (2003)):

$$J(k) = \sum_{p=N_1}^{N} \|\mathbf{y}^{sp}(k+p|k) - \mathbf{y}(k+p|k)\|^2 + \lambda \sum_{p=0}^{N_u-1} \|\triangle\mathbf{u}(k+p|k)\|^2, \quad (6.2)$$

where the scalar $\lambda \geq 0$ defines a weight attributed to damping the input moves versus the reduction of the control errors. Let us emphasize that assuming $\lambda = 0$ is possible, but when there are no constraints on amplitudes and rates of change of the process inputs, this often leads to practically unacceptable controller properties, particularly to huge input changes and a lack of robustness against modeling errors.

To calculate the values $\mathbf{y}(k+p|k)$ in the prediction horizon, $p = N_1, ..., N$, it is necessary to have a *process model*. Generally, it can be a non-linear model. So far, MPC algorithms with *linear process models* have been of the greatest importance.

First of all, the range of direct applications of these algorithms is fairly wide. Secondly, they constitute a basis for constructing relatively simple and often very efficient non-linear algorithms with linearized models.

In a linear case, applying the principle of superposition, it is possible to present the trajectory of predicted outputs $\mathbf{y}(k+p|k)$ in the form of a sum of a *free trajectory* $\mathbf{y}^0(k+p|k)$ dependent only on the realized (past) process inputs and a trajectory $\triangle\mathbf{y}(k+p|k)$ dependent only on the decision variables (current and future moves of the process inputs $\triangle\mathbf{u}(k+p|k)$). Thus, the trajectory $\triangle\mathbf{y}(k+p|k)$, $p=N_1,...,N$ is called a *forced output trajectory* (precisely, it is a forced component of the predicted output trajectory). Thus we have

$$\mathbf{y}(k+p|k)=\mathbf{y}^0(k+p|k)+\triangle\mathbf{y}(k+p|k),\quad p=N_1,...,N. \tag{6.3}$$

The above partition, although not necessary for the realization of a predictive control algorithm, is computationally convenient, because the values $\mathbf{y}^0(k+p|k)$, as dependent only on the past of the process, are calculated by the control algorithm *only once* for the current sampling instant k and remain then *fixed parameters* in further dynamic optimization of the future input changes.

One of the appealing properties of the MPC technique is the ability to take various constraints explicitly into account. The following constraints are important:

- constraints on *values (amplitudes) and increments of process control inputs*:

$$\mathbf{u}_{\min}\le\mathbf{u}(k+p|k)\le\mathbf{u}_{\max},\quad p=0,1,...,N_u-1, \tag{6.4}$$

$$-\triangle\mathbf{u}_{\max}\le\triangle\mathbf{u}(k+p|k)\le\triangle\mathbf{u}_{\max},\quad p=0,1,...,N_u-1; \tag{6.5}$$

- constraints on *values (amplitudes) of process outputs*:

$$\mathbf{y}(k+p|k)\le\mathbf{y}_{\max},\quad p=N_1,N_1+1,...,N, \tag{6.6}$$

$$\mathbf{y}(k+p|k)\ge\mathbf{y}_{\min},\quad p=N_1,N_1+1,...,N. \tag{6.7}$$

The controlled variables, for which desired (set-point) values are given, are in general not the only process output variables directly influenced by the MPC controller. There may be also components of the process output vector for which desired (reference) values are not given, but only the constraints are prescribed (one- or two-sided). A concentration of the chlorine in a municipal water distribution system may serve here as an example; its value should usually be only within prescribed minimal and maximal limits. Such process outputs are often called *constraint variables* (as opposed to controlled variables), see, e.g., the work of Blevins et al. (2003). The above-formulated basic elements of the MPC algorithm, i.e., the cost function and the constraints on outputs, also cover this case. One should only assign zero values to diagonal elements of the matrix $\mathbf{Q}$ corresponding to the constraint variables and formulate the constraints for these variables in the form (6.6) and/or (6.7). Therefore, we shall not distinguish controlled and constraint variables in the process output vector until necessary, in order to avoid unnecessary complication of the resulting notation.

The reader interested in a detailed presentation of the history and basics of predictive control is referred to review, e.g., the papers and books by Mayne et al. (2000), Maciejowski (2002), Qin and Badgwell (2003), Rossiter (2003), and Tatjewski (2007). In further sections, two most popular MPC algorithms with linear process models will be presented—the DMC and GPC algorithms. These techniques will then be a starting point for the description of computationally relatively simple and efficient MPC algorithms for non-linear process models, including fuzzy models (in the Takagi–Sugeno structure) and neural models.

6.1.2 Dynamic Matrix Control Algorithm

In the DMC algorithm the process dynamics are modeled by discrete step responses (Cutler and Ramaker, 1980; Garcia and Morshedi, 1986). Denoting subsequent coefficients of the unit step response of a SISO plant by s_j, we have

$$y(k+1) = y(0) + \sum_{j=0}^{k} s_j \triangle u(k-j). \tag{6.8}$$

Assuming that the unmeasured disturbances affect the process output, at current sampling instant k they are modeled as a difference between the current measured output value $y(k)$ and the value of the output predicted at the previous sampling instant, i.e.,

$$d(k) = y(k) - [y(0) + \sum_{j=1}^{k} s_j \triangle u(k-j)]. \tag{6.9}$$

In the DMC algorithm a lack of knowledge about future changes of the disturbances on a prediction horizon is assumed, therefore the following model is used (called the *constant output disturbance model*, or the DMC disturbance type model, see, e.g., the work of Maciejowski (2002)):

$$d(k+1|k) = d(k+2|k) = \cdots = d(k+N|k) = d(k). \tag{6.10}$$

For asymptotically stable processes (without integrated or runaway responses), the output stabilizes, after a step change in the input, at a certain value s_∞, where $\lim_{k\to\infty} s_k = s_\infty$. Therefore, it is enough to know a finite number, say D, of coefficients of the step response, i.e., the number of steps after which the value of the step response can be treated as constant, equal to the static process gain $k_m = s_\infty$ (D can be called the horizon of the process dynamics). The estimation $D \cong (T_0 + (3 \div 4)T)/T_p$ usually proves correct, where T_0 is a process time delay and T is a process dominant time constant.

Using the process model (6.8), together with the disturbance model (6.9), (6.10), it is easy to derive (Tatjewski, 2007) the following formula for the prediction of the process output at sample k, which is the sum of the forced component $\triangle \mathbf{y}(k+p|k)$ and the free component $y^0(k+p|k)$ of the predicted output trajectory:

$$y(k+p|k) = \triangle y(k+p|k) + y^0(k+p|k), \tag{6.11}$$

$$\triangle y(k+p|k) = \sum_{j=1}^{p} s_j \triangle u(k+p-j|k), \;\; p = 1,2,...,N, \tag{6.12}$$

$$y^0(k+p|k) = y(k) + \sum_{j=1}^{D-1} (s_{j+p} - s_j) \triangle u(k-j), \;\; p = 1,...,N. \tag{6.13}$$

Consider now a multivariable, *multi-input multi-output* process with n_y controlled outputs and n_u control inputs. We have the process model in the form of a set of $n_y \cdot n_u$ finite step responses $\{s_l^{ij},\, l = 1,2,...,D\}$, where i indexes controlled outputs, $i = 1,2,...,n_y$, and j indexes process control inputs, $j = 1,2,...,n_u$. Therefore, components of the vector $s^{ij} = [s_1^{ij}\; s_2^{ij}\; ...\; s_D^{ij}]$ are elements of a step response of the i-th output to a unit step on the j-th input (when all other inputs are kept constant). The dynamics horizon D is taken as common for all responses, i.e., it can be assumed that $s_l^{ij} = \text{const.}$ for $l \geq D$.

Let us define the *multivariable step response* $\{\mathbf{S}_1, \mathbf{S}_2, \mathbf{S}_3, ...\}$ as consisting of matrices:

$$\mathbf{S}_l = \begin{bmatrix} s_l^{11} & s_l^{12} & s_l^{13} & \cdots & s_l^{1n_u} \\ s_l^{21} & s_l^{22} & s_l^{23} & \cdots & s_l^{2n_u} \\ s_l^{31} & s_l^{32} & s_l^{33} & \cdots & s_l^{3n_u} \\ \vdots & \vdots & \vdots & \ddots & \vdots \\ s_l^{n_y1} & s_l^{n_y2} & s_l^{n_y3} & \cdots & s_l^{n_yn_u} \end{bmatrix}, \quad l = 1,2,...,D, \tag{6.14}$$

each of them further consisting of coefficients s_l^{ij} of all step responses, $i = 1,2,...,n_y$, $j = 1,2,...,n_u$, corresponding to the sampling instant l. Therefore, the process is represented by D matrices $\mathbf{S}_l$ of dimension $n_y \times n_u$, instead of $n_y \cdot n_u$ vectors of dimension D. This model representation makes it possible *to directly apply all the formulae for SISO processes to the case of a MIMO process*; only the matrices $\mathbf{S}_l$ must be inserted in places of the scalar coefficients s_l of a single-step response, $l = 1,2,...,D$. The formula describing predicted outputs is now as follows:

$$\begin{aligned} \mathbf{y}(k+p|k) & \\ &= \mathbf{y}^0(k+p|k) + \triangle \mathbf{y}(k+p|k) \\ &= \sum_{j=1}^{p} \mathbf{S}_j \triangle \mathbf{u}(k+p-j|k) + \mathbf{y}(k) \\ &\quad + \sum_{j=1}^{D-1} (\mathbf{S}_{j+p} - \mathbf{S}_j) \triangle \mathbf{u}(k-j). \end{aligned} \tag{6.15}$$

Define the vectors

$$\mathscr{Y}^{sp}(k) = \begin{bmatrix} \mathbf{y}^{sp}(k+N_1|k) \\ \vdots \\ \mathbf{y}^{sp}(k+N|k) \end{bmatrix}, \quad \mathscr{Y}^{0}(k) = \begin{bmatrix} \mathbf{y}^{0}(k+N_1|k) \\ \vdots \\ \mathbf{y}^{0}(k+N|k) \end{bmatrix},$$

$$\Delta\mathscr{Y}(k) = \begin{bmatrix} \triangle\mathbf{y}(k+N_1|k) \\ \vdots \\ \triangle\mathbf{y}(k+N|k) \end{bmatrix}, \quad \Delta\mathscr{U}(k) = \begin{bmatrix} \triangle\mathbf{u}(k|k) \\ \vdots \\ \triangle\mathbf{u}(k+N_u-1|k) \end{bmatrix},$$

$$\mathscr{Y}^{pred}(k) = \mathscr{Y}^{0}(k) + \Delta\mathscr{Y}(k) = [\mathbf{y}(k+N_1|k)^T \cdots \mathbf{y}(k+N|k)^T]^T,$$

and the matrices $\underline{\mathbf{Q}}=\mathrm{diag}\{\mathbf{Q},...,\mathbf{Q}\}$, $\underline{\mathbf{R}}=\mathrm{diag}\{\mathbf{R},...,\mathbf{R}\}$. Then the cost function can be written in the form

$$J(k) = \left\| [\mathscr{Y}^{sp}(k) - \mathscr{Y}^{0}(k)] - \Delta\mathscr{Y}(k) \right\|^2_{\underline{\mathbf{Q}}} + \|\Delta\mathscr{U}(k)\|^2_{\underline{\mathbf{R}}} \tag{6.16}$$

or, if additionally $\mathbf{Q} = \mathbf{I}$ and $\mathbf{R} = \lambda\mathbf{I}$, as

$$J(k) = \left\| [\mathscr{Y}^{sp}(k) - \mathscr{Y}^{0}(k)] - \Delta\mathscr{Y}(k) \right\|^2 + \lambda \|\Delta\mathscr{U}(k)\|^2. \tag{6.17}$$

Define also the vectors

$$\mathscr{Y}(k) = \begin{bmatrix} \mathbf{y}(k) \\ \vdots \\ \mathbf{y}(k) \end{bmatrix}, \quad \Delta\mathscr{U}^{P}(k) = \begin{bmatrix} \triangle\mathbf{u}(k-1) \\ \vdots \\ \triangle\mathbf{u}(k-(D-1)) \end{bmatrix},$$

where $\dim\mathscr{Y}(k) = n_{\mathscr{Y}} = n_y \cdot (N - N_1 + 1)$, $\dim\triangle\mathscr{U}^{P}(k) = n_u \cdot (D-1)$. Now, the vectors of free and forced output responses can be presented in the form

$$\mathscr{Y}^{0}(k) = \mathscr{Y}(k) + \mathbf{M}^{P}\triangle\mathscr{U}^{P}(k), \tag{6.18}$$

$$\Delta\mathscr{Y}(k) = \mathbf{M}\Delta\mathscr{U}(k), \tag{6.19}$$

where

$$\mathbf{M}^{P} = \begin{bmatrix} \mathbf{S}_{1+N_1} - \mathbf{S}_1 & \mathbf{S}_{2+N_1} - \mathbf{S}_2 & \mathbf{S}_{3+N_1} - \mathbf{S}_3 & \cdots & \mathbf{S}_{D-1+N_1} - \mathbf{S}_{D-1} \\ \mathbf{S}_{2+N_1} - \mathbf{S}_1 & \mathbf{S}_{3+N_1} - \mathbf{S}_2 & \mathbf{S}_{4+N_1} - \mathbf{S}_3 & \cdots & \mathbf{S}_{D+N_1} - \mathbf{S}_{D-1} \\ \vdots & \vdots & \vdots & \ddots & \vdots \\ \mathbf{S}_{N+1} - \mathbf{S}_1 & \mathbf{S}_{N+2} - \mathbf{S}_2 & \mathbf{S}_{N+3} - \mathbf{S}_3 & \cdots & \mathbf{S}_{N+D-1} - \mathbf{S}_{D-1} \end{bmatrix}, \tag{6.20}$$

$$\mathbf{M} = \begin{bmatrix} \mathbf{S}_{N_1} & \mathbf{S}_{N_1-1} & \cdots & \mathbf{S}_1 & \mathbf{0} & \cdots & \mathbf{0} \\ \mathbf{S}_{N_1+1} & \mathbf{S}_{N_1} & \cdots & \mathbf{S}_2 & \mathbf{S}_1 & \cdots & \mathbf{0} \\ \vdots & \vdots & \vdots & \vdots & \vdots & \ddots & \vdots \\ \mathbf{S}_{N_u} & \mathbf{S}_{N_u-1} & \cdots & \mathbf{S}_{N_u-N_1+1} & \mathbf{S}_{N_u-N_1} & \cdots & \mathbf{S}_1 \\ \mathbf{S}_{N_u+1} & \mathbf{S}_{N_u} & \cdots & \mathbf{S}_{N_u-N_1+2} & \mathbf{S}_{N_u-N_1+1} & \cdots & \mathbf{S}_2 \\ \vdots & \vdots & \vdots & \vdots & \vdots & \ddots & \vdots \\ \mathbf{S}_N & \mathbf{S}_{N-1} & \cdots & \mathbf{S}_{N-N_1+1} & \mathbf{S}_{N-N_1} & \cdots & \mathbf{S}_{N-N_u+1} \end{bmatrix}. \tag{6.21}$$

According to definitions of the vectors $\mathscr{Y}^0(k)$, $\Delta\mathscr{Y}(k)$, $\Delta\mathscr{U}^P(k)$ and $\Delta\mathscr{U}(k)$, the dimension of the matrix $\mathbf{M}^P$ is $n_{\mathscr{Y}} \times n_{\Delta\mathscr{U}^P} = n_y \cdot (N - N_1 + 1) \times n_u \cdot (D - 1)$, while the dynamic matrix $\mathbf{M}$ is of dimension $n_{\mathscr{Y}} \times n_{\Delta\mathscr{U}} = n_y \cdot (N - N_1 + 1) \times n_u \cdot N_u$.

Using (6.19), the cost function can be written in the form

$$J(k) = \left\| [\mathscr{Y}^{sp}(k) - \mathscr{Y}^0(k)] - \mathbf{M}\Delta\mathscr{U}(k) \right\|^2_{\underline{\mathbf{Q}}} + \|\Delta\mathscr{U}(k)\|^2_{\underline{\mathbf{R}}}, \tag{6.22}$$

which is a strictly convex function, provided $\mathbf{R} > 0$.

6.1.2.1 Analytic (Explicit, Unconstrained) DMC Algorithm

If there are no constraints (or the constraints are neglected), then the *vector of optimal input moves* minimizing the cost function follows from the necessary optimality conditions; it is easy to calculate that (Tatjewski, 2007)

$$\Delta\widehat{\mathscr{U}}(k) = [\mathbf{M}^T\underline{\mathbf{Q}}\mathbf{M} + \underline{\mathbf{R}}]^{-1}\mathbf{M}^T\underline{\mathbf{Q}}[\mathscr{Y}^{sp}(k) - \mathscr{Y}^0(k)] = \mathbf{K}[\mathscr{Y}^{sp}(k) - \mathscr{Y}^0(k)], \tag{6.23}$$

where

$$\mathbf{K} = [\mathbf{M}^T\underline{\mathbf{Q}}\mathbf{M} + \underline{\mathbf{R}}]^{-1}\mathbf{M}^T\underline{\mathbf{Q}}. \tag{6.24}$$

Denote

$$\mathbf{K} = \begin{bmatrix} \overline{\mathbf{K}}_1 \\ \overline{\mathbf{K}}_2 \\ \vdots \\ \overline{\mathbf{K}}_{N_u} \end{bmatrix} = \begin{bmatrix} \mathbf{K}_{1,1} & \mathbf{K}_{1,2} & \cdots & \mathbf{K}_{1,N-N_1+1} \\ \mathbf{K}_{2,1} & \mathbf{K}_{2,2} & \cdots & \mathbf{K}_{2,N-N_1+1} \\ \vdots & \vdots & \ddots & \vdots \\ \mathbf{K}_{N_u,1} & \mathbf{K}_{N_u,2} & \cdots & \mathbf{K}_{N_u,N-N_1+1} \end{bmatrix}, \tag{6.25}$$

where each submatrix $\overline{\mathbf{K}}_i$ is of dimension $n_u \times n_{\mathscr{Y}} = n_u \times n_y(N - N_1 + 1)$, and each submatrix $\mathbf{K}_{i,j}$ is of dimension $n_u \times n_y$. The optimal control input increments corresponding to the current sampling instant are applied to the process only, i.e., the vector $\triangle\hat{\mathbf{u}}(k|k)$ consisting of first n_u elements of the vector of the optimal solution (6.23):

$$\triangle\hat{\mathbf{u}}(k) = \triangle\hat{\mathbf{u}}(k|k) = \overline{\mathbf{K}}_1[\mathscr{Y}^{sp}(k) - \mathscr{Y}^0(k)]. \tag{6.26}$$

Therefore, the obtained *control law* is linear feedback from the difference between the set-point trajectory and the predicted free trajectory.

The DMC control law can be presented in a form showing a direct dependency of the optimal controller output on the past process inputs (i.e., past controller outputs) and on the process outputs. Presenting the matrix $\mathbf{M}^P$ in the form

$$\mathbf{M}^P = \left[\mathbf{M}_1^P\, \mathbf{M}_2^P \cdots \mathbf{M}_{D-1}^P\right], \tag{6.27}$$

where each submatrix $\mathbf{M}_j^P$ is of dimension $n_{\mathscr{Y}} \times n_u = n_y(N - N_1 + 1) \times n_u$, and using the structure (6.25) of the matrix $\mathbf{K}$, we have

$$\begin{aligned}\triangle\hat{\mathbf{u}}(k) &= \overline{\mathbf{K}}_1[\mathscr{Y}^{sp}(k) - \mathscr{Y}(k) - \mathbf{M}^P \triangle \mathscr{U}^P(k)] \\ &= \sum_{p=N_1}^{N} \mathbf{K}_{1,p-N_1+1}[\mathbf{y}^{sp}(k+p|k) - \mathbf{y}(k)] - \sum_{j=1}^{D-1} \mathbf{K}_j^u \triangle \mathbf{u}(k-j),\end{aligned} \tag{6.28}$$

where

$$\mathbf{K}_j^u = \overline{\mathbf{K}}_1 \mathbf{M}_j^P, \quad j = 1, 2, ..., D-1. \tag{6.29}$$

The formula (6.28) presents the structure of the DMC control law designed analytically in the case without inequality constraints, or when the existence of these constraints has been consciously neglected during the design. Thus, this controller can be referred to as an *unconstrained, explicit DMC controller*; the description *analytical DMC controller* can also be encountered (Tatjewski, 2007).

In continuous process control, one usually does not know when changes in the set-point values will occur in the future, i.e., in the prediction horizon. Therefore, it is common practice in the design of MPC algorithms for stabilization tasks to assume that the set-points are constant over the prediction horizon and equal to the current values,

$$\mathbf{y}^{sp}(k+N_1|k) = \mathbf{y}^{sp}(k+N_1+1|k) = \cdots = \mathbf{y}^{sp}(k+N|k) = \mathbf{y}^{sp}(k). \tag{6.30}$$

Then the control law (6.28) simplifies to the form

$$\begin{aligned}\triangle\hat{\mathbf{u}}(k) &= \sum_{p=N_1}^{N} \mathbf{K}_{1,p-N_1+1}[\mathbf{y}^{sp}(k) - \mathbf{y}(k)] - \sum_{j=1}^{D-1} \mathbf{K}_j^u \triangle \mathbf{u}(k-j) \\ &= \mathbf{K}^e[\mathbf{y}^{sp}(k) - \mathbf{y}(k)] - \sum_{j=1}^{D-1} \mathbf{K}_j^u \triangle \mathbf{u}(k-j),\end{aligned} \tag{6.31}$$

where

$$\mathbf{K}^e = \sum_{p=N_1}^{N} \mathbf{K}_{1,p-N_1+1}. \tag{6.32}$$

The structure of the linear control law (6.31) is illustrated in Fig. 6.2, where the box with a diagonal matrix of discrete transfer functions $\frac{1}{1-z^{-1}}\mathbf{I}$ implements discrete vector integration—a summation of consecutive increments of the process input vector in order to transform the control input increments into the control input values. Further, the box in the internal feedback loop fed by consecutive values of the controller output increments $\triangle\mathbf{u}(k)$ should be treated as performing all that is

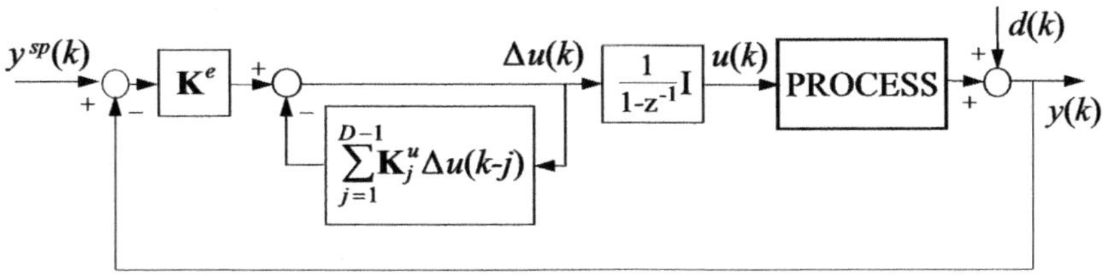

Fig. 6.2 Structure of the analytic (unconstrained) DMC controller, with constant set-points over the prediction horizon

necessary to implement the formula written in there, i.e., also keeping in its memory $D-1$ last control input increments.

In practice, there are always constraints on process inputs resulting from physical limitations of actuators. If these constraints are active, then feeding the block of internal feedback of the MPC controller with previous control increments resulting from the control law formula (6.28) or (6.31) would lead to significant deterioration of the control performance. The reason is that the control increments $\Delta\mathbf{u}(k-j)$ corresponding to the control inputs $\mathbf{u}(k-j)$, which really affected the process, should be used in the controller feedback loop. This is indicated in the structure shown in Fig. 6.3, where $z^{-1}\mathbf{I}$ denotes a diagonal matrix with unit delays on the diagonal, and the non-linear blocks should be also understood as implementing vector constraints, having vectors as inputs and outputs. The structure from Fig. 6.3 implements constraints on both amplitudes and rates of change, it also contains an additional feedback loop constraining integration of the controller signal (correcting the state of the integrator) in a way which is similar to an anti-windup scheme used in structures of PID controllers, (see, e.g., the works of Åström and Wittemark (1997) or Goodwin et al. (2001)). Applying the additional anti-windup loop introduces *full correction of the state of the integrator*, with a unit delay—at a sampling instant k the controller output signal $\Delta\hat{\mathbf{u}}(k)$ is added to the signal $\mathbf{u}(k-1)$ which *actually affected the process* at the previous sampling instant, i.e., to the signal $\mathbf{u}_{\text{int}}(k-1)$, after passing through the amplitude constraining element (the actuator output signal can also be directly used here).

6.1.2.2 DMC Algorithm in a Numeric Version

Treating the constraints on the controller output signal as presented in the previous section, i.e., after the calculation of the unconstrained control law, is generally suboptimal though often leads to acceptable results. Moreover, it is much more difficult to take into account in this way constraints on the process outputs, controlled and/or uncontrolled. Taking the constraints into account explicitly during the calculation of the optimal control signal leads to the necessity to solve a dynamic optimization problem at each sampling instant. This is a quadratic programming problem minimizing the cost function (6.22) under the constraints (6.4), (6.5), (6.6) and (6.7). Reformulating this problem to a standard form required by classic QP procedures,

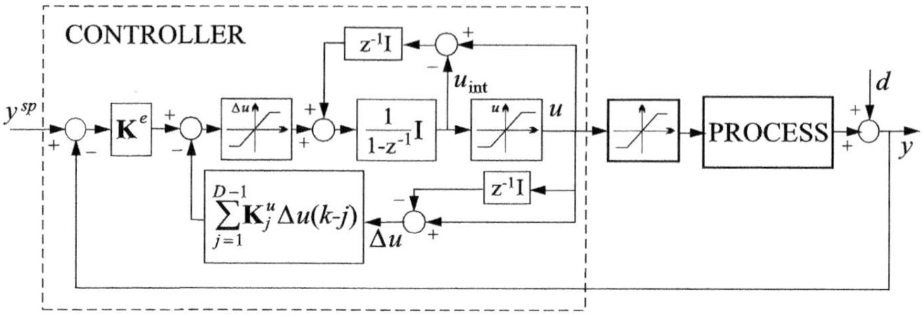

Fig. 6.3 Analytic DMC controller in the structure taking into account constraints on amplitudes and rates of change of the process control inputs

e.g., *QUADPROG* from the *Matlab* package, is simple, (see, e.g., the work of Tatjewski (2007)). The standard QP procedures are effective and reliable, thus enabling efficient implementations of predictive control algorithms.

The quadratic programming problem can be precisely and effectively solved; however, under the condition that there is a solution to this problem, namely, that the feasible set defined by all inequality constraints is not empty (the constraints are not contradictory). Let us recall here only that the feasible set can happen to be empty when there exist constraints on process output variables (controlled and/or constraint variables). Because these constraints are usually soft, i.e., they may be occasionally violated, the simplest and commonly applied technique assuring the feasibility of the QP problem is to treat these constraints via penalty functions. That is, there are additional variables $v_{\min} \in R^{n_y}$, $v_{\max} \in R^{n_y}$ introduced in the optimization problem, and the output constraints are relaxed on the prediction horizon to the form

$$\mathbf{y}(k+p|k) \leq \mathbf{y}_{\max} + \mathbf{v}_{\max}, \tag{6.33}$$

$$-\mathbf{y}(k+p|k) \leq -\mathbf{y}_{\min} + \mathbf{v}_{\min}. \tag{6.34}$$

Further, penalty terms are added to the cost function:

$$\rho_{\max}(\mathbf{v}_{\max})^T \mathbf{v}_{\max} + \rho_{\min}(\mathbf{v}_{\min})^T \mathbf{v}_{\min}, \tag{6.35}$$

and additional inequality constraints $v_{\min} \geq 0$, $v_{\max} \geq 0$ must be added, too. The above formulation is not the most general one possible. It describes the case when there are lower- and upper-bound constraints on all process outputs, which may not always be the case. Further, it does not exploit the possibility to use different penalty coefficients for different components of the process output vector, to simplify the notation.

6.1.3 Generalized Predictive Control Algorithm

In the GPC algorithm, a discrete process model represented by difference equations (or, equivalently, discrete transfer functions) is used (Clarke et al., 1987). The only difference between the DMC and GPC algorithms is that the dynamic matrix $\mathbf{M}$ and the free trajectory of predicted outputs are calculated using this model, and not the step response one. Therefore, techniques showing the ways these quantities are calculated will only be shown. In the literature, (e.g., the research of Camacho and Bordons (2004) or Tatjewski (2007)), a general procedure based on the use of the Bezout identity can be found, with integrated white noise as the disturbance model. However, formulae defining the GPC algorithm can be derived in a simpler way, provided the same disturbance model as in the DMC algorithm is assumed (i.e., constant output disturbance model).

Let us consider first a SISO model of the ARX type in the form

$$y(k) = -a_1 y(k-1) - \cdots - a_{n_A} y(k-n_A) + b_0 u(k-1) + \cdots + b_{n_B} u(k-n_B-1). \tag{6.36}$$

Assuming the input signal in the form of the unity step, the above model directly yields the sequence of step response coefficients $\{s_1, s_2, \ldots\}$, resulting from the equation

$$s_k = -\sum_{i=1}^{\min\{k-1,n_A\}} a_i s_{k-i} + \sum_{i=0}^{\min\{k-1,n_B\}} b_i. \tag{6.37}$$

Finally, these coefficients directly define the dynamic matrix $\mathbf{M}$.

To calculate the elements $y^0(k+p|k)$ of the predicted free output trajectory, the constant process input signal is assumed over the prediction horizon, equal to the value $u(k-1)$ calculated at the previous sampling instant (as in the DMC algorithm). Under this input signal, the predicted values $y^0(k+p|k)$ are calculated sequentially, for $p = 1, \ldots, N$, from the equation

$$y^0(k+p|k) = y(k+p) + d(k), \tag{6.38}$$

where the values $y(k+p)$ are calculated from the equation (6.36) applied for $k = k+p$, but for $k+p-j > k$ inserting in place of yet unknown values $y(k+p-j)$ the just calculated predictions $y^0(k+p-j|k)$. To the free outputs predicted in this way for every sampling instant over the prediction horizon, the value of the disturbance estimate $d(k)$ is added, calculated at sampling instant k and treated as constant over the prediction horizon (assumed constant output disturbance model):

$$d(k) = y(k) - y(k|k-1) = y(k) - [-\sum_{i=1}^{n_A} a_i y(k-i) + \sum_{i=0}^{n_B} b_i u(k-1-i)]. \tag{6.39}$$

The presented technique leads to the following final formula for elements of the free components of the predicted output trajectory (Tatjewski, 2007):

$$y^0(k+p|k) = -\sum_{i=1}^{\min\{n_A,p-1\}} a_i y^0(k+p-i|k) - \sum_{i=\min\{n_A,p-1\}+1}^{n_A} a_i y(k+p-i) + $$
$$+ \sum_{i=0}^{\min\{n_B,p\}} b_i u(k-1) + \sum_{i=\min\{n_B,p\}+1}^{n_B} b_i u(k-1+p-i) + d(k),$$
$$p = 1,2,\ldots,N, \quad (6.40)$$

where values of the disturbance estimate $d(k)$ are defined by (6.39).

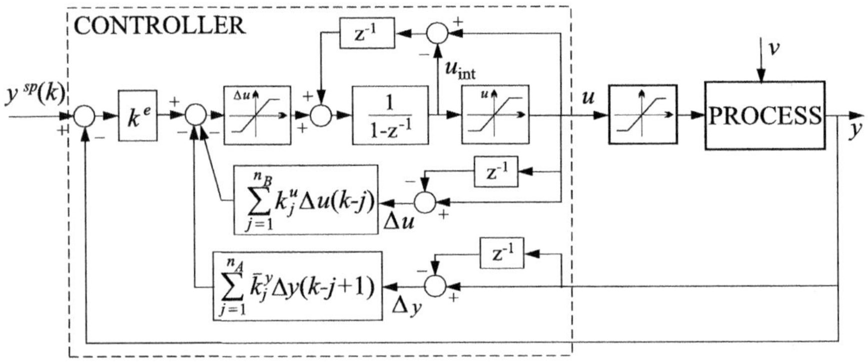

Fig. 6.4 Structure of the analytic GPC controller, with constraints on amplitudes and rates of change of the control inputs taken (a posteriori) into account, supplemented with an anti-windup loop

The formula (6.40) is recurrent and can be directly used for implementing the numerical version of the GPC algorithm, where at each sampling instant first the free output trajectory is calculated and then the quadratic programming problem is solved. Certainly, the derivation of an explicit formula describing the dependency of the free output trajectory on past process outputs and control signals is possible, recurrently using the formula (6.40). In the unconstrained case (or ignoring the constraints when solving the QP problem), this leads to an analytic GPC controller having the structure shown in Fig. 6.4 (Tatjewski, 2007). Comparing the structures of the analytic DMC and GPC controllers, it can be observed that the GPC controller is generally described by a significantly smaller number of feedback coefficients, because usually $D \gg n_A$ and $D \gg n_B$.

The generalization of the model (6.36) to the case of a MIMO plant, with n_u inputs and n_y outputs, is of the form

$$\mathbf{A}(z^{-1})\mathbf{y}(k) = \mathbf{B}(z^{-1})\mathbf{u}(k-1), \quad (6.41)$$

where $\mathbf{A}$ and $\mathbf{B}$ are polynomial matrices:

$$\mathbf{A}(z^{-1}) = \mathbf{1} + \mathbf{A}_1 z^{-1} + \mathbf{A}_2 z^{-2} + \cdots + \mathbf{A}_{n_a} z^{-n_A}, \tag{6.42}$$

$$\mathbf{B}(z^{-1}) = \mathbf{B}_0 + \mathbf{B}_1 z^{-1} + \mathbf{B}_2 z^{-2} + \cdots + \mathbf{B}_{n_b} z^{-n_B}, \tag{6.43}$$

and z^{-1} denotes the unit time delay operator.

Assuming the DMC type disturbance model (constant output disturbance model) for every output,we can easily generalize the equations (6.37) and (6.40) to the MIMO case, using directly the process model. Assuming that the matrix $\mathbf{A}(z^{-1})$ is diagonal, denote by $A^m(z^{-1})$ and $B^{m,j}(z^{-1})$ polynomials (elements of the matrices $\mathbf{A}(z^{-1})$ and $\mathbf{B}(z^{-1})$) defining the response of the m-th output to the change of the n_u inputs:

$$y_m(k) = -\sum_{i=1}^{n_A} a_i^m y_m(k-i) + \sum_{j=1}^{n_u} \sum_{i=0}^{n_B} b_i^{m,j} u_j(k-1-i), \tag{6.44}$$

$m = 1,...,n_y$. (It was assumed here, without loss of generality, that all polynomials are of the same degree, $n_A^m = n_A$, $n_B^{m,j} = n_B$.) Using (6.44), formulae for elements of step responses can be easily obtained, for every pair (j-th) input–(m-th) output:

$$s_k^{m,j} = -\sum_{i=1}^{\min\{k-1,n_A\}} a_i^m s_{k-i}^{m,j} + \sum_{i=0}^{\min\{k-1,n_B\}} b_i^{m,j}, \tag{6.45}$$

$m = 1,...,n_y$, $j = 1,...,n_u$ (compare with (6.37)). Knowing elements of the multivariable step-response, elements of the dynamic matrix can be directly calculated (as was shown in the DMC case), and thus elements of the forced trajectory of the predicted outputs.

Further, reasoning analogously as in the SISO case, we can easily obtain recurrent formulae for elements of free trajectories of the predicted outputs over the prediction horizon, which are equivalents of (6.40) for the MIMO case, in the form

$$y_m^0(k+p|k) = -\sum_{i=1}^{\min\{n_A,p-1\}} a_i^m y_m^0(k+p-i|k) - \sum_{i=\min\{n_A,p-1\}+1}^{n_A} a_i^m y_m(k+p-i) +$$
$$+ \sum_{j=1}^{n_u} \left[\sum_{i=0}^{\min\{n_B,p\}} b_i^{m,j} u_j(k-1) + \sum_{i=\min\{n_B,p\}+1}^{n_B} b_i^{m,j} u_j(k-1+p-i) \right] + d_m(k),$$
$$p = 1,...,N, \tag{6.46}$$

where

$$d_m(k) = y_m(k) - \left[-\sum_{i=1}^{n_A} a_i^m y_m(k-i) + \sum_{j=1}^{n_u} \sum_{i=0}^{n_B} b_i^{m,j} u_j(k-1-i) \right] \tag{6.47}$$

are estimates of additive disturbances calculated at the sampling instant k, $m = 1,...,n_y$.

6.1.4 Non-linear Predictive Control

6.1.4.1 MPC Algorithms with Optimization Using Directly a Non-linear Model

Implementations of MPC algorithms using directly a non-linear process model are significantly more difficult than when linear models are used. The main reason is that the dynamic optimization problem is then a non-linear programming problem, generally a non-convex one, due to the non-linearity of process models. Certainly, there are many non-linear programming procedures available now, but they are in general not capable to assure properties required for on-line applications—to find the true solution with assumed accuracy not exceeding a strictly prescribed time limit. This can be particularly critical if a numerically complex non-linear process model is used (e.g., having internal iterative procedures). Therefore, if the MPC algorithm with dynamic optimization using directly a non-linear process model is designed, which can be reasonable for processes with slower dynamics, then it is recommended to use a model with possibly good numerical features in the optimization procedure. First of all, the model should be sufficiently smooth, enabling the application of gradient optimization procedures, it should assure sufficiently quick calculation of the process outputs. If it is not the case with the original process model available, then an often recommended procedure is to design numerically more efficient approximation of this model for direct use in the MPC controller, e.g., in the form of a fuzzy model (Marusak, 2008; 2009; Marusak and Tatjewski, 2009) or a neural network model (Ławryńczuk, 2007; 2008; Ławryńczuk and Tatjewski, 2008). These models can provide the required approximation quality and have, as mentioned, good numerical properties for effective implementation of non-linear optimization, although the optimization still remains non-linear and generally non-convex, thus not guaranteeing the properties required for on-line applications.

6.1.4.2 MPC Algorithms with Linearizations of the Non-linear Model

A sound solution in practical non-linear MPC design is the tendency to apply linearizations of the non-linear model (Morari and Lee, 1999). The reason is that this results in the formulation of predictive algorithms in such a way that at each sampling instant a strictly convex quadratic programming problem is solved. This QP problem is successively modified, every few steps or even at each step (sampling instant) of the MPC algorithm, to be consistent with the changes of the non-linear process working point. In this way, generally suboptimal but usually efficient algorithms are created (suboptimal with respect to the MPC algorithm solving the true non-linear optimization problem).

The simplest solution which enables a natural generalization of well-known good practical properties of predictive algorithms with linear process models to non-linear predictive control are algorithms which at each sampling instant perform the *linearization of the non-linear model*, at a current process state, and then calculate the control inputs using a linear MPC algorithm, e.g., DMC or GPC, with the linearized model (see, e.g., the work of Tatjewski (2007) and the extensive list of references given therein). This is a *suboptimal approach*, but the one which allows keeping the fundamental, for practical applications, feature of *control reliability*, i.e., a guarantee that at each of the successive algorithm steps an optimal solution of the (quadratic) optimization problem will be found, and always within a predefined time. The discussed structure of suboptimal non-linear MPC control with model linearizations will be called MPC-NSL (*MPC Non-linear with Successive Linearizations*). It is presented in Fig. 6.5.

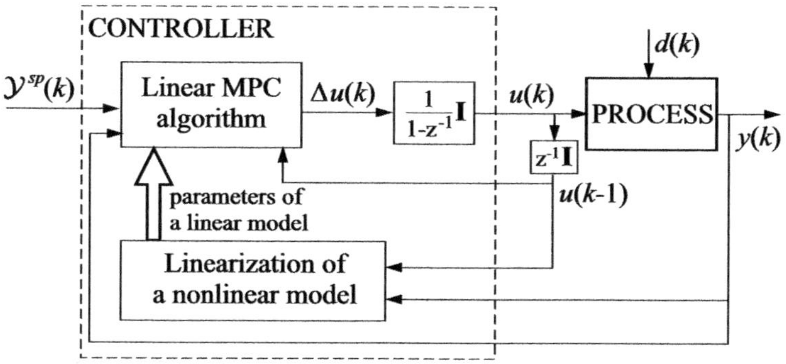

Fig. 6.5 Structure of a non-linear MPC algorithm with successive linearizations of the process model (MPC-NSL algorithm)

For weakly non-linear processes, or operating close to certain equilibrium points during longer time periods and under slowly varying disturbances, the linearization may not be necessary at each sampling instant. It may then be sufficient to perform it more rarely, e.g., only after every predefined number of samples. Such an example is given by Maciejowski (2002).

A structurally more precise solution than the MPC-NSL algorithm, and at the same time still simple in implementation and preserving the required good properties of a QP optimization problem, is to use a linearized model only in the optimization problem (only for the calculation of the forced trajectory of predicted outputs), but to evaluate the free trajectory from a non-linear model, at each sampling instant, see the research by Tatjewski (2007), where a list of references on the subject is given. This structure will be called *MPC with non-linear prediction and linearization*, and it is presented in Fig. 6.6.

In fact, there is no reason to give up a non-linear prediction of a free output trajectory having a non-linear process model. This is a relatively easy task, performed only once at each sampling instant of the MPC controller. Thus, it does not significantly affect the total amount of calculations performed at each sampling instant, which is first of all determined by a numerical solution of the optimization problem, even if it is a quadratic programming one. Thus, MPC-NPL algorithms should be preferred to MPC-NSL algorithms.

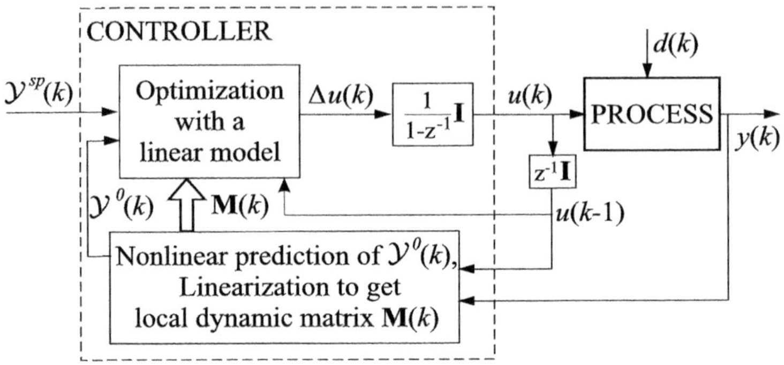

Fig. 6.6 Structure of the MPC-NPL algorithm

To calculate the free component $\mathscr{Y}^0(k)$ of a trajectory of predicted outputs on a prediction horizon, it is necessary to have not only values of the already applied (past) process input and output signals, but also to assume certain (initial) values of future process inputs in the prediction horizon, i.e., an initial trajectory of control inputs $\mathscr{U}^0(k)$. Proceeding identically as in the case of MPC algorithms with linear models, in order to evaluate the free output trajectory we can assume zero increments of process input signals in a prediction horizon, i.e., process inputs constant and equal to $\mathbf{u}(k-1)$. Another, in non-linear cases often better, approach is to apply an appropriately adopted optimal control trajectory evaluated at a previous sampling instant as the current initial control trajectory (Tatjewski, 2007). In the dynamic optimization task of the MPC-NPL controller, future control increments are calculated with respect to the initial trajectory $\mathscr{U}^0(k)$, and the dynamic matrix $\mathbf{M}$ needed for these calculations is derived from the linearized model. Thus, the cost function for the QP dynamic optimization problem will be

$$\left\|\mathscr{Y}^{zad}(k) - \mathscr{Y}^0(k) - \mathbf{M}(k)\Delta\mathscr{U}(k)\right\|^2_{\underline{\mathbf{Q}}} + \|\Delta\mathscr{U}(k)\|^2_{\underline{\mathbf{R}}}, \tag{6.48}$$

which is optimized subject to constraints on amplitudes and rates of change of control increments, and subject to constraints of process outputs, which are constructed also with non-linear prediction and linearization:

$$\mathscr{Y}_{\min} \leq \mathscr{Y}^0(k) + \mathbf{M}(k)\Delta\mathscr{U}(k) \leq \mathscr{Y}_{\max}. \tag{6.49}$$

The smaller the optimal control input increments evaluated as a solution of the QP problem, the better the MPC-NPL algorithm should operate. Therefore, it may deliver almost optimal behavior even for strongly non-linear processes and for transitions to distant operating points, as long as the trajectories are smooth and realized with a sequence of relatively small process input changes.

6.1.4.3 Fuzzy and Neural Models in Non-linear MPC Algorithms

As mentioned earlier, a non-linear model used in non-linear MPC with on-line non-linear optimization should have good numerical properties. In MPC approaches which are based on linear approximations of the non-linear model of the process (the MPC-NSL and MPC-NPL algorithms) calculated on-line, linearization should be calculated in a relatively uncomplicated and numerically reliable manner. Thanks to their advantages, two kinds of non-linear models, i.e., *fuzzy models*, in particular *Takagi–Sugeno (TS) structures*, and *neural networks*, are important. It is a well-known fact that both of the mentioned model types are good approximators of non-linear relations (Pedrycz, 1993; Haykin, 1999). Moreover, they have simple structures, which is particularly important from the numerical point of view.

In non-linear fuzzy modeling, taking into account applications of models in MPC, Takagi–Sugeno fuzzy models play a vital role. In such models, consequents of fuzzy rules (THEN-parts) are functions. Usually, in consequents, linear models are used. These models constitute local linear approximations of the original non-linear models. These local models are switched using the fuzzy approach. The fuzzy combination of local linear models constitutes a non-linear fuzzy model (Pedrycz, 1993; Jang et al., 1997; Piegat, 2001; Tatjewski, 2007).

Let us assume that the fuzzy model consists of r rules, and that all consequents are linear models in the form of discrete difference equations corresponding to the linear model (6.36) used in the GPC algorithm. The fuzzy model can then be formulated as consisting of the following rules:

$$\begin{aligned} \mathrm{R}^i:\ \text{IF}\quad & y(k) \text{ is } A_0^i \text{ and } y(k-1) \text{ is } A_1^i \text{ and } \cdots \text{ and } y(k-n_R) \text{ is } A_{n_R}^i \\ & \text{and } u(k) \text{ is } B_0^i \text{ and } u(k-1) \text{ is} B_1^i \text{ and } \cdots \text{ and } u(k-m_R) \text{ is } B_{m_R}^i \\ \text{THEN}\ & y^i(k+1) = -a_1^i y(k) - \cdots - a_{n_A}^i y(k-n_A+1) \\ & + b_0^i u(k) + \cdots + b_{n_B}^i u(k-n_B), \end{aligned} \tag{6.50}$$

where i indexes rules ($i = 1, \ldots r$) and at the same time subdomains of the fuzzy TS model, $A_j^i \in \mathbb{Y}_j$, $B_j^i \in \mathbb{U}_j$, while a_j^i and b_j^i are coefficients of functions in rule consequents.

Elements of each of the sets $\mathbb{Y}_j$ are fuzzy sets covering the area of the variable $y(k-j)$, $j = 0, \ldots, n_R$; analogously, $\mathbb{U}_j$ for $u(k-j)$, $j = 1, \ldots, m_R$. Usually, partitions of the domain of $y(k)$ are the same, independent of time delays, i.e., $\mathbb{Y}_0 = \ldots = \mathbb{Y}_{n_R} = \mathbb{Y}$; analogously, $\mathbb{U}_1 = \ldots = \mathbb{U}_{m_R} = \mathbb{U}$ (Tatjewski, 2007).

The set of rules is complemented by the standard formula for a final conclusion, namely, for the model output

$$y(k+1) = \sum_{i=1}^{r} \tilde{w}^i(k) y^i(k+1), \tag{6.51}$$

where $\tilde{w}^i(k)$ are normalized activation levels of the rules (6.50). Activation levels of the rules depend at the sampling instant k on n_R+1 values of outputs (i.e., $y(k), y(k-1), \ldots y(k-n_R)$) and m_R+1 values of the output (i.e., $u(k), u(k-1), \ldots, u(k-m_R)$).

Precisely, they depend on grades of the membership of these outputs and inputs to the fuzzy sets A^i_j and B^i_j, where, in general, $n_R \neq n_A$ and $m_R \neq n_B$.

A concise description of the fuzzy model (6.50), (6.51), compliant with the model used in the GPC algorithm (6.36), is

$$y(k+1) = -a_1(k)y(k) - \cdots - a_{n_A}(k)y(k-n_A+1) + b_0(k)u(k) + \cdots + b_{n_B}(k)u(k-n_B), \tag{6.52}$$

where

$$a_j(k) = \sum_{i=1}^{r} \tilde{w}^i(k)a^i_j, \quad b_j(k) = \sum_{i=1}^{r} \tilde{w}^i(k)b^i_j. \tag{6.53}$$

Thus, the non-linear TS fuzzy model has the same structure as the linear model used in the GPC algorithm, but *coefficients of the fuzzy model depend on the current operating point of the process*. It is necessary to emphasize the fact that, thanks to this property, the linearization of the non-linear TS model is straightforward and can be performed very efficiently during on-line control. Details are given in the works of Tatjewski and Ławryńczuk, (2006), Tatjewski, (2002; 2007).

Thanks to their properties, neural networks constitute the second class of non-linear models which are recommended for MPC. Let us consider a non-linear NARX model of a SISO dynamic process:

$$y(k) = f(u(k-\tau), \ldots, u(k-n_B-1), y(k-1), \ldots, y(k-n_A)). \tag{6.54}$$

Such a model corresponds to the linear model (6.36) used in the GPC algorithm. The structure of a feedforward neural model with two layers (Haykin, 1999) which realizes the NARX model is depicted in Fig. 6.7. The output signal of the model is calculated from

$$y(k) = w^2_0 + \sum_{i=1}^{K} w^2_i \varphi(z_i(k)), \tag{6.55}$$

where $z_i(k)$ is the sum of inputs of the i-th hidden node, $\varphi : \mathbb{R} \to \mathbb{R}$ denotes a non-linear transfer function used in the hidden layer (usually the tanh function is used for this purpose), K is the number of hidden nodes. From (6.54), one has

$$z_i(k) = w^1_{i,0} + \sum_{j=1}^{I_u} w^1_{i,j} u(k-\tau+1-j) + \sum_{j=1}^{n_A} w^1_{i,I_u+j} y(k-j). \tag{6.56}$$

Weights of the first layer are denoted by $w^1_{i,j}$, $i = 1, \ldots, K$, $j = 0, \ldots, n_A + n_B - \tau + 2$, while w^2_i, $i = 0, \ldots, K$, denotes weights of the second layer, $I_u = n_B - \tau + 2$. When the process considered has as many as n_u inputs and n_y outputs (MIMO process), the neural model usually consists of n_y neural networks, and each network has one output:

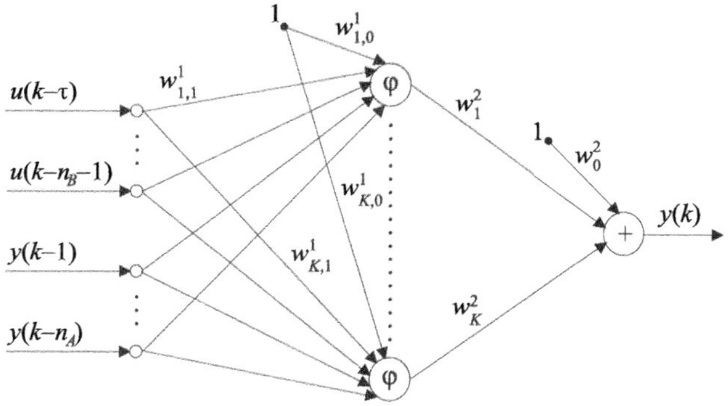

Fig. 6.7 Structure of the neural model of a SISO dynamic process

$$\begin{aligned} y_1(k) = f_1(&u_1(k-\tau^{1,1}),\ldots,u_1(k-n_B^{1,1}),\ldots, \\ &u_{n_u}(k-\tau^{1,n_u}),\ldots,u_{n_u}(k-n_B^{1,n_u}-1), \\ &y_1(k-1),\ldots,y_1(k-n_A^1)) \end{aligned} \tag{6.57}$$

$$\vdots$$

$$\begin{aligned} y_{n_y}(k) = f_{n_y}(&u_1(k-\tau^{n_y,1}),\ldots,u_1(k-n_B^{n_y,1}),\ldots, \\ &u_{n_u}(k-\tau^{n_y,n_u}),\ldots,u_{n_u}(k-n_B^{n_y,n_u}-1), \\ &y_{n_y}(k-1),\ldots,y_{n_y}(k-n_A^{n_y})). \end{aligned} \tag{6.58}$$

Functions f_m, $m = 1,\ldots,n_y$ are realized by separate neural networks. Their structures are similar to those of a neural network model of a SISO dynamic model shown in Fig. 6.7. Alternatively, it is possible to use only one neural network as a MIMO dynamic model, then this network has n_y output nodes (i.e., $y_1(k),\ldots,y_{n_y}(k)$). However, such a model is usually more complicated than that which consists of n_y networks with one output. Moreover, training neural networks with many outputs is more difficult because all parameters (weights) are optimized at the same time.

Taking into account the very specific role of a dynamic model in MPC algorithms with a long-range horizon, it is necessary to stress the advantages of neural models. First of all, two-layered neural networks with one non-linear hidden layer are universal approximators which are able to approximate any continuous non-linear function with arbitrary accuracy. Thanks to this property, it is possible to use neural networks as sufficiently precise but simpler models of full scale complex dynamic models. Neural networks have a very simple and regular structure, their number of parameters (weights) is usually moderate. As a result, neural models of processes can be used efficiently in non-linear MPC algorithms. Non-linear prediction, linearization, calculation of the non-linear free trajectory as well as unmeasured disturbance estimation are calculated in a straightforward way. Implementation details of the

MPC-NSL and MPC-NPL algorithms in which neural models are used are given by Ławryńczuk (2007), Tatjewski and Ławryńczuk (2006) and Tatjewski (2007).

6.1.5 Optimization of Set-Points

The basic multi-layer control structure with an MPC supervisory feedback control layer is shown in Fig. 6.8 (Tatjewski, 2007; 2008). The upper most layer represents

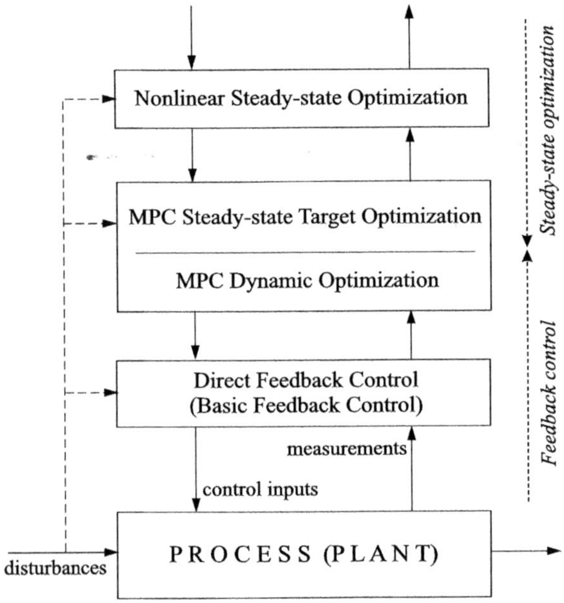

Fig. 6.8 Structure of multi-layer control with an MPC supervisory feedback control layer

economic Steady-state Set-point Optimization (SSO), usually performed using a non-linear, comprehensive model of the underlying process, appropriately adapted to changing environment conditions (slowly varying disturbances, production requirements) with frequency significantly smaller than the intervention frequency of the underlying MPC controller. A typical form of the SSO problem involves a linear economic cost function with

$$\max_{\mathbf{u}^{ss},\mathbf{y}^{ss}} \{\mathbf{q}^T \mathbf{y}^{ss} - \mathbf{p}^T \mathbf{u}^{ss}\}, \tag{6.59a}$$

$$\begin{cases} \mathbf{u}_{\min} \leq \mathbf{u}^{ss} \leq \mathbf{u}_{\max}, \\ \mathbf{y}_{\min} \leq \mathbf{y}^{ss} \leq \mathbf{y}_{\max}, \\ \mathbf{y}^{ss} = F(\mathbf{u}^{ss}, \mathbf{w}), \end{cases} \tag{6.59b}$$

where the operator F represents a non-linear static process model, for current values of its parameters w (estimates of disturbances, tunable parameters). Algorithms used to solve the SSO problem are specialized (when needed) or standard non-linear programming routines, e.g., Sequential Quadratic Programming (SQP) procedures, which are regarded now as very efficient (Fletcher, 1987; Bertsekas, 1995). Let us denote the optimal solution point of (6.59b) by $(\hat{\mathbf{u}}^{ss}, \hat{\mathbf{y}}^{ss})$.

Since the MPC controller is executed significantly more frequently than the SSO optimizer, controlling the process under disturbances and constraints, then it occurred reasonable to adopt additionally the set-point, before every MPC dynamic optimization (see, e.g., the research by Rao and Rawlings (1999), Kassman et al. (2000), Blevins et al. (2003) or Tatjewski (2007)). Such an algorithm of auxiliary set-point optimization (recalculation) can be called Steady-State Target Optimization (SSTO) (Kassman et al., 2000).

The simplest way to define the SSTO problem is to apply the statics of the dynamic process model used in MPC as a steady-state process model in SSTO. In the case of a linear model, used, e.g., in the DMC or GPC algorithms, this is simply the gain matrix $\mathbf{G}$ corresponding to this model. Two alternative formulations of the SSTO problem can be found, differing in the optimization goal:

SSTO1: To find steady-state targets $(\mathbf{u}^{ss}, \mathbf{y}^{ss})$ as close as possible to the set-points $(\hat{\mathbf{u}}^{ss}, \hat{\mathbf{y}}^{ss})$, despite static process model limitations and input and output constraints, solving an appropriate optimization problem, e.g., the one with a quadratic penalty function:

$$\begin{aligned} &\max_{\Delta\mathbf{u}^{ss}} \{||\mathbf{y}^{ss} - \hat{\mathbf{y}}^{ss}||^2 + ||\mathbf{u}^{ss} - \hat{\mathbf{u}}^{ss}||^2\} \\ &\qquad \mathbf{u}_{\min} \leq \mathbf{u}^{ss} \leq \mathbf{u}_{\max} \\ &\qquad \mathbf{y}_{\min} \leq \mathbf{y}^{ss} \leq \mathbf{y}_{\max} \\ &\qquad \mathbf{y}^{ss} = \mathbf{y}^0(k+N|k) + \mathbf{G}\Delta\mathbf{u}^{ss} \\ &\qquad \mathbf{u}^{ss} = \mathbf{u}(k-1) + \Delta\mathbf{u}^{ss}, \end{aligned} \tag{6.60}$$

where $\mathbf{y}^0(k+N|k)$ denotes values of the outputs at the end of the prediction horizon calculated provided the control signal is constant over the whole prediction horizon and equal to the last (at a previous sampling instant, $k-1$-th) value $\mathbf{u}(k-1)$ applied (Rao and Rawlings, 1999);

SSTO2: To calculate the steady-state targets $(\mathbf{u}^{ss}, \mathbf{y}^{ss})$ on the basis of the original economic performance function (as in SSO), but using a simple steady-state model and all constraints, i.e., solving at the current sampling instant the optimization problem of the form (Kassman et al., 2000; Blevins et al., 2003)

$$\max_{\Delta \mathbf{u}^{ss}}\{\mathbf{q}^T \Delta \mathbf{y}^{ss} - \mathbf{p}^T \Delta \mathbf{u}^{ss}\}$$

$$\begin{aligned} &\mathbf{u}_{\min} \le \mathbf{u}^{ss} \le \mathbf{u}_{\max} \\ &\mathbf{y}_{\min} \le \mathbf{y}^{ss} \le \mathbf{y}_{\max} \\ &\mathbf{y}^{ss} = \mathbf{y}^0(k+N|k) + \mathbf{G}\Delta \mathbf{u}^{ss} \\ &\mathbf{u}^{ss} = \mathbf{u}(k-1) + \Delta \mathbf{u}^{ss}. \end{aligned} \tag{6.61}$$

Notice that, in both cases, the modification of the set-points is relative to the current "steady-state" $(\mathbf{u}(k-1), \mathbf{y}^0(k+N|k))$. This means that, before solving the problem (6.60) or (6.61), the MPC controller calculates the free trajectory of predicted outputs, transfers the last calculated value $\mathbf{y}^0(k+N|k)$ to the SSTO problem, which then calculates the corrected set-point $(\mathbf{u}^{ss}, \mathbf{y}^{ss})$ and sends it back to MPC. Finally, the MPC Dynamic Optimization (MPC-DO) problem is solved.

The desired set-point $(\hat{\mathbf{u}}^{ss}, \hat{\mathbf{y}}^{ss})$ may be calculated in the SSO problem, but it may also stem directly from certain requirements, like, e.g., norms in the case of desired water purity or allowed air pollution, in drinking water distribution or energy production systems—this is the case when SSTO1 can be a reasonable formulation.

The quality of the target set-points $(\mathbf{u}^{ss}, \mathbf{y}^{ss})$ generated in SSTO2 depends on that of the static process model used, i.e., the gain matrix $\mathbf{G}$. If the process is truly non-linear, then its static characteristics change with the changes of the set-point. The MPC controller designed as a robust one should then work properly, but with differing performance. However, the values of the constant matrix $\mathbf{G}$ can be, for many set-points, far from correct and thus the results of SSTO may be far from optimal. The solution is then the adaptation of $\mathbf{G}$, based on the non-linear model used in the higher SSO layer (Ławryńczuk et al., 2007; 2008; Tatjewski, 2007; 2008). The adaptation would be best if preformed at every sampling instant, but this may be unrealistic or unnecessary—it may be then performed more rarely, e.g., after every predefined number of samples. The SSTO2 problem with an adaptively changing gain matrix would have the form

$$\max_{\Delta \mathbf{u}^{ss}}\{\mathbf{q}^T \Delta \mathbf{y}^{ss} - \mathbf{p}^T \Delta \mathbf{u}^{ss}\} \tag{6.62a}$$

$$\begin{cases} \mathbf{u}_{\min} \le \mathbf{u}^{ss} \le \mathbf{u}_{\max} \\ \mathbf{y}_{\min} \le \mathbf{y}^{ss} \le \mathbf{y}_{\max} \\ \mathbf{y}^{ss} = F(\mathbf{u}(k-1), \mathbf{w}) + \mathbf{G}(k)\Delta \mathbf{u}^{ss} \\ \mathbf{u}^{ss} = \mathbf{u}(k-1) + \Delta \mathbf{u}^{ss}, \end{cases} \tag{6.62b}$$

where $F(\mathbf{u}(k-1), \mathbf{w})$ is the process output calculated from the non-linear SSO model, for current values of the input $\mathbf{u}(k-1)$ and disturbance w. The corresponding control structure is shown in Fig. 6.9.

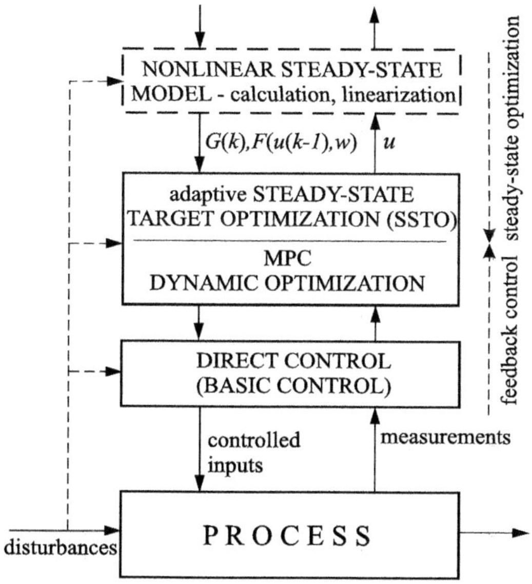

Fig. 6.9 Multi-layer structure with adaptive SSTO based on linearizations of the SSO model

6.1.6 Examples

6.1.6.1 CSTR with the Van de Vusse Reaction

The example control plant is a non-linear chemical reactor in which the van de Vusse reaction is carried out (Fig. 6.10). The equations describing the process are as follows (Doyle III et al., 1995):

$$\begin{aligned} \frac{dC_A}{dt} &= -50 \cdot C_A - 10 \cdot C_A^2 + F(C_{Af} - C_A) \\ \frac{dC_B}{dt} &= 50 \cdot C_A - 100 \cdot C_B - F \cdot C_B, \end{aligned} \tag{6.63}$$

where C_A, C_B are concentrations of substances A and B, F is the raw flow rate (it is assumed that the volume in which the reactions takes place is constant), C_{Af} is the concentration of substance A in the inlet flow. The output of the process is the concentration C_B and the manipulated variable is the flow rate of the raw substance F measured in l/h.

The control plant has an inverse response which clearly manifests itself in the step response obtained near the operating point $C_B = 1.12\ mol/l$, $F = 34.3\ l/h$ (Fig. 6.11). Thus, the reactor is hard to control and it is advisable to apply a predictive control algorithm for it.

The simulation experiments were carried out using *PExSim*—an element of the *DiaSter* package. The DMC controller was designed for the control plant. In order to do so, first the control plant was simulated using the *Non-linear dynamic* block.

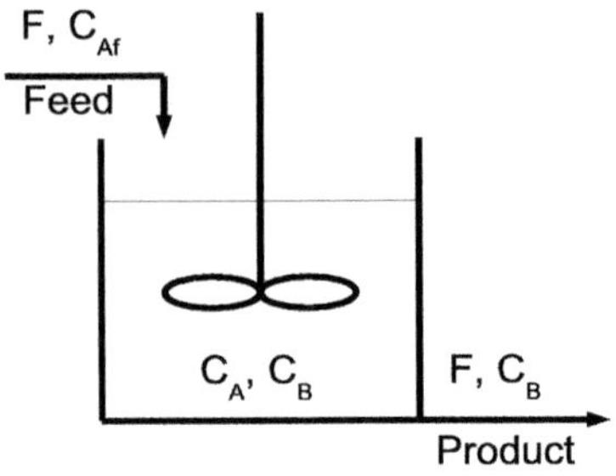

Fig. 6.10 Diagram of the CSTR with the van de Vusse reaction

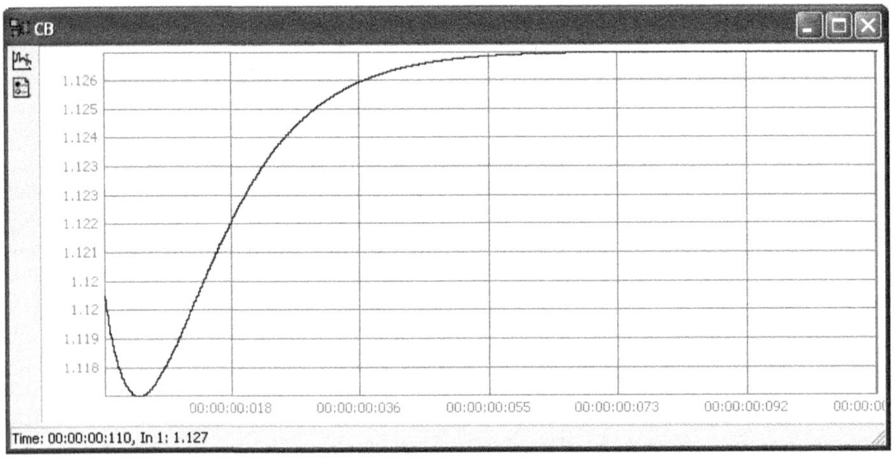

Fig. 6.11 Step response of the CSTR

The normalized step response was registered using the *Text file output* block. The diagram of the system used to perform the experiment is shown in Fig. 6.12.

Next, the *DMC algorithm* block, containing the implementation of the DMC algorithm, was used. The sampling period $T_s = 0.12$ *min* and the dynamics horizon $D = 40$ were assumed along with the following parameters of the controller: prediction horizon $N = 40$, control horizon $N_u = 20$, $\lambda = 0.001$.

Responses obtained after changes of the set-point value are shown in Figs. 6.13 and 6.14. The obtained results are satisfactory for both kinds of set-point changes (to lower as well as upper values). The overshoot is in both cases small. The change of the set-point value to $C_{B,sp} = 1.2\ l/mol$ is slower because of the non-linearity of the control plant.

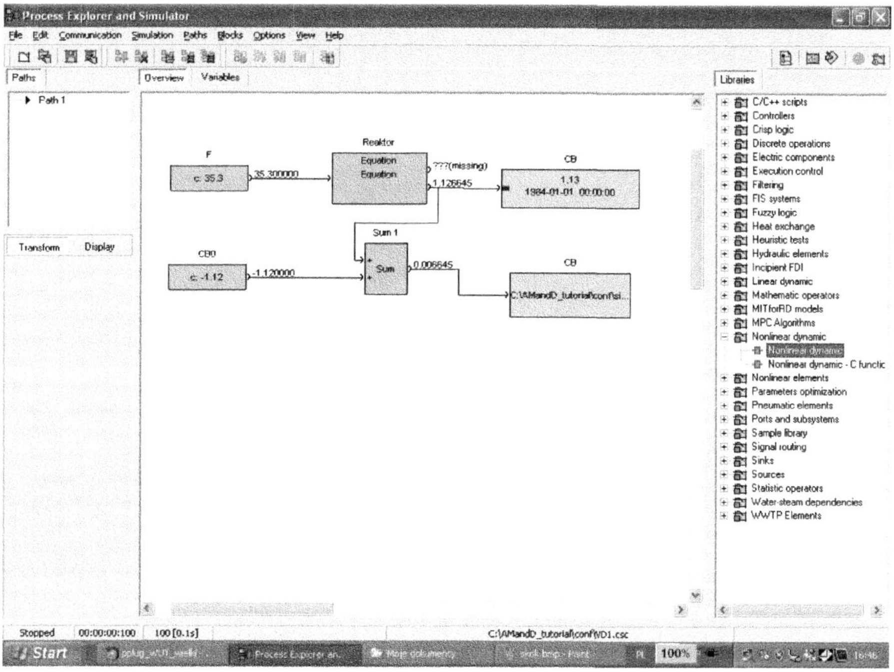

Fig. 6.12 System of connections used to obtain the step response

6.1.6.2 MIMO Chemical Reactor

In order to demonstrate the efficiency of the GPC algorithm in controlling multivariable processes and handle constraints, a chemical reactor, shown in Fig. 6.15 (Camacho and Bordons, 2004), is considered. The process has two inputs and two outputs, while its continuous-time transfer function model is (time constants in minutes)

$$\begin{bmatrix} Y_1(s) \\ Y_2(s) \end{bmatrix} = \begin{bmatrix} \dfrac{1}{1+0.7s} & \dfrac{5}{1+0.3s} \\ \dfrac{1}{1+0.5s} & \dfrac{2}{1+0.4s} \end{bmatrix} \begin{bmatrix} U_1(s) \\ U_2(s) \end{bmatrix}, \tag{6.64}$$

where dimensionless manipulated variables include U_1: the flow rate of the feed, and U_2: the flow rate of the cooling substance; dimensionless controlled variables are Y_1: the concentration of the product, and Y_2: the temperature in the reactor. Using the sampling period 0.03 min, one obtains the discrete-time dynamic model

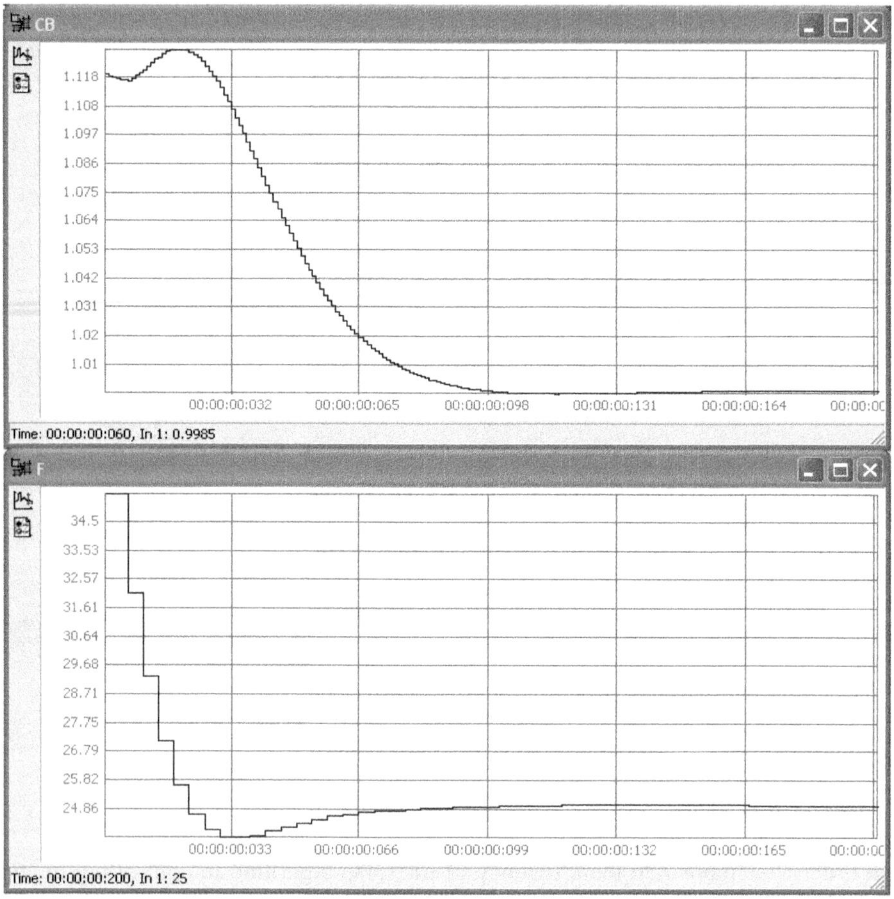

Fig. 6.13 Response of the control system to the change of the set-point value to $C_{B,sp} = 1$ l/mol

$$\begin{aligned} y_1(k) =& 0.041951u_1(k-1) - 0.037959u_1(k-2) \\ &+ 0.475812u_2(k-1) - 0.455851u_2(k-2) \\ &+ 1.862885y_1(k-1) - 0.866877y_1(k-2), \end{aligned} \tag{6.65}$$

$$\begin{aligned} y_2(k) =& 0.058235u_1(k-1) - 0.054027u_1(k-2) \\ &+ 0.144513u_2(k-1) - 0.136097u_2(k-2) \\ &+ 1.869508y_2(k-1) - 0.873715y_2(k-2). \end{aligned} \tag{6.66}$$

All experiments are carried out in *PExSim*. At first the simulated process is implemented using the elementary block *1st order inertia*, which belongs to the *Linear*

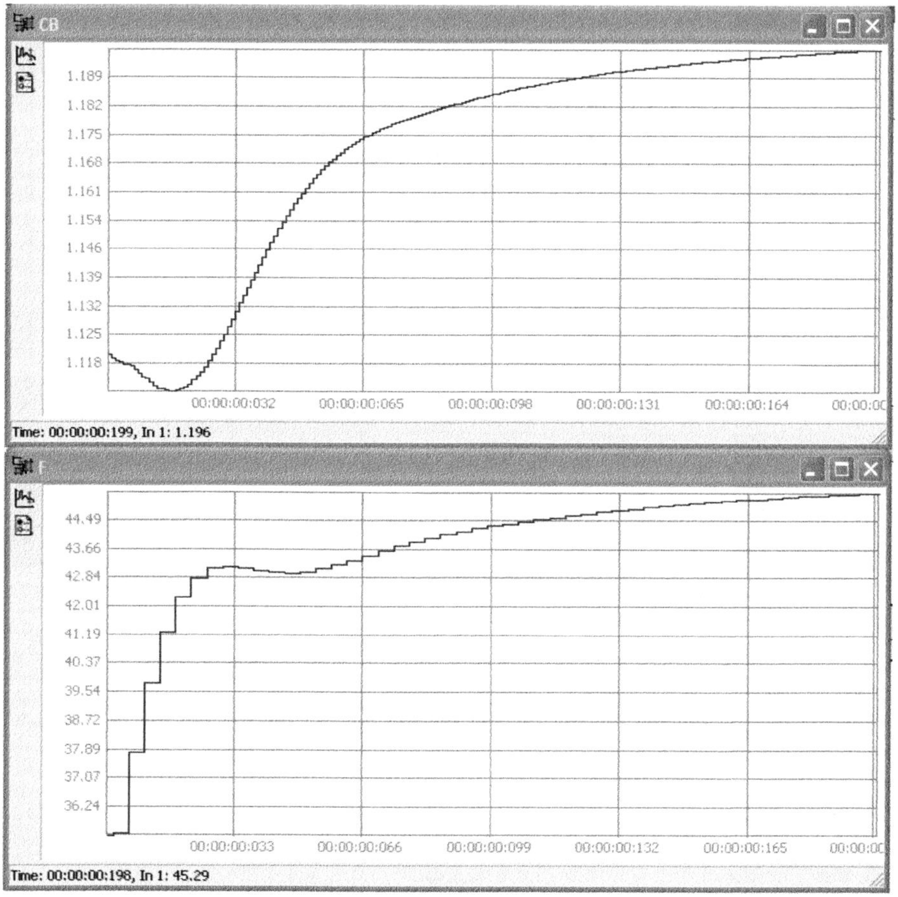

Fig. 6.14 Response of the control system to the change of the set-point value to $C_{B,sp} = 1.2$ l/mol

dynamic group. The implementation of the simulated chemical reactor is shown in Fig. 6.16.

Two versions of the GPC algorithm are compared: the explicit one and the numerical one. In both algorithms the same parameters are used: the prediction horizon $N = 10$, the control horizon $N_u = 2$, weighting matrices are $\mathbf{Q} = \mathbf{I}_{2\times 2}$ and $\mathbf{R} = 0.1\mathbf{I}_{2\times 2}$, where $\mathbf{I}$ is an identity matrix of appropriate dimensionality.

At first the explicit (unconstrained) GPC algorithm is considered. The algorithm is realized by the block *GPC algorithm* (from the *MPC algorithms* group). Figure 6.17 shows the implementation of this algorithm for the MIMO chemical reactor in *PExSim*. The obtained simulation results are depicted in Fig. 6.18 (four top panels). It is assumed that set-points for both outputs (y_1, y_2) change from 0 to 1 at

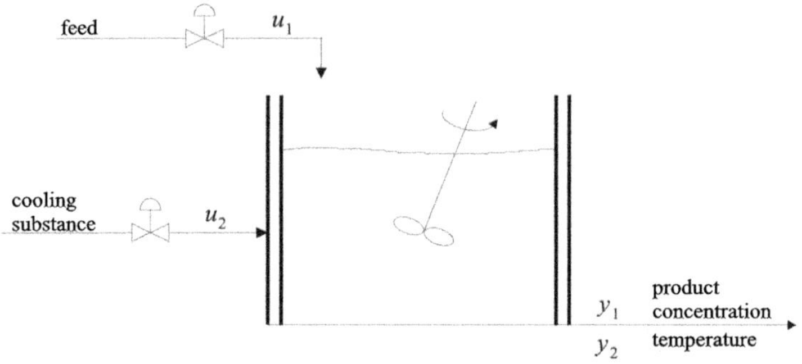

Fig. 6.15 Chemical reactor with two inputs and two outputs

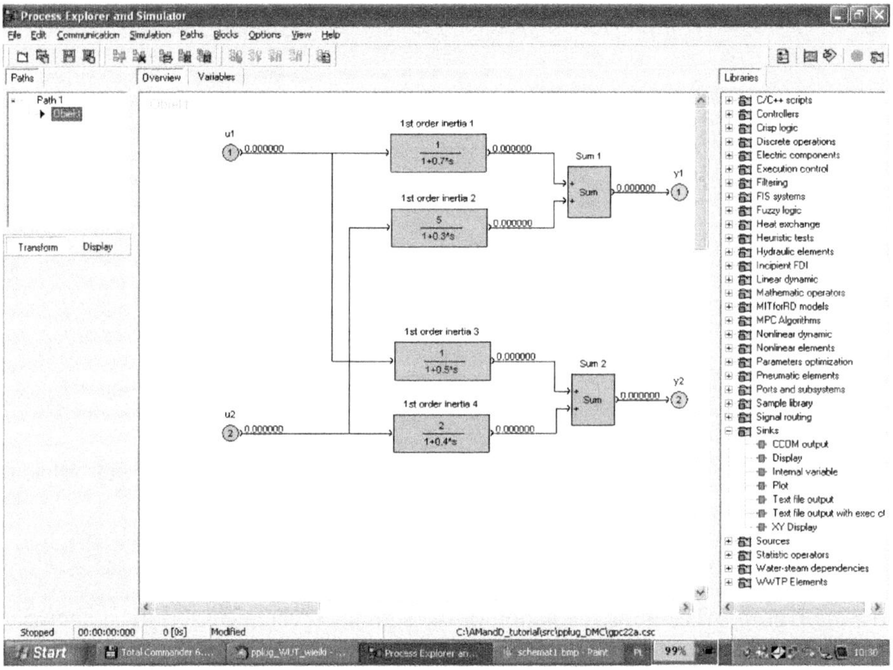

Fig. 6.16 Implementation of the MIMO chemical reactor in *PExSim* (simulated process)

the beginning of the simulation. For chosen weighting matrices **Q** and **R**, output trajectories are fast (i.e., the process converges to new set-points quickly), but changes of manipulated variables are very fast, particularly during the first three iterations of the algorithm. (In spite of the fact that, in the rudimentary explicit GPC

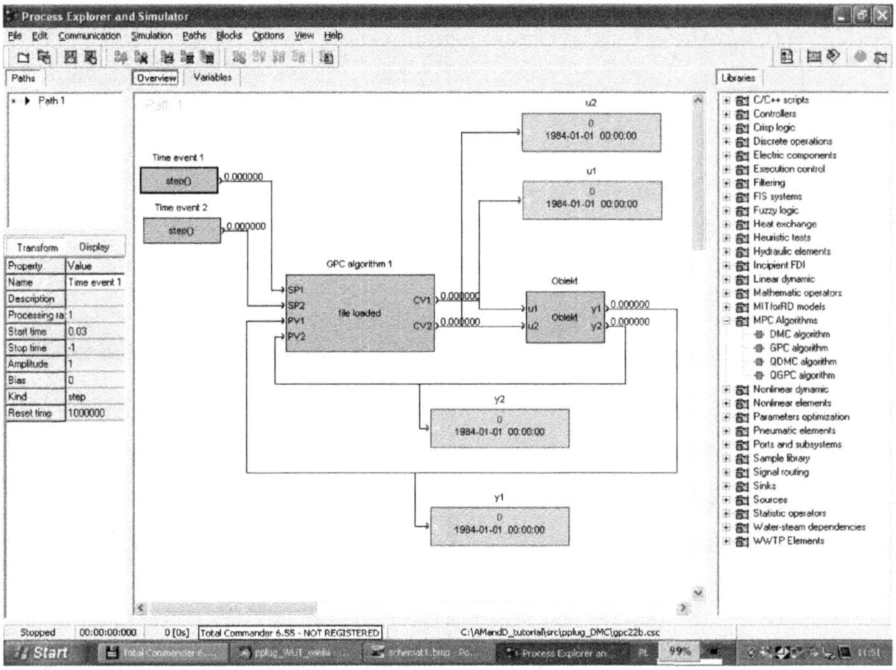

Fig. 6.17 Implementation of the explicit GPC algorithm for the MIMO chemical reactor

algorithm, constraints are not taken into account, in the algorithm available in *PExSim* it is possible to use a constraint projection mechanism, i.e., calculated manipulated variables for the current sampling instant are projected onto the admissible set of constraints.)

Next, in order to demonstrate the advantages of the numerical GPC algorithm, the explicit algorithm is replaced by its numerical version. The algorithm is realized by the *QGPC algorithm* block (from the *MPC algorithms* group). This algorithm takes into account all imposed constraints in the quadratic programming optimization problem solved at each sampling instant. Constraints imposed on values of manipulated variables are $u_{1\,\min} = -2$, $u_{1\,\max} = 2$, $u_{2\min} = -0.5$, $u_{2\max} = 0.5$, constraints imposed on increments of manipulated variables are $\Delta u_{1\,\max} = 0.5$, $\Delta u_{2\max} = 0.2$. Figure 6.18 (four bottom panels) depicts simulation results. In comparison with simulations obtained in the explicit GPC algorithm, output trajectories are slightly slower, but all constraints are satisfied in consecutive sampling instants (iterations of the algorithm).

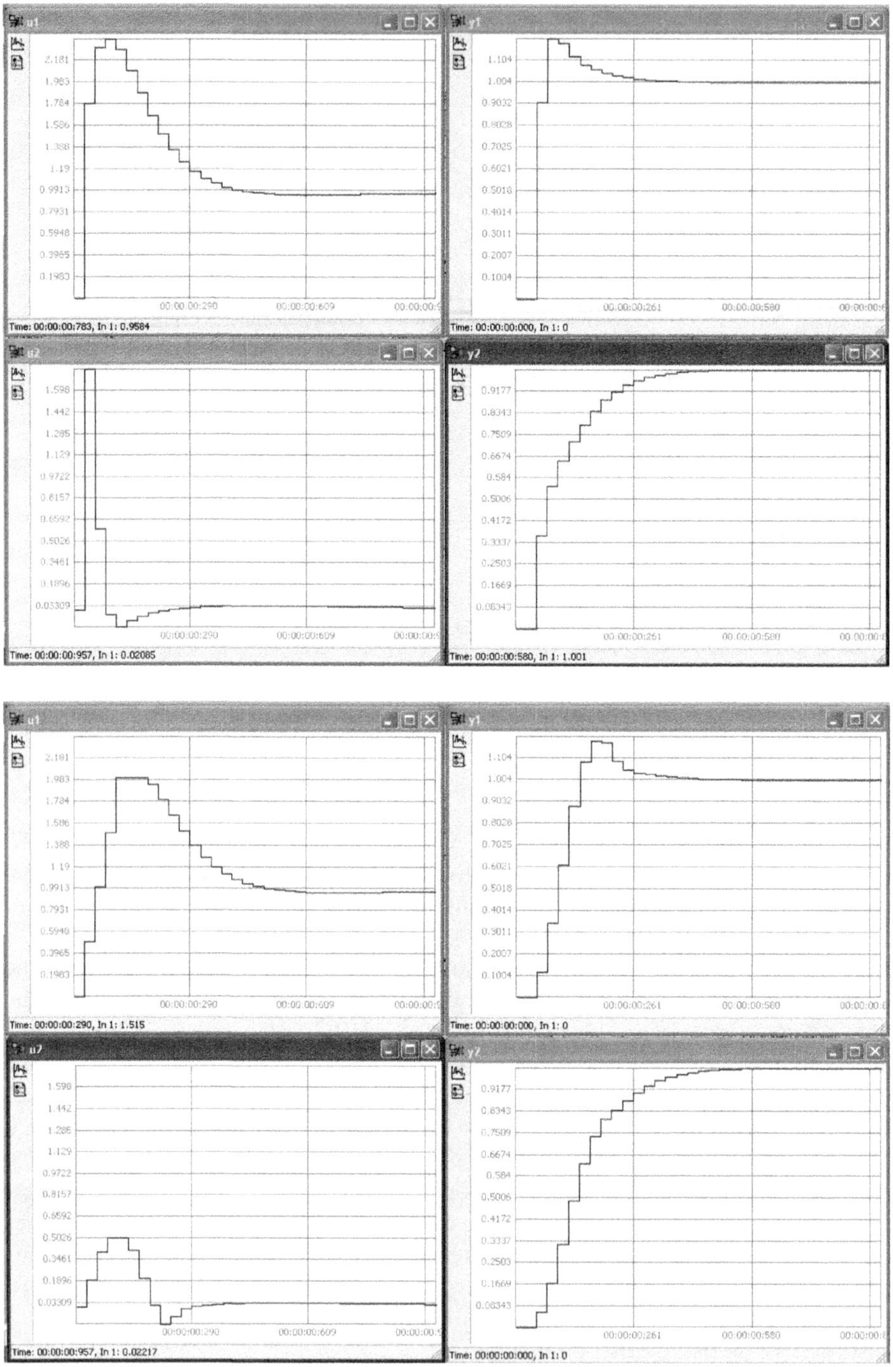

Fig. 6.18 Simulation results of the MIMO chemical reactor: explicit (unconstrained) GPC algorithm (four top panels), numerical GPC algorithm (four bottom panels)

6.2 Self-tuning and Adaptation of Control Loops

The purpose of this section is to present of practical algorithms for self-tuning and adaptation of PID loops implemented in the *DiaSter* package (also in some instrument controllers and PC-based soft controllers). Self-tuning can be executed by means of the step response or relay control (Åström et al., 1993). The step response is applied if the same steady-state follows the same input, i.e., when the plant remains relatively isolated. Relay self-tuning is less demanding, as it tolerates plants with interactions, load disturbances, etc. If the closed loop response does not appear precisely specified, some corrections of controller settings, called fine-tuning, are needed. The standard version of relay self-tuning is extended here by making the tuning condition dependent on plant dynamics. Overshoot is avoided by choosing an appropriate controller structure. Relay control does not have to be symmetric.

Adaptation means automatic adjustment of controller settings after a major change of the process dynamics. Comparing several adaptive controllers revealed that the EXACT algorithm from Foxboro (now in Invensys) is particularly useful (Kraus and Myron, 1984). Details of EXACT have not been disclosed, however. An alternative algorithm presented here uses the same specifications as EXACT. During operation it employs three two-dimensional surfaces expressing overshoot, damping and frequency in terms of loop gain and controller zero. The algorithm executes steps on those surfaces to get to the target given as the crossing of some contour lines.

The last part deals with the input/output structure of PID, self-tuning and adaptation blocks, and explains how they cooperate. The internal or the external set-point can be chosen. The latter comes from MPC, optimization or other *DiaSter* components. Remarks on bumpless switching are also given.

The algorithms presented here are tested on two benchmark plants:

$$\mathrm{A}: \frac{1}{(20s+1)(0.5s+1)^4}, \qquad \mathrm{B}: \frac{\mathrm{e}^{-2s}}{(1.2s+1)^5}, \tag{6.67}$$

used originally by Kraus and Myron (1984) to demonstrate the convergence of EXACT. Such plants are considered *bracketing extremes* for most processes encountered in practice. The plant A is "easy", the plant B "difficult". So one can hope that reasonable results obtained for A and B will be also available elsewhere.

Unlike in the earlier section, here we apply a continuous approach to simplify explanations, and because loop control is usually executed much faster than prediction and optimization.

6.2.1 Step Response Method

As indicated above, step response self-tuning is feasible when the plant exhibits a repeatable steady-state. Besides, the plant must remain in the steady-state before the step input is applied. The algorithm implemented in *DiaSter* constructs two models of the plant at the same time, i.e.,

$$G_1(s) = \frac{k_o}{Ts+1}\,\mathrm{e}^{-\tau s}\,, \qquad G_2(s) = \frac{k_o}{(Ts+1)^2}\,\mathrm{e}^{-\tau s}. \tag{6.68}$$

The one that provides better approximation is selected. If the first model turns out better, the PI controller is chosen. PID is suitable for the second one. The convergence of identification is fast due to the simplicity of the models. Integral time and derivative time (for PID) are set to cancel time constants. Then, after the 1-st order Padé approximation of the delay term $\mathrm{e}^{-\tau s}$, the closed loop transfer function assumes the standard 2-nd order form. This in turn allows us to develop explicit rules for controller gain. Note that the extensive set of PI and PID tuning rules was collected by O'Dwyer (2003).

6.2.1.1 Plant Identification

Figure 6.19(a),(b) shows step responses of the benchmark plants A, B, supplemented with some noise. The input step is 1.0 (100%), standard deviation of the noise is 0.015 (1.5%). Before the step is applied, the controller monitors the output for some time and determines the initial mean value. After applying the step, the controller records the response in some time horizon T_H, until the steady-state is reached. For the responses of Fig. 6.19(a),(b) we have A: $T_H = 100$, B: $T_H = 25$ (counting from the moment the step is applied).

After recording the response, the controller calculates

- delay τ from the initial part,
- moment T_T at which the dynamic part of the response terminates,
- the steady-state mean value in the interval from T_T to T_H,
- plant gain k_o from the input step and initial and steady-state means,
- time constants T of the two models by least-squares approximation of the dynamic part,
- identification errors of the two approximations.

The delay τ is determined as the last moment at which the response does not exceed the initial mean. This yields $\tau = 2.1$ and $\tau = 3.7$ for the plants A, B, respectively (Fig. 6.19(a),(b)). Likewise, T_T is the first moment at which the response exceeds the final mean. So $T_T = 70.3$ and $T_T = 14.1$, respectively. It should be clear, however, that such numerical values characterize a particular realization of the noise. Any other realization would give slightly different results. The gains k_o computed from the means are both 0.99.

Identification of T. Let us begin with the 1-st order model of (6.68). It can be written as

$$\dot{y} + ay = ak_o u\,, \qquad a = \frac{1}{T}, \tag{6.69}$$

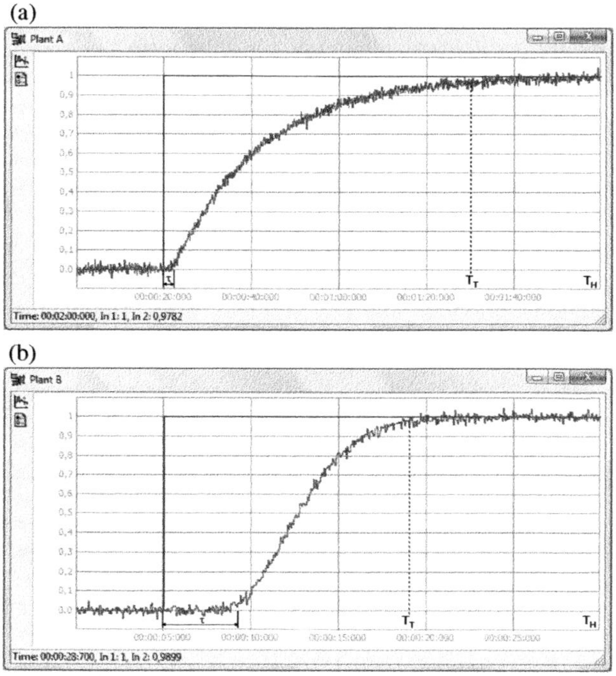

Fig. 6.19 Step response: plant A (a), plant B (b)

for $t > \tau$, where the input u is constant. We shall identify the coefficient a by minimizing the criterion

$$I(a) = \int_{\tau}^{T_T} [y(t,a) - z(t)]^2 D_t, \tag{6.70}$$

where $z(t)$ denotes the step response of the plant and $y(t,a)$ is a solution of (6.69). Let $\eta(t,a) = \partial y(t,a)/\partial a$ be the sensitivity function of y with respect to a. Differentiating both sides of (6.69) with respect to a yields the equation

$$\dot{\eta} + a\eta = k_o - y. \tag{6.71}$$

The gradient and the Hessian of the criterion I are given by

$$\frac{\partial I}{\partial a} = 2\int_{\tau}^{T_T} (y-z)\frac{\partial y}{\partial a} D_t = 2\int_{\tau}^{T_T} (y-z)\eta D_t = I_a, \tag{6.72}$$

$$\frac{\partial^2 I}{\partial a^2} = 2\int_{\tau}^{T_T} [\eta\eta + (y-z)\frac{\partial \eta}{\partial a}] D_t \cong 2\int_{\tau}^{T_T} \eta^2 D_t = I_{aa}. \tag{6.73}$$

Simplification in the second integral above is justified for small errors $(y - z)$. The Newton formula for iterative calculation of a has the form

$$a_{i+1} = a_i - \frac{I_a}{I_{aa}}. \tag{6.74}$$

The time constant T is an inverse of the resulting a. Note that $\dot{y}$, I, η, $\partial I/\partial a$ and $\partial^2 I/\partial a^2$ can be computed in parallel.

The 1-st order transfer functions that approximate the responses are given in the first row of Table 6.1, together with values of the criterion I. Convergence is reached in just two iterations, provided that $(2...3)/(T_T - \tau)$ is taken as the starting value for a.

Table 6.1 Models of the benchmark plants A, B

Model Ident. criterion	Plant A $\frac{1}{(20s+1)(0.5s+1)^4}$	Plant B $\frac{e^{-2s}}{(1.2s+1)^5}$
$\frac{k_o}{Ts+1}e^{-\tau s}$	$\frac{0.99}{19.5s+1}e^{-2.1s}$	$\frac{0.99}{4.7s+1}e^{-3.7s}$
I	$1.1 \cdot 10^{-4}$	$2.7 \cdot 10^{-4}$
$\frac{k_o}{(Ts+1)^2}e^{-\tau s}$	$\frac{0.99}{(9.3s+1)^2}e^{-2.1s}$	$\frac{0.99}{(2.2s+1)^2}e^{-3.7s}$
I	$6.9 \cdot 10^{-4}$	$3.9 \cdot 10^{-5}$

Now consider the 2-nd order model of (6.68). The model and sensitivity equations become

$$\ddot{y} + 2a\dot{y} + a^2 y = a^2 k_o u\,, \qquad a = \frac{1}{T}, \tag{6.75}$$

$$\ddot{\eta} + 2a\dot{\eta} + a^2\eta = 2a(k_o u - y) - 2\dot{y}. \tag{6.76}$$

Note that both y and $\dot{y}$ are needed here to calculate η. Results are given in the second row of Table 6.1. Two iterations suffice if a starts from $4/(T_T - \tau)$.

Values of the criteria I indicate that the 1-st order model is suitable for the plant A and the 2-nd order one for the plant B. Thus

$$\mathrm{A_{model}}: \ \frac{0.99}{19.5s+1}e^{-2.1s}, \qquad \mathrm{B_{model}}: \ \frac{0.99}{(2.2s+1)^2}e^{-3.7s}. \tag{6.77}$$

The plant and model responses are shown in Fig. 6.20(a),(b) (noise removed).

We remark that the increase of the noise makes the delay τ a little larger at the expense of the time constant T.

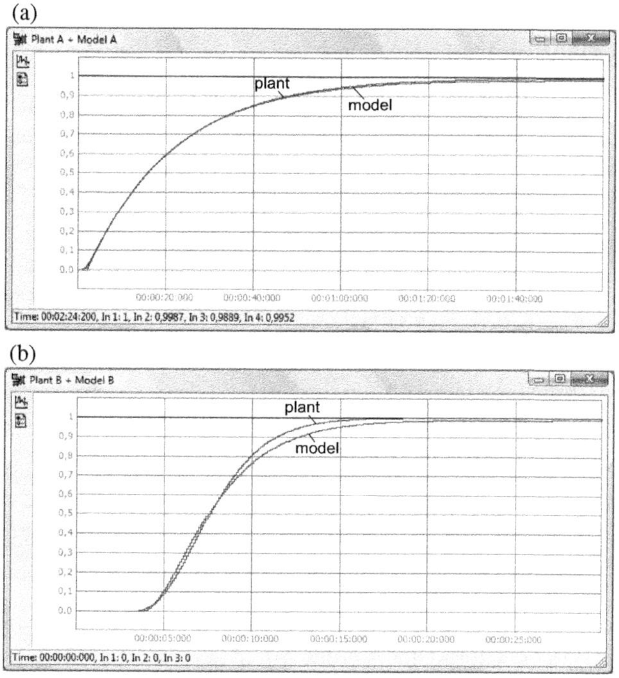

Fig. 6.20 Step responses of the plant and the model: plant A (a), plant B (b)

6.2.1.2 Controller Settings

The typical tuning objective is to get a closed loop response with a short setting time and a shape similar to that of the standard 2-nd order transfer function:

$$G_{II}(s) = \frac{\omega_n^2}{s^2 + 2\xi\omega_n s + \omega_n^2}. \tag{6.78}$$

The shape is determined by the damping factor ξ, i.e., critical damping for $\xi = 1$, oscillation for $\xi < 1$ and overdamping for $\xi > 1$. The percentage overshoot $p_{\%}$ and the settling time t_s are related to ξ and ω_n by textbook formulae:

$$\xi = |\ln(\frac{p_{\%}}{100})| / \sqrt{\pi^2 + \ln^2 \frac{p_{\%}}{100}}, \qquad t_s = \frac{4}{\xi\omega_n}. \tag{6.79}$$

As indicated at the beginning, the controller is set to cancel the time constant of the plant, since cancellation makes the loop quite fast. Hence the tuning objective is reduced to getting a response with a specified overshoot $p_{\%}$ or, equivalently, with a specified damping factor ξ. However, if the resulting controller turns out too sensitive (large gain), the objective must be replaced by the specification of the settling time t_s for the overdamped response.

PI controller. Consider the controller $k_p(1+1/(T_i s))$ for the model $k_o e^{-\tau s}/(Ts+1)$. The time constant is canceled by setting $T_i = T$. The 1-st order Padé approximation of $e^{-\tau s}$ yields the following open loop transfer function:

$$\begin{aligned} G_{open}(s) &= \frac{k_o}{Ts+1} e^{-\tau s}\, k_p\left(1+\frac{1}{Ts}\right) = \frac{k_p k_o}{Ts} \cong \frac{k_p k_o}{Ts} \frac{\frac{-\tau}{2}s+1}{\frac{\tau}{2}s+1} = \\ &= k' \frac{-s'+1}{s'(s'+1)} = k' G'(s') \,, \qquad k' = \frac{k_p k_o}{2} \frac{\tau}{T} \,, \qquad s' = \frac{\tau}{2} s. \end{aligned} \tag{6.80}$$

The closed loop transfer function becomes

$$G_{closed}(s') = \frac{k' G'(s')}{1 + k' G'(s')} = \frac{k'(-s'+1)}{s'^2 + (1-k')s' + k'} . \tag{6.81}$$

Now the goal is to find an expression for the relative gain k', so that the characteristic polynomial $s'^2 + (1-k')s' + k'$ would have a specified damping factor ξ.

Critical damping ($\xi = 1$). k' must solve the equation $(1-k')^2 - 4k' = 0$. Hence $k' = 3 - 2\sqrt{2} \cong 0.17$. Closed loop poles are $s'_{1,2} = (k'-1)/2 = 1 - \sqrt{2}$, so the relative settling time t'_s may be evaluated as $4/(\sqrt{2}-1) \cong 9.6$ (four time constants).

Oscillation ($\xi < 1$). Expressions for k' and t'_s in terms of specified ξ are as follows:

$$k' = \frac{(R^2 + R - t^2 R^2)(R-1) + t^2 R^2 (2R+1)}{(R-1)^2 + t^2 R^2} \,, \qquad t'_s = \frac{4}{|R|} , \tag{6.82}$$

where $t = \frac{\sqrt{1-\xi^2}}{\xi}$, $R = \frac{1-\sqrt{t^2+2}}{t^2+1}$. They were developed by Trybus (2005) using root locus. Several sets of ξ, k', t'_s for typical $p_\%$ are given in Table 6.2 for reference.

Table 6.2 Overshoot, damping factor, relative gain, settling time

$p_\%$	0	5	10	15	20	25	30
ξ	1	0.69	0.59	0.52	0.46	0.40	0.36
k'	0.17	0.27	0.32	0.37	0.41	0.45	0.49
t'_s	9.6	11	11.9	12.7	13.6	14.7	15.8

Overdamping ($\xi > 1$). As seen from Table 6.2 and k' in (6.80), the smallest gain k_p corresponding to critical damping equals $0.34\,(T/\tau)/k_o$. If such gain is too large because of practical limitations, the specification of the overshoot $p_\%$ must be replaced by the specification of the settling time t_s. This often happens for small delay τ. Specified t_s may be expressed in terms of the time constant T, i.e.,

$$t_s = \alpha T, \tag{6.83}$$

with some design parameter α. For $\alpha = 1$ we have $t_s = T$, so the loop will be four times faster than the plant (four time constants). The dominating pole of $G_{closed}(s')$ in (6.81) is $s'_1 = (k' - 1 + \sqrt{1 - 6k' + k'^2})/2$, so the relative settling time may be evaluated as $t'_s = 4/|s'_1|$. Since $t_s = t'_s \tau/2$, from (6.83) and the expression for s'_1 we get the following equation:

$$|k' - 1 + \sqrt{1 - 6k' + k'^2}| = \frac{4}{\alpha} \frac{\tau}{T} \tag{6.84}$$

to compute k'. A two-dimensional look-up table with α and τ/T coordinates is used in *DiaSter* to get k'.

The rules for absolute settings and the settling time (needed for specified $p_{\%}$) are as follows:

$$\text{PI}: \qquad k_p = 2k' \frac{T}{\tau} \frac{1}{k_o}, \qquad T_i = T, \qquad t_s = t'_s \frac{\tau}{2}. \tag{6.85}$$

PID controller. Consider now the controller $k_p(1 + 1/(T_i s) + T_d s)$ for the plant model $k_o e^{-\tau s}/(Ts + 1)^2$. Let $T_d = T_i/4$; thus the controller becomes $k_p(T_i/(2s) + 1)^2/(T_i s)$. The time constant T is canceled by $T_i = 2T$. The open loop transfer function is almost the same as (6.80), so

$$G_{open}(s) = k_p k_o \frac{1}{2T} \mathrm{e}^{-\tau s}, \tag{6.86}$$

except for $2T$ instead of T. This, however, does not affect the expressions for k' and t'_s. Absolute settings and the settling time are given by

$$\text{PID}: \qquad k_p = 4k' \frac{T}{\tau} \frac{1}{k_o}, \qquad T_i = 2T, \qquad T_d = \frac{T_i}{4}, \qquad t_s = t'_s \frac{\tau}{2}. \tag{6.87}$$

Noise filter. Since the dynamics of the loop are known, we can design a low-pass filter for the plant output. The filter is particularly useful for the PID controller. Recall that $\tau/2$ is a "time constant" of the open loop (see (6.80)). The dynamics of the loop designed for specified $p_{\%}$ (using Table 6.2) will not be affected if we take the time constant of the filter a few times less than $\tau/2$, say $\tau/(8...10)$. For the loop designed for t_s (using (6.84)), the filter time constant can be about $t_s/20$.

6.2.1.3 Initial Response and Gain Fine-Tuning

Plant A. For the model $0.99\mathrm{e}^{-2.1s}/(19.5s + 1)$, from (6.77) the rules (6.85) yield PI: $k_p = 18.7k'$, $T_i = 19.5$, $t_s = 1.05t'_s$. We shall consider two cases:

- critical damping with $\xi = 1$ ($p_{\%} = 0$),
- oscillation with the overshoot $p_{\%} = 20$.

From Table 6.2 we have $k' = 0.17, 0.41$ and $t'_s = 9.6, 13.6$, respectively, for the two cases. The absolute values are $k_p = 3.2, 7.7$ and $t_s = 10, 14$. The time constant of the filter is $2.1/10 = 0.21$. Filtered closed loop responses are shown in Fig. 6.21(a). The response y_1 for the first case appears satisfactory. However, the 32% overshoot of y_2 exceeds 20% too much, so some correction is needed. The actual settling times are somewhat longer than expected.

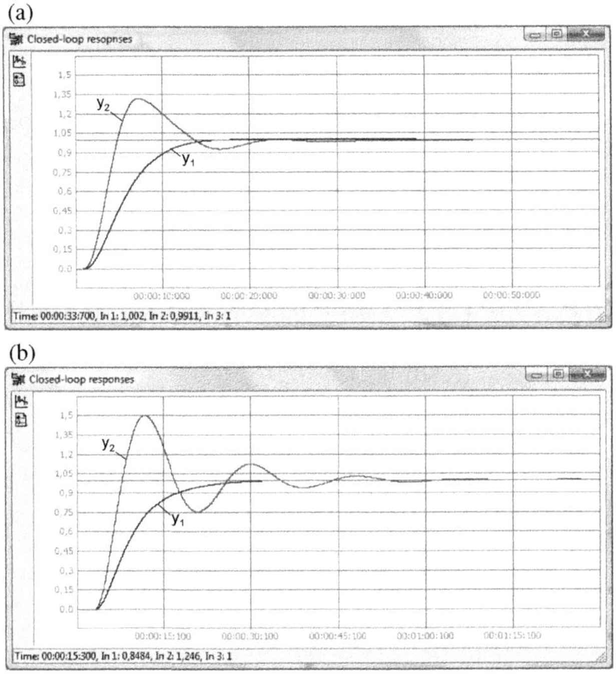

Fig. 6.21 Closed loop responses for preliminary tuning: plant A (a), plant B (b)

Plant B. For the model $0.99\mathrm{e}^{-3.7s}/(2.2s+1)^2$ and the rules (6.87), we have PID: $k_p = 2.4k'$, $T_i = 4.4$, $T_d = 1.1$ and $t_s = 1.05t'_s$. Hence $k_p = 0.41, 0.98$ and $t_s = 18, 27$ for the two cases ($p_{\%} = 0, 20$). The filter time constant is $3.7/10 = 0.37$. The PID transfer function $k_p(1 + 1/(T_i s) + (T_d s/D + 1))$ with a typical divisor $D = 5$ is implemented. The filtered responses are shown in Fig. 6.21(b). As before, we may be satisfied with the damped response y_1. However, the 55% overshoot of y_2 is far too large than the required 20%.

The two examples indicate that if oscillatory responses are needed, due to the difference between the dynamics of the plant and its model, the closed loop response may differ considerably from specification, so controller settings must be corrected.

Such corrections are usually called fine-tuning. However, fine-tuning is rarely dealt with in the literature, being rather left to engineering experience.

Fine-tuning applied here is restricted to the correction of the gain k_p, based on the measurement of the overshoot $p_\%$. T_i and T_d are left unchanged. The underlying theory is explained below.

Let $p_\%$ be the overshoot obtained initially and ξ the corresponding damping factor. For $p_\% = 32,55$ of the y_2 plots in Fig. 6.21(a),(b), we have $\xi = 0.34, 0.19$, respectively. Using $s' = s\,\tau/2$ in the close-loop transfer function $G_{II}(s)$ of (6.78), we write its denominator as $(s'^2 + 2\xi\,\omega_n' s' + \omega_n'^2)$. However, the denominator may be also written in the form $(s'^2 + (1-k')s' + k')$ from (6.81). Such k' corresponds to the response with the actual damping ξ. By comparing coefficients of the two representations we obtain

$$k' = \left(\sqrt{\xi^2+1} - \xi\right)^2. \tag{6.88}$$

This value will be interpreted as relative gain of the loop, corresponding to a real response.

To develop an iterative procedure, assume that at a step i, for some controller gain $k_{p,i}$, a closed loop response with the damping ξ_i is obtained. We are looking for a new gain $k_{p,i+1}$ that will provide a response with a specified damping ξ_{spec}. Since k_p is proportional to k' (see (6.85), (6.87)), by using (6.88) we may write

$$k_{p,i+1} = k_{p,i}\left(\frac{\sqrt{\xi_{spec}^2+1} - \xi_{spec}}{\sqrt{\xi_i^2+1} - \xi_i}\right)^2. \tag{6.89}$$

This formula will be applied in fine-tuning steps. We take $p_{\%,spec} = 20$ as before, thus $\xi_{spec} = 0.456$.

Plant A. Let $i = 0$ denote the initial tuning, hence $k_{p,0} = 7.7$, $\xi_0 = 0.34$ (see above). For $\xi_{spec} = 0.456$, the formula (6.89) yields $k_{p,1} = 6.2$. The response y_1 shown in Fig. 6.22(a) has $p_\% = 19.3$. Another tuning step is not needed.

Plant B. Now $k_{p,0} = 0.98$ and $\xi_0 = 0.19$ (for $p_\% = 55$). $k_{p,1} = 0.59$ results in y_1 with $p_\% = 11$ (Fig. 6.22(b)), so $\xi_1 = 0.57$. Now we need to increase the gain to get $p_{\%,spec} = 20\%$. $k_{p,2} = 0.73$ yields y_2 with $p_\% = 27$ ($\xi = 0.38$). In the third step we apply $k_{p,3} = 0.64$ and get y_3 with $p_\% = 16.4$. It seems reasonably close to 20%, so fine-tuning may be terminated.

Finally we recall that, due to a small τ/T ratio $(2.1/19.5 = 0.11)$, the PI controller designed as above, i.e., for $p_\%$ specification, is quite sensitive. In Section 7.4.2 we consider a plant with similar τ/T, but with the PI controller designed for a specified settling time t_s.

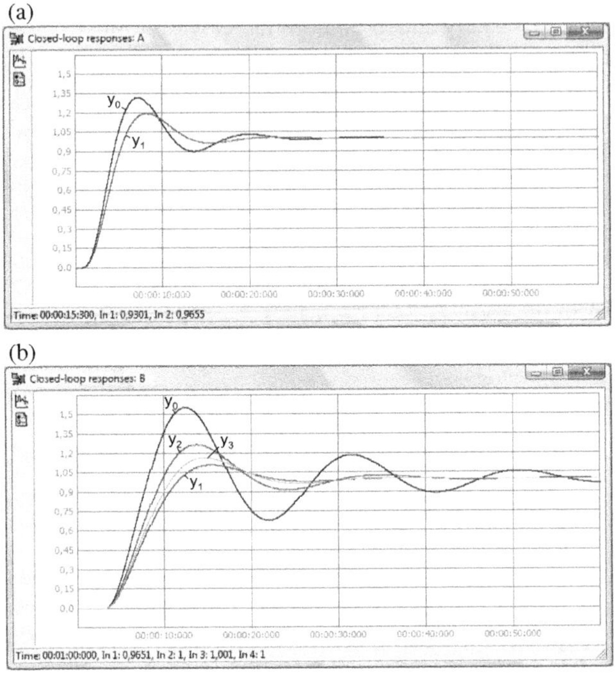

Fig. 6.22 Close loop responses during fine-tuning: plant A (a), plant B (b)

6.2.2 Relay Self-tuning

Self-tuning based on relay oscillations is a method for automatic execution of the well known Ziegler–Nichols experiment. Requirements on the steady-state are not as strict as for the step response. However, one cannot expect particularly good results, since relay oscillations provide information on the plant dynamics only, and not on the steady-state gain.

6.2.2.1 Åström and Hägglund Method

In 1943, Ziegler and Nichols presented a limit cycle method for experimental tuning of PID controllers, based on bringing the loop with a P controller to the stability limit. Having critical gain k_{cr} and oscillation period T_{cr}, the settings of the P, PI or PID controller, whichever preferred, are calculated from simple rules, e.g., P: $k_p = 0.5\,k_{cr}$, PI: $k_p = 0.45\,k_{cr}$, $T_i = 0.85\,T_{cr}$ (O'Dwyer, 2003). As can be seen, the loop with the P controller has the gain margin $GM = 2\ (= 1/0.5)$.

The Ziegler–Nichols experiment is somewhat cumbersome due to step-by-step adjustments of the gain while approaching the stability limit. Hence broad interest, 25 years ago, in the simple idea of Åström and Hägglund (1984), who proposed automatic realization of the experiment by replacing the P controller with relay, as shown in Fig. 6.23(a). The relay generates on-off control with some magnitude U, so the loop exhibits sustained oscillation, whose amplitude A_{cr} and period T_{cr} are measured (Fig. 6.23(b)). By employing the describing function of the relay, one can calculate controller settings using frequency methods.

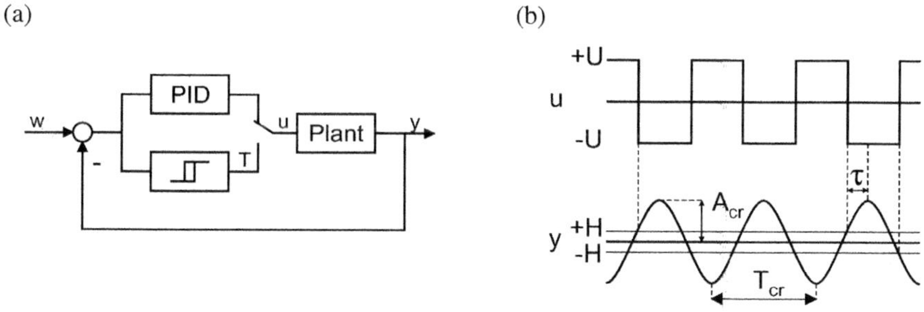

Fig. 6.23 Loop with relay self-tuning (a), typical time diagrams (b)

Plant magnitude and angle. The describing function of the relay with hysteresis H is given by

$$N(A) = \frac{4U}{\pi A}\left(\sqrt{1-\left(\frac{H}{A}\right)^2} - j\frac{H}{A}\right). \tag{6.90}$$

Let R and I denote absolute values of real and imaginary parts of the plant transfer function at frequency $\omega_{cr} = 2\pi/T_{cr}$. Hence

$$\begin{aligned} G_o(j\omega_{cr}) &= -R - jI = M\,\mathrm{e}^{-j\phi}\,, \\ M &= \sqrt{R^2+I^2}\,, \qquad \phi = \pi - \arctan\tfrac{I}{R}. \end{aligned} \tag{6.91}$$

The Nyquist condition $1 + N(A_{cr})G_o(j\omega_{cr}) = 0$ for the limit cycle yields

$$R = \frac{\pi}{4U}\sqrt{A_{cr}^2 - H^2}\,, \qquad I = \frac{\pi H}{4U}. \tag{6.92}$$

So we have a single point of the plant frequency characteristic (Fig. 6.24(a)). This is merely a single point, hence one cannot expect particularly good final results.

Design condition. To calculate controller settings, the following condition was applied by Hägglund and Åström (1991):

$$G_o(j\omega_{cr})G_{PID}(j\omega_{cr}) = \frac{1}{GM}\,\mathrm{e}^{-j(\pi-PM)}, \tag{6.93}$$

with the gain margin $GM = 2$ and the phase margin $PM = \pi/4$ (45°). Figure 6.24(b) shows how a single point of the open loop characteristic is affected by this condition, when the plant angle ϕ exceeds $\frac{3}{4}\pi$ (145°). Such ϕ is common in practice, since A_{cr} is substantially grater than H for sufficiently large U. It is clear from (6.91) and (6.93) that the controller must provide phase shift $\phi - \pi + PM$ at ω_{cr}. Since $\phi > \frac{3}{4}\pi$, for $PM = \pi/4$ we have positive shift $\phi - \frac{3}{4}\pi$. Such shift can be introduced by the PID controller only, and not by PI (to design PI we would have to take $PM = 0$).

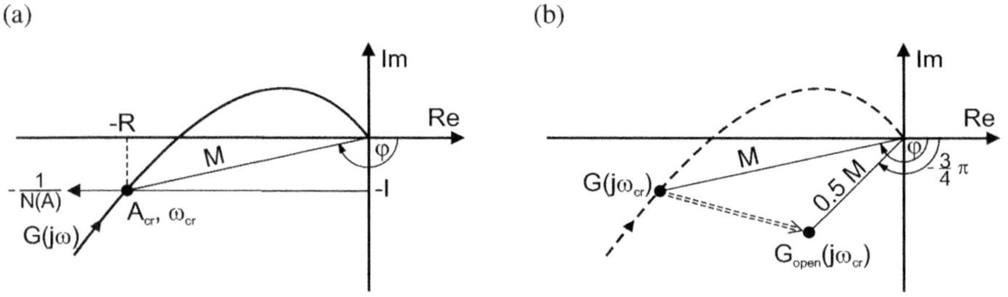

Fig. 6.24 Frequency characteristic: plant (a), single point of the open-loop for $GM = 2$, $PM = \pi/4$ (b)

PID settings. The condition (6.93) gives two relations for the calculation of three settings. From Ziegler–Nichols rules, we take $T_d = T_i/4$ as the third one. By using (6.91) and $k_p(1 + 1/(T_i s) + T_d s)$ in (6.93), one obtains

$$k_p = \frac{GM}{M} \frac{\omega_{cr} T_i}{(\omega_{cr} T_i)^2 + 4}, \quad T_i = \frac{2}{\omega_{cr}} \tan \frac{1}{2} \left(\phi - \frac{\pi}{2} + PM \right), \quad T_d = \frac{T_i}{4}. \tag{6.94}$$

For small or large delays, a PI controller with

$$\text{small delay}: \quad k_p = 0.5 \tfrac{1}{M}, \quad T_i = \frac{4}{\omega_{cr}} \tag{6.95}$$

$$\text{large delay}: \quad k_p = 0.25 \tfrac{1}{M}, \quad T_i = \frac{1.6}{\omega_{cr}} \tag{6.96}$$

is recommended by Hägglund and Åström (1991). They do not specify, however, when to switch from PID to PI.

We will look at loop responses for the margins $GM = 2$ and $PM = \pi/4$.

Plant A. Relay control simulated for the on-off magnitude $U = 0.22$ (22%) and the hysteresis $H = 0.005$ gives $A_{cr} = 0.016$ and $T_{cr} = 9.1$. From the expressions (6.91), (6.92), we get $R = 0.055$, $I = 0.018$, $M = 0.058$, $\phi = 2.83$ (162°), and from (6.94), the PID settings $k_p = 7.7$, $T_i = 4.7$, $T_d = 1.2$. The response of the standard unity feedback loop, i.e., with the PID controller driven by the control error $(w - y)$, is shown in Fig. 6.25(a). The overshoot is 31% and still remains, although reduced roughly by half, if we split the controller by leaving the I component driven by the error and driving PD by the negative output $-y$. The PI controller tuned according to (6.96) and split into I+P exhibits oscillations and a much longer settling time.

Plant B. Due to remarkable delay, the magnitude U must be much smaller than before to avoid excessive output changes. $U = 0.05$ (5%) and $H = 0.005$ give $A_{cr} = 0.038$ and $T_{cr} = 16$. Now $R = 0.59$, $I = 0.078$, $M = 0.60$, $\phi = 3.0$ (172°). PID settings are $k_p = 0.66$, $T_i = 10$, $T_d = 2.5$. The response of the standard loop shown in Fig. 6.25(b) seems quite poor. Much better is the response for the PI controller tuned according to (6.96), i.e, $k_p = 0.42$, $T_i = 4.1$.

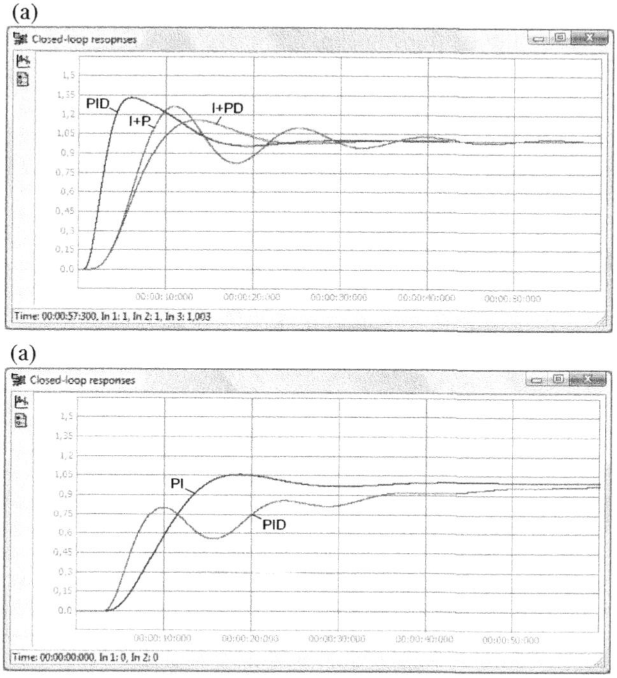

Fig. 6.25 Closed loop responses for different controllers: plant A (a), plant B (b)

6.2.2.2 Plant Dependent Margins

The examples reveal that the margins $GM = 2$ and $PM = \pi/4$ do not satisfy all cases. It would be better to relate them to plant dynamics, evaluated somehow by looking at relay oscillations more carefully.

Delay and period. Consider a typical plant $k_o e^{-\tau s}/(Ts + 1)$ whose dynamics may be characterized by relative delay τ/T. The equation $\tau\omega_{cr} + \arctan(T\omega_{cr}) = \phi$

describes relay oscillations (see (6.90)–(6.92)). Thus the ratio τ/T_{cr}, being a function of τ/T and ϕ, may also characterize the plant dynamics. The angle ϕ does not change much, assuming values in a fairly narrow interval (say 155...175°, Fig. 6.24). The delay τ can be read out from the oscillation plot, as the time interval between the control change and the extremum of the output, shown in the right hand side of Fig. 6.23(b). Simulations of relay control for the plants $k_o e^{-\tau s}/(Ts+1)$, $\tau/T \in (0.05, 2)$ and $k_o/(Ts+1)^n$, $n = 1,2,3...$ have indicated that the ratio τ/T_{cr} increases from about 0.1 for small τ/T or n up to about 0.3 for large τ/T or n. Therefore, τ/T_{cr} will be used here as an indicator what plant we are dealing with.

GM, PM vs. τ/T_{cr}. In our modification of relay self-tuning the margins *GM*, *PM* are not constant but related to τ/T_{cr} according to the nomograms of Fig. 6.26(a),(b). For τ/T_{cr} close to 0.2, we have $GM \cong 2$ and $PM \cong \pi/4$, as in the work of Hägglund and Åström (1991) (H–A dotted lines in Fig. 6.26). By adjusting *GM*, *PM* to the plant dynamics, roughly similar responses are obtained for delays in a typical range. The behavior of the PID controller for $\tau/T_{cr} \cong 0.1$ and 0.3 resembles that of PI, so there is no need to switch the controllers for small or large delays. The responses of the standard unity feedback loop exhibit modest overshoot of about 10...15%.

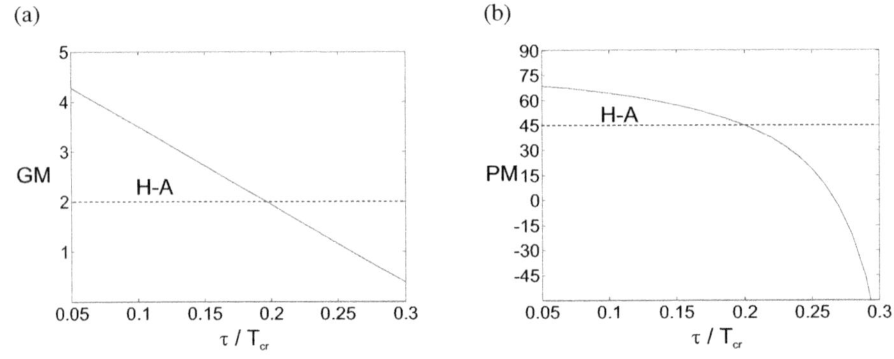

Fig. 6.26 Nomograms of design margins in terms of τ/T_{cr}: GM (a), PM (b)

Overshoot elimination. A number of applications, e.g., temperature control, do not tolerate overshoot after the set-point change. It turns out, however, that controller settings obtained from relay self-tuning very often produce overshoots, even if PID is split into I and PD components. Removing the overshoot in all cases is by no means easy, as indicated by (Åström et al., 1993). After numerous trials and comparison tests, the PID controller structure shown in Fig. 6.27 is recommended to avoid overshoot. The P component is driven by the control error $(w-y)$, D by the negative output $-y$, and I by modified error $w_f - y$, with a filtered set point w_f coming from a low pass filter $1/(T_i s/2+1)$. Such a structure has been favorably evaluated in a number of temperature loops with relay self-tuning. Sample tuning and response are shown in Section 7.4.2.

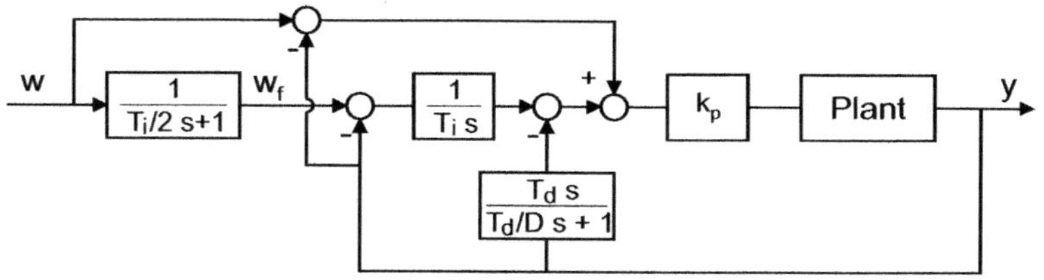

Fig. 6.27 No-overshoot loop with a PID controller of a dedicated structure

6.2.2.3 Relay Asymmetry

The method of Åström and Hägglund (1984; 1991) assumes the symmetry of the relay (Fig. 6.23(a)), which requires either the selection of the magnitude U or proper adjustment of the limits U_{min}, U_{max}. This sometimes does not happen, especially if the tuning algorithm must be service free.

Suppose the set-point w, control u and output y are all in the range 0...1 (0...100%), so $U_{min} = 0.0$ and $U_{max} = 1.0$. The symmetry of the relay is provided only for $w = 0.5$ and a linear plant with unity gain $k_o = 1$. Any deviation from these values causes asymmetry, extending the oscillation period T_{cr} and, if expressions for controller settings remain unchanged, deteriorating close loop responses.

Here the asymmetry is dealt with as follows:

- the switch-on time t_{on} is determined from relay control (Fig. 6.28(a)) and the ratio t_{on}/T_{cr} expressing the degree of asymmetry is calculated;
- using the ratio t_{on}/T_{cr} and the nomogram of Fig. 6.28(b), the value of the correcting divisor $f(t_{on}/T_{cr})$ is determined;
- the corrected period T^*_{cr} defined by

$$T^*_{cr} = \frac{T_{cr}}{f(t_{on}/T_{cr})} \tag{6.97}$$

  is used in *GM* and *PM* nomograms of Fig. 6.26(a),(b) instead of original T_{cr}.

If relay control is symmetric, i.e., $f(t_{on}/T_{cr} = 0.5) = 1$, then $T^*_{cr} = T_{cr}$. In other cases we have $T^*_{cr} < T_{cr}$, so the extension of the period T_{cr} due to asymmetry is compensated (better or worse). The nomogram of Fig. 6.28(b) has been developed under the assumption that settings obtained for the linear plant at any set-point w must be almost the same as those obtained for $w = 0.5$. This is also valid for the plant with the gain $k_o \neq 1$.

A compensation of relay asymmetry according to (6.97) gives good results for linear plants, where changes of settings for different w and k_o do not exceed 10%...15%. The results are worse for non-linear plants, but in general they are still better than those obtained while neglecting asymmetry altogether.

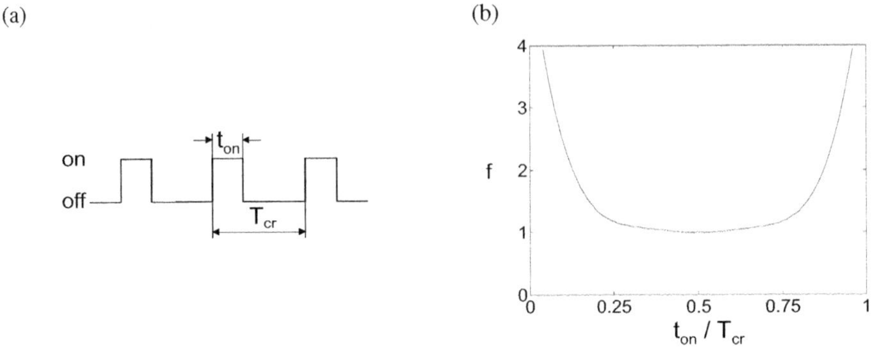

Fig. 6.28 Asymmetry of the relay: control (a), nomogram of the period correcting divisor (b)

6.2.3 Loop Adaptation

As has been mentioned at the beginning, the EXACTPID algorithm (Kraus and Myron, 1984) has been found particularly useful for industrial applications. Given an error transient after a set-point change or strong disturbance, the algorithm makes such adjustments of controller settings so as to get transients with specified overshoot and damping, and a short settling time. The simplicity of specifications and robustness even for a large change of plant properties are basic advantages of EXACT. However, algorithmic details are protected by a patent.

An alternative algorithm based on the same specifications and operating similarly is described below (Świder and Trybus, 2004). The algorithm employs three two-dimensional surfaces ("maps") that express overshoot, damping and frequency in terms of two settings of the controller. The maps are given in the form of look-up tables with contour lines. After detecting a transient, the algorithm determines a point where the loop operates, as the crossing of contours for overshoot and damping, or frequency. User specifications, i.e., the required overshoot and damping, determine the target point. Having the distance between the two points, the settings are adjusted accordingly.

6.2.3.1 Overshoot, Damping and Frequency vs. Settings

EXACT algorithm. Consider the standard control loop of Fig. 6.29 and assume that the controller has been pre-tuned initially. The expected error transients to load-step disturbance d are shown in Fig. 6.30. If the actual transient does not meet specifications, EXACT executes adaptation by adjusting PID settings in a number of steps. Each step follows a reference or disturbance excitation. Here, however, we restrict ourselves only to disturbances, since internal structures of PID controllers affecting reference responses may be different (see Fig. 6.27).

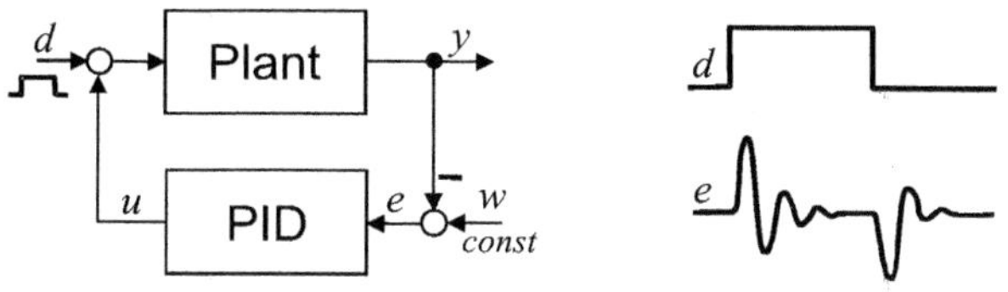

Fig. 6.29 Control loop structure during adaptation

The EXACT algorithm uses two indicators of the shape of the transient, namely, overshoot OVS and damping DMP, defined by

$$OVS = -\frac{E_2}{E_1}, \qquad DMP = \frac{E_3 - E_2}{E_1 - E_2}, \tag{6.98}$$

using the "peaks" E_1, E_2 and E_3 ($E_2 < 0$ for A, B in Fig. 6.30). The algorithm also employs the period T_p to determine the time-scale. Tuning specifications involve the required overshoot OVS^* and damping DMP^*. Both OVS^* and DMP^* must be non-negative, which means that a transient such as A, B or D may be specified as a target. The transient C is excluded due to rapid changes of control.

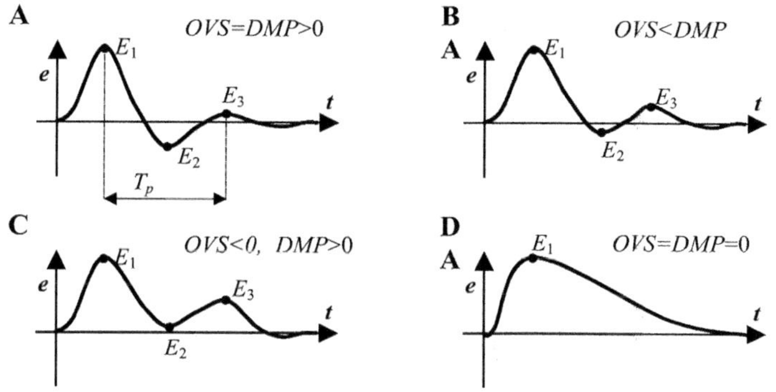

Fig. 6.30 Error transients for load-step disturbances

Default specifications of the overshoot and damping provided initially by EXACT are $OVS^* = DMP^* = 0.3$, which implies 2-nd order damped oscillatory responses. Notice, however, that the equality $OVS^* = DMP^*$ does not determine loop behavior uniquely, because the same overshoot and damping can be often obtained both for slow and fast responses. For obvious reasons, EXACT tunes for a fast response in such a case.

Relative frequency. The transients A, D, most interesting in practice, can be described by

$$E(s) = \frac{c}{s^2 + 2\xi\omega_n s + \omega_n^2}, \qquad \mathrm{A}: \ \xi < 1, \qquad \mathrm{D}: \ \xi \geq 1. \tag{6.99}$$

B and C may have other components in the numerator and the denominator. The natural frequency ω_n of the transient A is calculated from standard expressions:

$$\omega_n = \frac{\omega}{\sqrt{1-\xi^2}}, \qquad \xi = \frac{\sigma}{\sigma^2 + \omega^2}, \qquad \omega = \frac{2\pi}{T_p}, \qquad \sigma = \frac{-\ln|E_2/E_1|}{T_p}. \tag{6.100}$$

We will return to the transient D further on.

The PID transfer function $k_p + 1/(T_i s) + T_d s$ for $T_d = T_i/4$ may be written as

$$\mathrm{PID}: \ \frac{k_p}{2z}\frac{(s+z)^2}{s}, \qquad z = \frac{2}{T_i}. \tag{6.101}$$

The controller has two independent settings, gain k_p and zero z. Let e be a transient A with certain natural frequency ω_n, obtained for some k_p and z. To relate ω_n to z, we define relative natural frequency:

$$OMN = \frac{\omega_n}{z}. \tag{6.102}$$

It will be a third indicator used for adaptation.

Settings plane. Consider the following typical plant:

$$G(s) = \frac{k_o}{Ts+1}\mathrm{e}^{-\tau s}. \tag{6.103}$$

Taking the PID controller (6.101), one can simulate the closed loop response for some settings k_p, z. From the error transient, we measure E_1, E_2, E_3, determine OVS and DMP, and, for the transient A (or B close to A), calculate ω_n and, finally, OMN.

Define the relative zero of the controller as the product $z\tau$ and loop gain as $k_p k_o$. The plane with logarithmic coordinates,

$$a = \log(z\tau), \qquad b = \log(k_p k_o), \tag{6.104}$$

will be called the settings plane for the PID controller (6.101) and the plant (6.103).

Suppose now that an extensive number of simulations have been performed and the corresponding indicators OVS, DMP and OMN evaluated. Points at which $OVS = const_i$, $DMP = const_j$, $OMN = const_k$ can be connected together giving sets of contour lines of three surfaces $OVS(a,b)$, $DMP(a,b)$, $OMN(a,b)$ on the settings plane. The contours for four plants with $\tau/T = 0.05$, 0.2, 1 and 4 are shown in Fig. 6.31.

Areas A, B, C, D. The interval $\tau/T \in [0.05, 4]$ satisfies most industrial plants. The 1.5 decade range along both axes in Fig. 6.31 covers a fairly large span of the settings

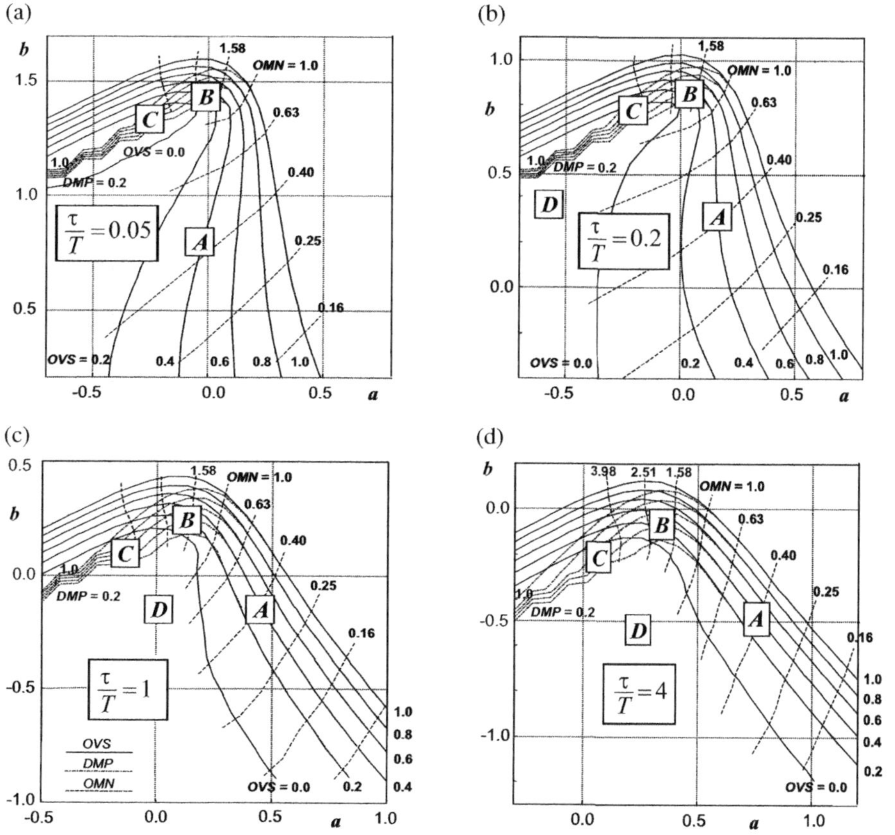

Fig. 6.31 Contour lines of OVS, DMP and OMN for $\tau/T = 0.05, 0.2, 1, 4$ on the settings plane

k_p, z ($10^{1.5} \cong 31.6$). The squares with characters A, B, C, D indicate areas where error transients look like in Fig. 6.30. In particular, the transient A occurs for large a (a strong I action in PID) and small or modest b. The transient B corresponds to modest a and large b (a strong P action). Instability occurs above the contour $OVS = 1.0$ on the right and $DMP = 1.0$ on the left. The relative frequency OMN increases while moving up from the bottom (increase of gain). The OMN contours are spaced by one fifth of decade ($\log 0.4 = -0.4$, $\log 0.63 = -0.2$, etc.). Hence a step from one OMN contour to another changes the settling time roughly by 1/3 ($t_s = 4/(\xi\omega_n)$; $1.0 - 0.63 = 0.37 \cong 1/3$ in Fig. 6.31).

6.2.3.2 Fine-Tuning for Known τ/T

Target point. The contours of Fig. 6.31 reveal to what target our algorithm will converge. Given some equality specifications $OVS^* = DMP^*$ (transient A), the target is a point at which the contours OVS^*, DMP^* begin to diverge (a black square in

Fig. 6.32(a)). If, however, $OVS^* < DMP^*$, the target is a crossing point of the contours OVS^*, DMP^* (Fig. 6.32(b)). The divergence point is particularly important, since it determines a limit transient A having the largest frequency OMN, hence the shortest settling time. Any increase in controller sensitivity beyond the divergence point results in a transient with the shape B (Fig. 6.30).

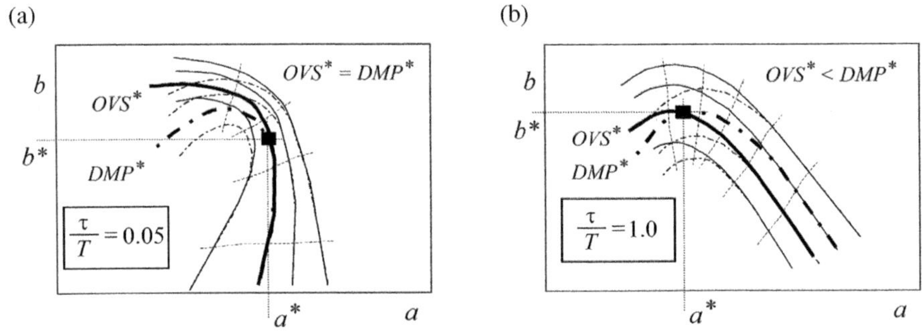

Fig. 6.32 OVS^* and DMP^* contours: divergence point (a), crossing point (b)

To demonstrate how the contours of Fig. 6.31 can be employed for controller tuning, assume temporarily that the relative delay τ/T of the plant (6.103) is known, e.g., from preliminary tuning. Having τ/T, we can create three surfaces $OVS(a,b)$, $DMP(a,b)$, $OMN(a,b)$ and, given a starting point, use them to get to the target. This is done as follows:

1. *the starting point:* suppose some initial settings k_p^0, z^0 yield a transient e^0 from which E_1^0, E_2^0, E_3^0 are measured and OVS^0, DMP^0 computed.
   - If $OVS^0 = DMP^0$, which means that e^0 is a transient A, ω_n^0 is calculated from (6.100) and OMN^0 from (6.102). The starting point (a^0, b^0) is the crossing of the contours OVS^0, OMN^0;
   - If $OVS^0 < DMP^0$ (both positive), e^0 represents the transient B, the crossing of the contours OVS^0, DMP^0 becomes the starting point (a^0, b^0). Frequency OMN^0 is not needed here;
2. *the target point:* coordinates (a^*, b^*) of the target are determined either as the divergence point ($OVS^* = DMP^*$) or the crossing ($OVS^* < DMP^*$). The first case is shown in Fig. 6.33(a),(b);
3. *new settings:* measure the distances

$$\Delta a = a^* - a^0 , \qquad \Delta b = b^* - b^0 \tag{6.105}$$

(Fig. 6.33(a),(b)) and compute the final settings

$$k_p^* = k_p^0\, 10^{\Delta b}, \qquad z_c^* = z_c^0\, 10^{\Delta a}. \tag{6.106}$$

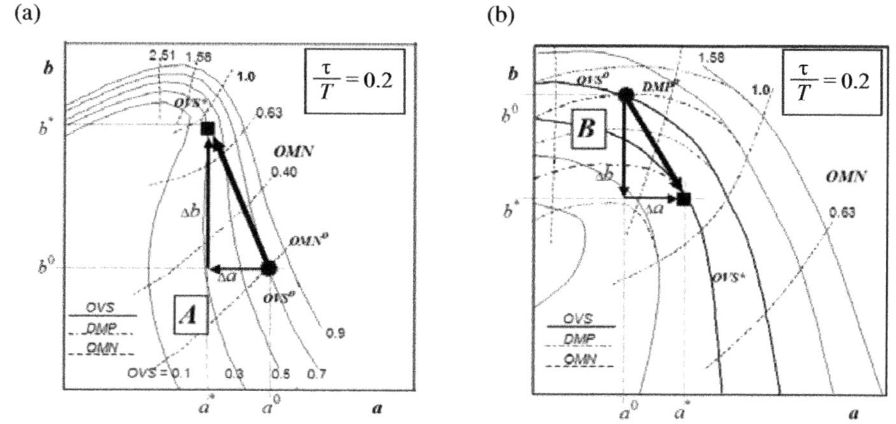

Fig. 6.33 Convergence for the plant with $\tau/T = 0.2$, the target point $OVS^* \cong DMP^* \cong 0.3$, and the starting point in A or B: area A (a), area B (b)

As seen, if τ/T is known, the target can be reached in one step.

Such an algorithm is able to operate in the areas A, B. If the starting point is in C or D (Fig. 6.31), we first have to get to A or B. Increasing a (stronger I action) moves the operating point from D to A, and decreasing b (weaker P) from B to D.

6.2.3.3 Adaptation for Unknown τ/T

Now consider the case of unknown τ/T. To keep our algorithm unchanged, a certain generally applicable template value of τ/T must be chosen, such that the corresponding three surfaces $OVS_{tmpl}(a,b)$, $DMP_{tmpl}(a,b)$, $OMN_{tmpl}(a,b)$ could be applied for all plants. Hence the question arises as to what value of τ/T can be used as a template. Before giving the answer, we recall that plants with large delay τ are considered "difficult". Thus it should not be surprising that the template we are looking for will involve relatively large delay.

Consequences of template surfaces. The problem is illustrated in Fig. 6.34 by the contours $OVS = 0.3$ for $\tau/T = 0.05$, 0.2, 1 and 4. The contours have been moved to have a common crossing at $OMN = 1.0$, near the divergence point of OVS and DMP. As seen, for the transients B (upper left part), the contours almost overlap, so τ/T does not really matter. The situation is different for the transients A (lower part). Suppose that the starting and target points both lie on the same contour, i.e., $OVS^0 = OVS^*$. This means that the initial transient has proper overshoot but is too slow. If we choose the template at, say $\tau/T = 1$, then only for points with $\tau = T$ the algorithm will be able to reach the target in one step (thickest arrow in Fig. 6.34). For $\tau < T$, as for a majority of industrial plants, the step made from $OMN^0 < 1.0$ along the contour for $\tau/T = 1$ will be oriented inwards, because the contours for $\tau/T < 1$ run more vertically than for $\tau/T = 1$ (compare $\tau/T = 0.05, 0.2$ with 1 in

Fig. 6.34). The step inwards means a decrease in the overshoot, so in the next step the algorithm will have to turn right, to return to OVS^*.

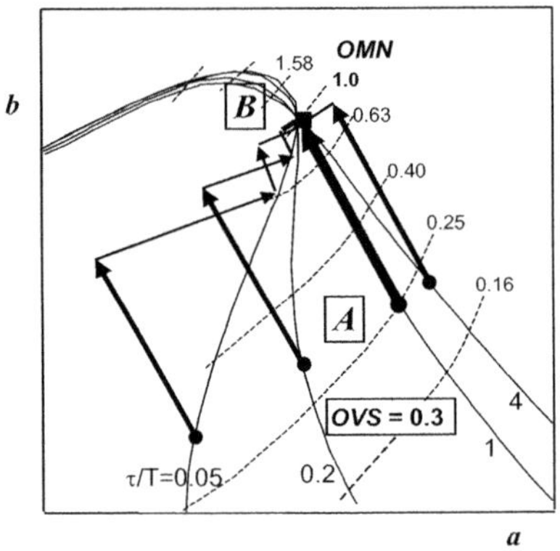

Fig. 6.34 Contours $OVS = 0.3$ for $\tau/T = 0.05$, 0.2, 1, 4 and steps towards the target made along the contour for $\tau/T = 1$

The situation is different for plants with $\tau > T$ (dominant delay). When we start from $OMN^0 < 1.0$, the step along the contour for $\tau/T = 1$ results in somewhat bigger overshoot than the initial $OVS^0 = OVS^*$. Fortunately, the increase in the overshoot is small because the slope of the contours for $\tau/T > 1$ is only slightly smaller than for $\tau/T = 1$ (compare $\tau/T = 1$ with 4 in Fig. 6.34). In the second step, the algorithm will decrease the overshoot by turning left.

Template surfaces. Taking above into account, $OVS_{tmpl}(a,b)$, $DMP_{tmpl}(a,b)$, $OMN_{tmpl}(a,b)$ for $\tau/T = 1$ are chosen as template surfaces, i.e., for the plant whose delay is the same as the time constant ("difficult" plant). Contour lines are obtained for

$$G_{tmpl} = \frac{e^{-s}}{s+1} \tag{6.107}$$

and have already been shown in Fig. 6.31 (lower left).

We must add, however, that the template $\tau/T = 1$ is suitable only if the target point is close to the "top" of the contours, as in Fig. 6.34. Then $OMN^0 < OMN^* \cong 1.0$, so the tuning will decrease the settling time. If we would like to slow down the loop having $OVS^* < OVS^0 < 1.0$, a step down should be made. By comparing the cases in Fig. 6.31, one can find that such a step has to go down almost vertically, not along the slope of $\tau/T = 1$, to avoid getting into the instability area (left side) for plants with small τ/T. So when we want to make the loop slower, $e^{-0.05s}/(s+1)$ is taken to create another three template surfaces.

6.2.3.4 Implementation Issues and Tests

Damped response. To tune the controller for the transient D (Fig. 6.30), the following negative overshoot is defined:

$$OVS = -\mathrm{e}^{-\pi(2-\xi)/\sqrt{1-(2-\xi)^2}} \qquad \text{for} \qquad \xi \in (1,2), \tag{6.108}$$

and the corresponding contour lines are developed. ξ is calculated from three values of time at which the transient overcomes the initial level (say 2%), reaches maximum E_1 and falls below the initial level. The limit contour $OVS = -0.1$ in Fig. 6.35 represents $\xi \cong 1.4$ (poles of $E(s) = c/(s^2 + 2\xi\omega_n s + \omega_n^2)$ differ six times). The transient D is further classified according to OVS and OMN as

$$\begin{array}{lll} \mathrm{D_A}: & \text{close to A} & - \ OVS \in [0,-0.1], \ OMN \leq 1.0 \\ \mathrm{D_B}: & \text{close to B} & - \ OVS \in [0,-0.1], \ OMN > 1.0 \\ \mathrm{D_D}: & \text{"deep" in D} - & \xi \geq 1.4. \end{array} \tag{6.109}$$

The notion of negative overshoot has also been used in EXACT (Kraus and Myron, 1984).

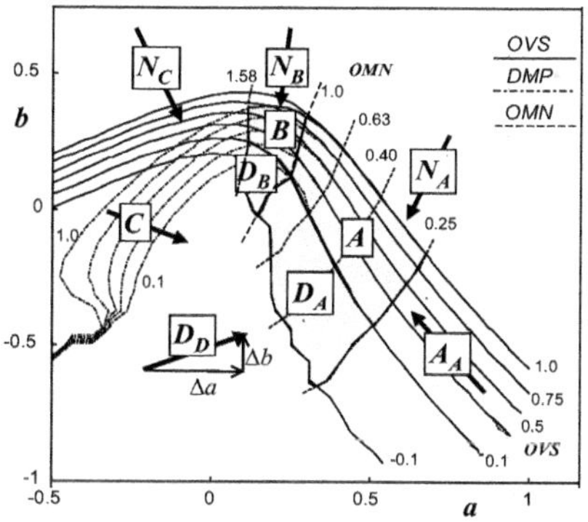

Fig. 6.35 Template contours and specific areas distinguished by the adaptation algorithm

Range of contours. Look-up tables with template contours must be restricted for implementation reasons. So only the most important area of the settings plane restricted from the bottom by $OMN = 0.25$ and from above by $OMN = 1.58$ is covered by the surfaces $OVS_{tmpl}(a,b)$, $DMP_{tmpl}(a,b)$, $OMN_{tmpl}(a,b)$. The low limit $OMN = 0.25$ provides transients up to four times slower than the divergence point $OVS = DMP$ (or $OMN = 1.0$). If the starting point belongs to this area, the settings are computed according to (6.106).

Fixed steps. If the starting point lies outside the area between $OMN = 0.25$ and $OMN = 1.58$, then certain fixed steps Δa, Δb associated with the particular location are applied in (6.106). Such steps are indicated in Fig. 6.35 by arrows with area symbols in squares. Lengths and orientations of these arrows vary from area to area.

Instability. A large, rapid change of the process dynamics may cause instability indicated by the overshoot $OVS > 1.0$. Based on such OVS and also on DMP, the corresponding point is assigned to one of the areas N_A, N_B, N_C. Fixed steps indicated by the arrows (Fig. 6.35) are applied to bring the loop back into the stability area.

Band-pass filter. Following Hägglund and Åström (1991), the error transient, before being processed by the adaptation algorithm, is first fed into the band-pass filter. This protects the algorithm from being triggered by slow or fast periodic disturbances, which could eventually de-tune the loop. The medium frequency of the filter is $2/T_i$.

Tests. For the standard plant (6.103), the algorithm converges in steps shown in Fig. 6.34. Convergence for the benchmark plants A, B of (6.68) is illustrated in Fig. 6.36. As can be seen, no matter where the starting point is, the adaptation converges in several up to about ten steps. We remark that a somewhat simplified version of the algorithm has been implemented in the RF-537 instrument controller manufactured by the ZPDA company from Ostrów Wlkp. in Poland.

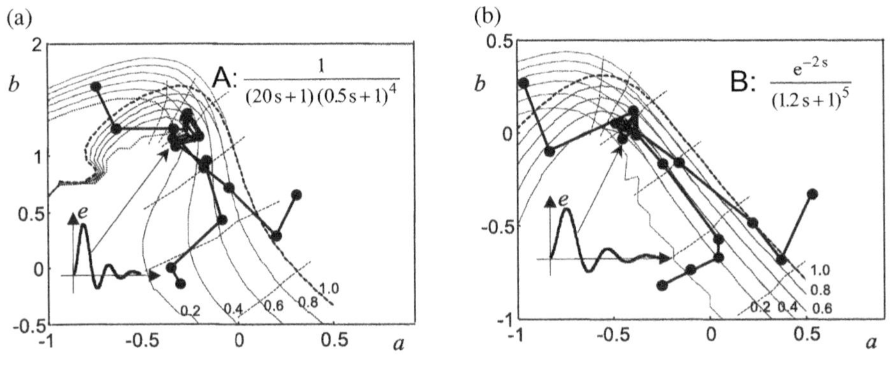

Fig. 6.36 Plant contours and adaptation convergence: plant A (a), plant B (b)

6.2.4 *Function Blocks*

This section presents input-output structures and principles of cooperation between basic function blocks of control loops, i.e., PID controllers, self-tuning blocks and the adaptive control block. Providing a standard control loop with self-tuning or adaptation functionality requires placing an appropriate block in front of the PID controller. The higher level of the *DiaSter* software is interfaced to the loop by means of a set-point generator that selects either an internal or an external set-point. The latter may come from DMC, GPC or optimization algorithms described in the first part of the chapter. The cascade loop can also be built in this way. Brief indications on bumpless switchings are given.

6.2.4.1 PID, SELF and ADAPT

Two PID controllers, SELF_step, SELF_relay and ADAPT blocks, together with typical connections, are shown in Fig. 6.37(a),(b), respectively. The first PID is standard, the second one structured according to Fig. 6.27.

(a)

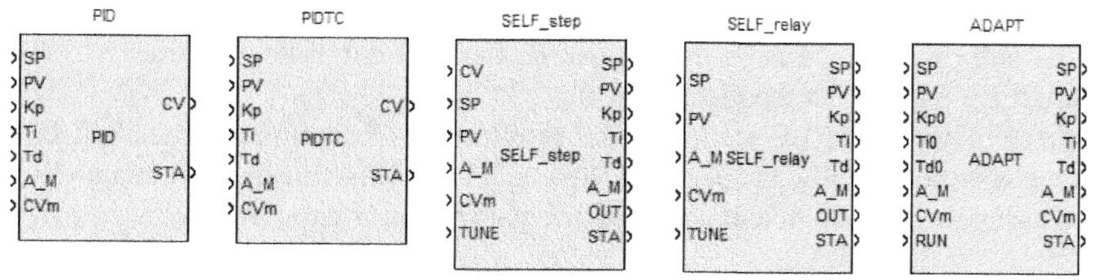

(b)

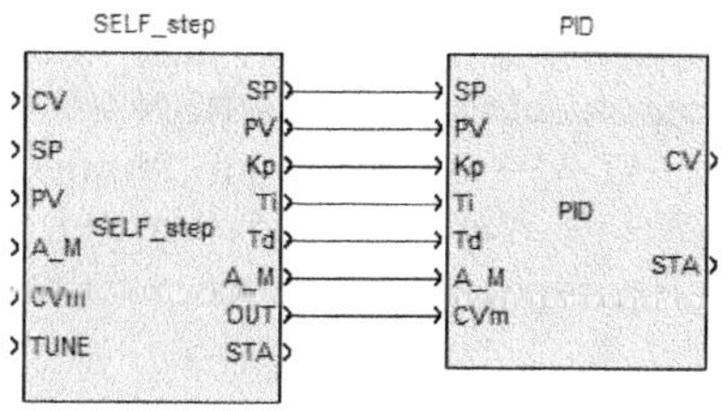

Fig. 6.37 Controller, self-tuning and adaptive control blocks (a), typical connections (b)

PID. The inputs denote, respectively, SP: set-point, PV: process variable, K_p, T_i, T_d: controller settings, A_M: *Automatic/Manual* mode, CVm: control variable for *Manual*. The outputs include CV: control variable, STA: block status (earlier in this section the variables SP, PV and CV have been denoted by w, y and u, respectively). A_M = 1 sets the controller to *Automatic*, A_M = 0 to *Manual*. CV equals CVm in *Manual*. The status STA can be 0: *ready*, another value: error (e.g., settings out of range). Ranges of the settings K_p, T_i, T_d and other limits are set in the block properties window, but we will not show it here. After switching from *Manual* to *Automatic*, CV begins from CVm to provide bumpless transfer. It is recommended, however, to keep SP close to PV before switching. Bumpless transfer from *Automatic* to *Manual* requires setting CVm close to CV.

SELF. Before self-tuning begins, the controller is usually in the *Manual* mode (A_M = 0; although it is not necessary). Setting TUNE to 1 activates the procedure. The SELF block sets the A_M output to 0, placing PID into *Manual* (if not in it already). Then, depending whether SELF_step or SELF_relay is used, a control step or relay control appears at the output OUT connected to CVm in PID. In the case of SELF_step, the step is made from the level at the CV input (usually connected to the CV output of PID). SELF_relay operates as an on-off controller with a set-point SP and a process output PV. The size of the control step, relay limits, hysteresis, etc. is set in block properties. While the experiment goes on, SELF monitors the inputs and executes corresponding calculations. Termination is indicated by the output status STA = 2: *success*. Other types of status denote, respectively, 0: *ready*, 1: *experiment on*, 3,4...: errors. When STA = *success*, PID settings appear at the outputs K_p, T_i, T_d. Now the user may deactivate self-tuning by setting TUNE to 0. The SELF block becomes transparent, i.e., the inputs SP, PV, A_M and CVm are directly transferred to the outputs. If A_M is set to *Automatic*, the PID controller acquires K_p, T_i, T_d from SELF and begins feedback control.

We remark that SELF_step does not provide gain fine-tuning if the closed loop response is not exactly as specified (Sec. 6.2.1.3). This can be done manually by measuring the overshoot and adjusting the gain according to (6.89).

ADAPT. If RUN is 1: *on*, ADAPT executes the "walking on the map" algorithm described in Section 6.2.3. So when a transient (at least medium size) settles down, new settings K_p, T_i, T_d are output and acquired by the PID controller. The status STA may be 1: *off* (RUN = 0), 0: *waiting* or tracking a transient, 1: *new settings ready*, 3,4,...: errors. STA = 1 is a pulse, so it appears only briefly. For RUN = 0, the inputs K_{p0}, T_{i0}, T_{d0} are transferred to the outputs (ADAPT is transparent). Normally, before ADAPT is switched on, K_{p0}, T_{i0}, T_{d0} are set to values received from SELF.

6.2.4.2 Set-Point Control

The connection of the control loop to a higher level is provided by the SP generator, whose structure is shown in Fig. 6.38. The output SP is a set-point for PID, SELF and ADAPT. By means of Switch and SPC (set-point control), the generator switches between internal (upper) and external (lower) set-points.

Options for the internal set-point include SP manual, Sinusoid, and a time scheduler composed of Time line (time base), Lookup table and Gain (scaling). Mode = 1 selects SP manual in Selector. The External SP block at the bottom acquires a value from a buffer, written by DMC, GPC, an optimum operating point or by the *InView* environment. A *DiaSter* subpath of External SP transfers the value to the SP Ext

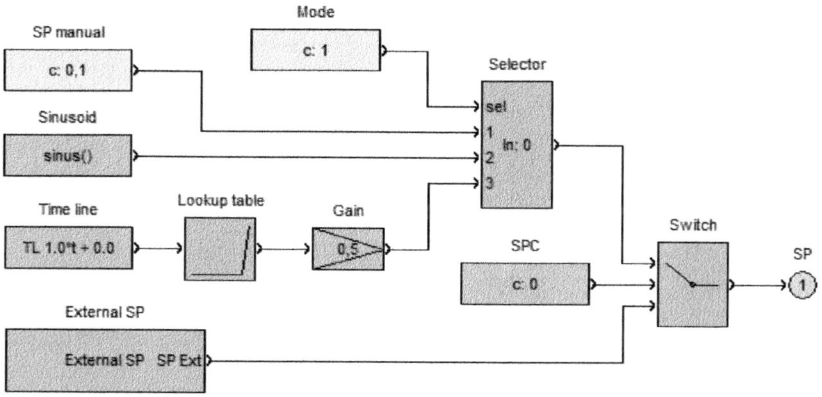

Fig. 6.38 Structure of the SP generator for a set-point

output. External SP is selected for SPC = 1 (Switch in lower position). The internal and external set-points must be equal for bumpless switching.

Chapter 7
Application of the DiaSter System

Michał Syfert, Paweł Chrzanowski, Bartłomiej Fajdek, Maciej Ławryńczuk, Piotr Marusak, Krzysztof Patan, Tomasz Rogala, Andrzej Stec, Robert Szulim, Piotr Tomasik, Dominik Wachla, and Marcin Witczak

7.1 Introduction

The functions of the *DiaSter* system are realized by means of specialized user packages. These packages cooperate by means of the system software platform. Particular packages can be developed as specialized libraries of modeling, system

Michał Syfert · Bartłomiej Fajdek
Institute of Automatic Control and Robotics, Warsaw University of Technology, ul. Św. Andrzeja Boboli 8, 02–525 Warsaw, Poland
e-mail: {m.syfert,b.fajdek}@mchtr.pw.edu.pl

Paweł Chrzanowski · Tomasz Rogala · Piotr Tomasik · Dominik Wachla
Department of Fundamentals of Machinery Design, Silesian University of Technology, ul. Konarskiego 18A, 44–100 Gliwice, Poland
e-mail: {pawel.chrzanowski,tomasz.rogala,piotr.tomasik, dominik.wachla}@polsl.pl

Maciej Ławryńczuk · Piotr Marusak
Institute of Control and Computation Engineering, Warsaw University of Technology, ul. Nowowiejska 15/19, 00-665 Warsaw, Poland
e-mail: {m.lawrynczuk,p.marusak}@ia.pw.edu.pl

Krzysztof Patan · Marcin Witczak
Institute of Control and Computation Engineering, University of Zielona Góra, ul. Podgórna 50, 65–246 Zielona Góra, Poland
e-mail: {k.patan,m.witczak}@issi.uz.zgora.pl

Andrzej Stec
Department of Computer and Control Engineering, Rzeszów University of Technology, W. Pola 2, 35–959 Rzeszów, Poland
e-mail: astec@prz-rzeszow.pl

Robert Szulim
Institute of Electrical Metrology, University of Zielona Góra
ul. Podgórna 50, 65–246 Zielona Góra, Poland
e-mail: r.szulim@ime.uz.zgora.pl

variables processing and visualization modules or as stand-alone software modules. Individual packages or a specified set of packages can be used to implement particular tasks connected with process modeling, supervisory control, optimization and diagnostics. The functions of individual packages complete each other, giving the possibility to realize even more complex tasks.

Simple, unrelated examples were presented in previous chapters. They were used to illustrate the presented theory and functioning rules of particular algorithms of the *DiaSter* system. The aim of this chapter is to present a complete example where different tasks are realized by different packages and the *DiaSter* platform for a chosen object of automatic control, monitoring and diagnostics. Due to educational purposes and clarity of presentation, the relatively simple set-up of a three tank system was chosen.

7.2 System of Automatic Control and Diagnostics

The application of selected *DiaSter* modules and packages is presented with the use of an example of the realization of control and diagnostic tasks for a Three Tank (TT) system. This system consists of

- three serially connected tanks,
- a water supply system composed of the pump and the control valve together with the positioner.

The water is supplied through the control valve to Tank 1. From the outlet of Tank 3, the water freely pours out due to gravity. A diagram of the system with indicated process variable measurements and control loops is presented in Fig. 7.1.

The control of the water level in Tank 3 is realized in the basic configuration. The water level of 0.25 m is selected as a default working point. An additional (nominally constant and not measured) water inflow to Tank 3 is considered. The fluctuations of that flow are chosen as the main, unmeasured process disturbance. The level of that flow changes randomly round a fixed constant value equal to about 2,7 l/min.

Process variables. The set of process variables, together with their descriptions and value ranges, is presented in Table 7.1. These variables are the main process variables that would be available (measured) in a real process of a similar kind.

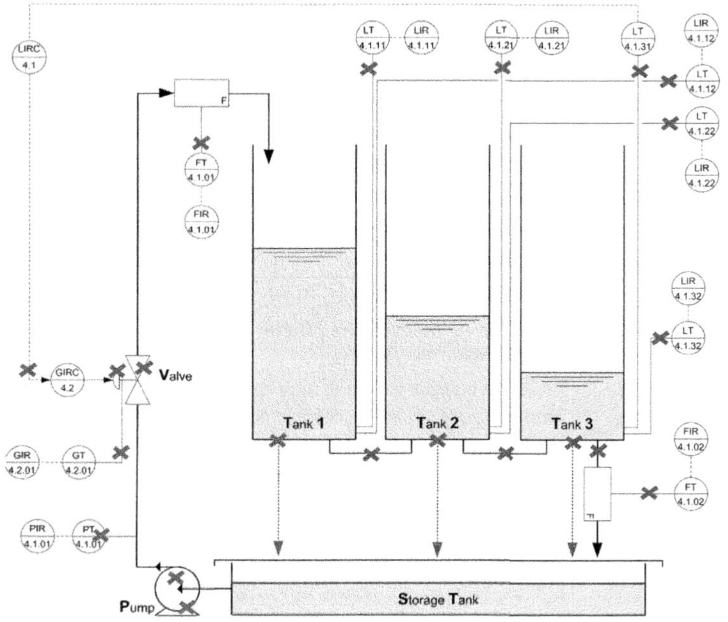

Fig. 7.1 Process and instrumentation diagram of the TT system together with symbolic designation of faults

Table 7.1 Set of process variables for the TT system

Symbol	Description	Value range
$SP(LIC.4.1.SP)$	Control value of the system (desired position of the valve plug on the water inlet to Tank 1)	$0...0.5m$
$CV(LIC.4.1.CV)$	Feedback signal of the position of the valve plug on the water inlet to Tank 1	$0...100\%$
$G(GT.4.2.01)$	Water outflow from Tank 1	$0...100\%$
$F_1(FT.4.1.01)$	Water inflow to Tank 1	$0...28l/min$
$L_{11}(LT.4.1.11)$ $L_{12}(LT.4.1.12)$	Water level in Tank 1 (two redundant sensors)	$0...1m$
$L_{21}(LT.4.1.21)$ $L_{22}(LT.4.1.22)$	Water level in Tank 2 (two redundant sensors)	$0...1m$
$L_{31}(LIC.4.1.PV)$ $L_{32}(LT.4.1.32)$	Water level in Tank 3 (two redundant sensors)	$0...1m$
$F_2(FT.4.1.02)$	Water outflow from Tank 3	$0...33l/min$
$P_1(PT.4.1.01)$	Pressure in the inlet to the control valve (pressure of the pump)	$80...260kPa$

Faults. The set of examined faults for the TT system consists of the faults of instruments, actuators and process components. Symbolic locations of fault introduction are shown in Fig. 7.1, while fault specification is given in Table 7.2.

Table 7.2 Set of discussed faults for the TT system

f_k	Description
f_1	Water pump operation stoppage (switch-off)
f_2	Decrease of water pump efficiency
f_3	Fault in the *CV* signal path
f_4	Fault of the control valve servomotor
f_5	Fault of the *G* valve position measuring path
f_6	Fault of the control valve plug or seat
f_7	Fault of the F_1 flow measurement path
f_8	Fault of the level L_{11} measurement path
f_9	Leakage from Tank 1
f_{10}	Partial clogging of the pipe between Tanks 1 and 2
f_{11}	Fault of the level L_{21} measurement path
f_{12}	Leakage from Tank 2
f_{13}	Partial clogging of the pipe between Tanks 2 and 3
f_{14}	Fault of the level L_{31} measurement path
f_{15}	Leakage from Tank 3
f_{16}	Partial clogging of the outlet from Tank 3
f_{17}	Fault of the F_2 measurement path
f_{18}	Fault of the L_{12} measurement path
f_{19}	Fault of the L_{22} measurement path
f_{20}	Fault of the L_{32} measurement path
f_{21}	Pressure *P* measurement path fault

7.3 Process Information Model in the DiaSter Platform

Basic process components and their relations that reflect the process structure are defined in the *DiaSter* system configuration. These elements form the *information model* of the process and the whole system. Particular packages use this model to manage the processed information and the stored configuration.

Several subsystems and particular constituent components are distinguished in the process structure. They are called *assets*. The following basic types of assets are used: `process` (element symbolizing the whole TT system), `section` (separate subsystems consisting of other subsystems or individual components), `component` (representation of a single device), `measuring device` (component type), `actuator` (component type) and `pipeline` (component type: water transportation elements connecting particular devices). The additional `task` type is used for organizational purposes.

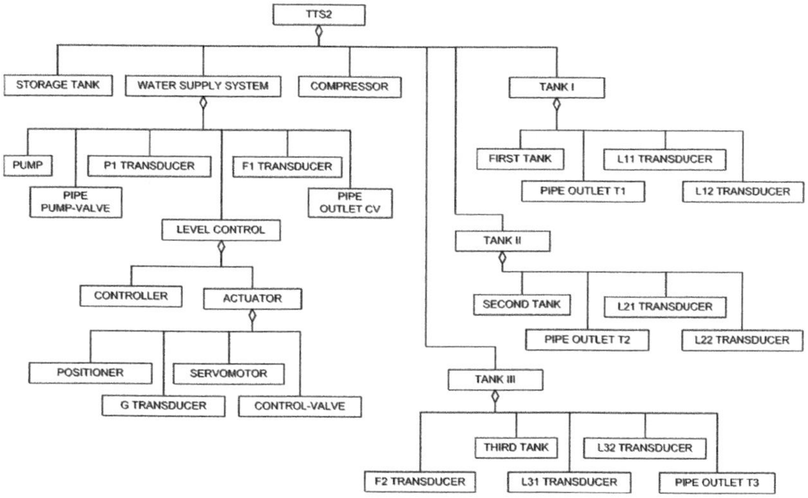

Fig. 7.2 Graphical representation of the component hierarchical structure

The set of defined assets is show in Fig. 7.2. The assets connected by the relation "belongs to" formulate a hierarchical structure corresponding to the process structure.

The relation "belongs to", shown in Fig. 7.2, is just one of the relations defining connections between process and system components that can be defined in the *DiaSter* platform information model. There is also a possibility to use several other built-in relations as well as user-defined ones. The following are defined and used at a general software platform level (generally available in all system packages):

- the relation "*connected as in/out*", which defines the direction of material/signals flow between assets;
- the relation "*delivered by*", which points out measuring devices that "produce" particular process variables;
- the relation "*of*", which determines the connection between assets and their faults.

The above relations were defined for all analyzed assets, process variables and faults. Their graphical representation is shown in Figs. 7.4 and 7.5.

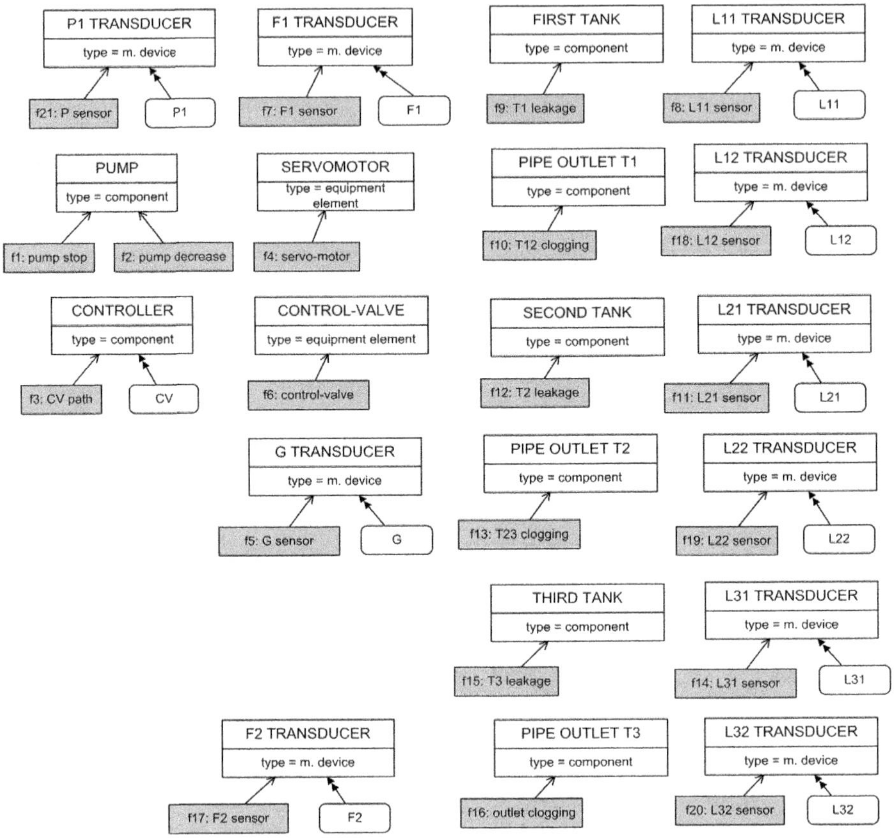

Fig. 7.3 Graphical representation of the relations "delivered by" and "of"

Figure 7.5 presents the user interface of the main *DiaSter* configuration module. The tree structure shown in the left hand window represents the hierarchical structure of the process and system components as well as their relations.

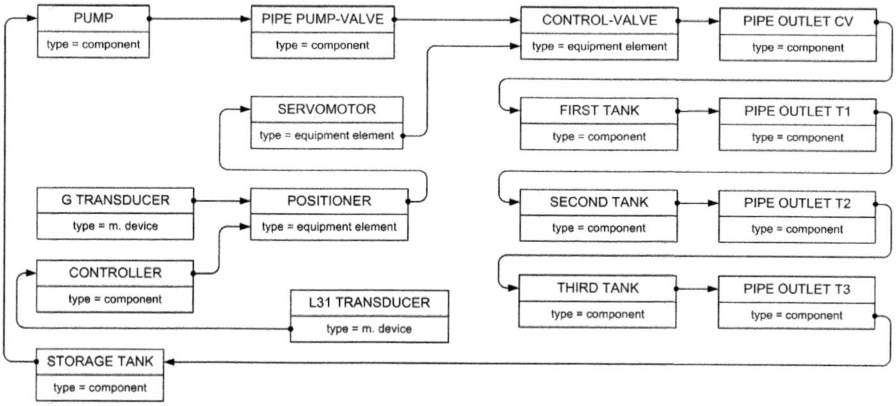

Fig. 7.4 Graphical representation of the component hierarchical structure

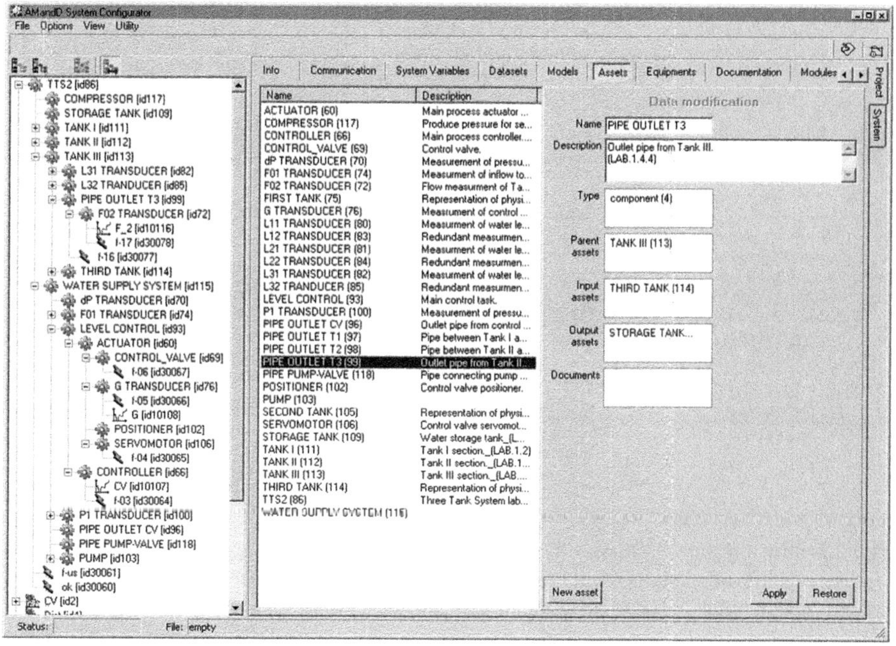

Fig. 7.5 Configuration user interface used to define process and system components as well as their relations

7.4 Applications of the DiaSter System Packages

Different modules and packages of the *DiaSter* system were applied to fulfill the tasks of simulation, control, advanced monitoring and diagnostics of the process presented in Section 7.2. The general structure of the system is presented in Fig.7.6.

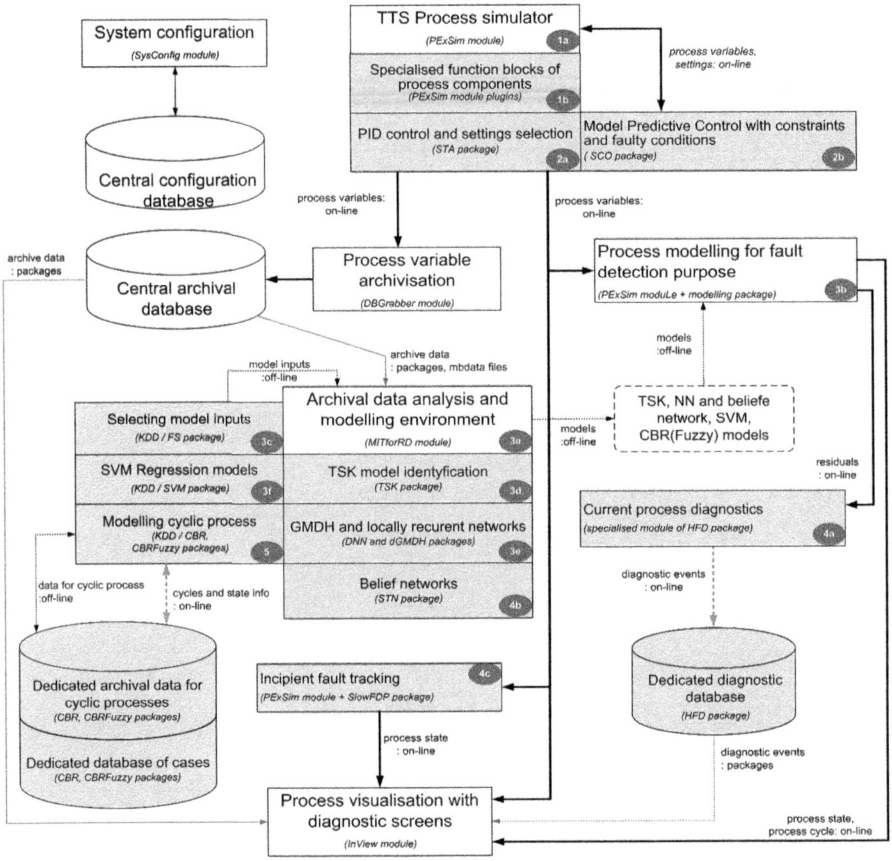

Fig. 7.6 Structure of the *DiaSter* system module for an example application. White blocks represent platform modules while grey ones represent different user packages (realized as system module plug-ins or stand alone modules).

The way of realizing particular tasks, together with examples of test results, is presented in the succeeding sections. The performed tasks are described in an order that corresponds to the logic of the implementation of the described system modules for a real application. The following groups of tasks are distinguished:

- *the way of realizing the process simulator with the use of tools available in the DiaSter system*: the simulator plays the role of a "pattern process"—other system features are presented with its use. The simulator is realized directly in the main system calculation module *PExSim* (Block 1a in Fig.7.6) with the use of a specialized function block representing process components (Block 1b in Fig. 7.6);
- *the example of control loop realizations that utilize the traditional PID controller (with the auto-tuning capability) as well as predictive control*: these tasks are fulfilled by specialized calculation module plug-ins of the *STA* package (Self-tuning and Adaptation, Block 2a in Fig. 7.6) and the *SCO* package (Supervisory Control and Optimization, Block 2b in Fig. 7.6);
- *the application of modeling based on collected archival data sets for the purpose of process variable reconstruction and fault detection*: TSK, SVM regression and neural network models (locally recurrent networks and GMDH dynamic networks) were applied for this purpose. Models are identified in an off-line mode with the use of the archival data analysis and modeling module *MITforRD* (block 3a in Fig. 7.6). The process of model input selection was supported by automatic features selection realized by the *KDD-FS* package (Knowledge Discovery in Database—Selection of Model Input Variables). Then, the models are used in the *PExSim* module in an on-line mode to estimate selected process variables (Block 3b in Fig. 7.6). Both kinds of models are realized as *MITforRD* plug-ins (Blocks 3d, 3e and 3f in Fig. 7.6) of the *TSK* (Takagi–Sugeno–Kanga) models, *DNN* (Locally Recurrent Neural Networks), *dGMDH* (Dynamic GMDH Neural Networks) and *SVM* (Support Vector Machines Regression Models);
- *the realization of diagnostic tasks with respect to current process diagnostics of abrupt faults as well as detection, tracking and development prediction of incipient faults (slow process degradation)*: two different approaches for current process diagnostics are presented. One utilizes detection based on the set of partial models and fuzzy reasoning, while the other is based on a belief network. The former is realized as a standalone system module (Block 4a in Fig. 7.6) of the *HFD* package (Fuzzy Diagnostics in the Hierarchical Structure) while the latter is a *MITforRD* plug-in (Block 4b in Fig. 7.6) of the *STN* package (Belief Networks Models). Incipient fault monitoring is realized by a specialized plug-in of the *PExSim* calculation module (block 4c in Fig. 7.6) in the *SlowFDP* package (Incipient Fault Tracking);
- *specialized tasks of cyclic process modeling and recognition of the process state* with the use of the base of cases realized by the *CBR* and *CBRFuzzy* (Fuzzy Case Base Reasoning Models for Cyclic Processes, Block 5 in Fig. 7.6) packages.

7.4.1 Process Simulator

The process simulator was realized in the variable processing module *PExSim*. Figure 7.7 shows a block diagram of the TT system simulator. The described blocks represent main simulator components:

- pump + pipe + control valve: a water supply subsystem (including the control valve and inflow pipes);
- tanks: a physical (balance) model of three tanks and pipelines;
- noise: a special subsystem responsible for generating measurement noise;
- disturbance: a special subsystem responsible for generating process disturbance.

The simulator has several parameters that influence its behavior. These parameters were tuned in a way that would reflect, with satisfactory accuracy, the operation of a real three tank system located at the Institute of Automatic Control and Robotics of the Warsaw University of Technology.

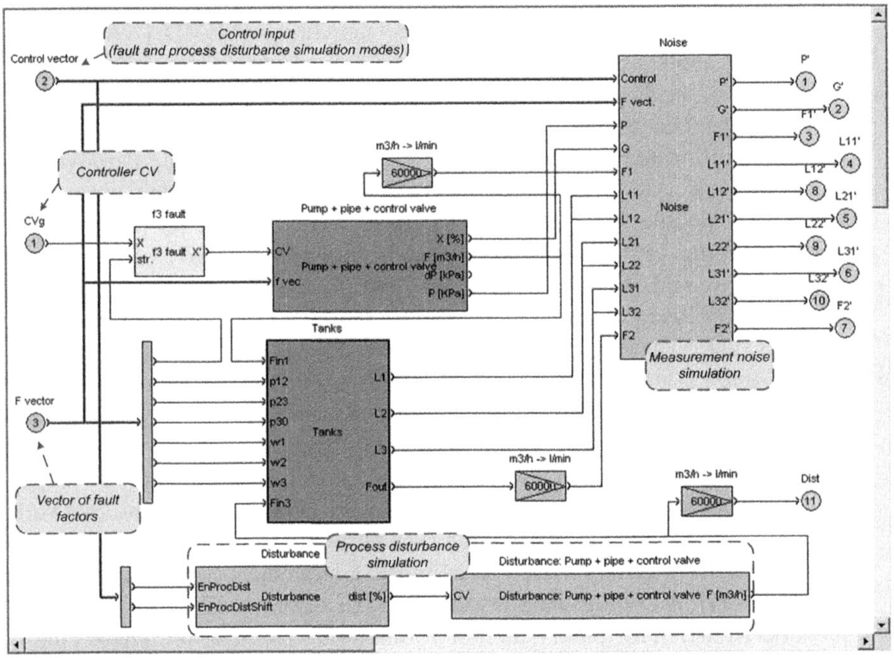

Fig. 7.7 Block diagram of the TT system simulator

The following techniques were used during simulator development: the modeling of selected process components based on real process data (pump model), theoretical characteristics describing the actuator (valve model), balance equations of the medium flows in consecutive tanks, and a theoretical physical law for the unbounded gravity outflow from Tank 3.

The model of the pump was realized as a set of soft switched (with the use of fuzzy logic) characteristics $P(F)$ for different pump control signals MP. Functional dependencies $P = f(F, MP)$ of the assembly: a pump with a pipe between the pump and the control valve) were identified from archival process data acquired during the operation of a real pump in a laboratory set-up. These characteristics are presented in Fig. 7.8.

Characteristics of the control valve were modeled as the composition of a linear part (in the range 0–10% of the opening degree) and a non-linear one (in the range 10–90% of the valve plug opening degree) calculated according to the formula

$$F = K_v\left(S_v\right)\sqrt{\frac{p_{1v} - p_{2v}}{\rho}}, \tag{7.1}$$

where p_1, p_2 denote pressure on the inlet and outlet of the pump, ρ denotes specific water density, K_v is a theoretical flow factor for the valve and S_v is a cross-section of the valve opening.

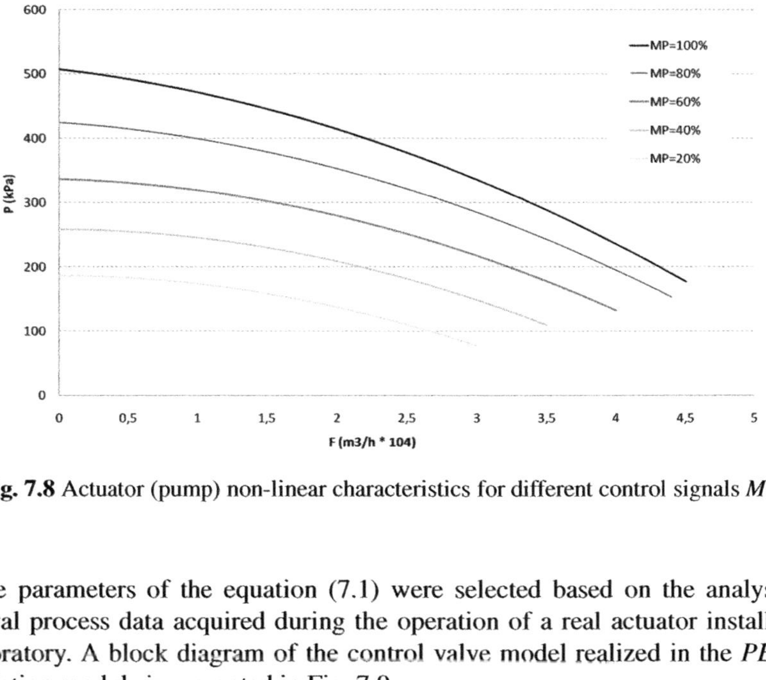

Fig. 7.8 Actuator (pump) non-linear characteristics for different control signals *MP*

The parameters of the equation (7.1) were selected based on the analysis of archival process data acquired during the operation of a real actuator installed at a laboratory. A block diagram of the control valve model realized in the *PExSim* calculation module is presented in Fig. 7.9.

The following balance equations were formulated for each tank (1,2 and 3):

$$S_1\frac{dL_{11}}{dt} = F - \alpha_{12}S_{12}\sqrt{2g(L_{11} - L_{21})}, \tag{7.2}$$

$$S_2\frac{dL_{21}}{dt} = \alpha_{12}S_{12}\sqrt{2g(L_{11} - L_{21})} - \alpha_{23}S_{23}\sqrt{2g(L_{21} - L_{31})}, \tag{7.3}$$

$$S_3\frac{dL_3}{dt} = \alpha_{23}S_{23}\sqrt{2g(L_{21} - L_{31})} - \alpha_{30}S_{30}\sqrt{2gL_{31}}, \tag{7.4}$$

where S_1, S_2, S_3 denote tank cross-sections, S_{12}, S_{23} stand for cross-sections of the pipes connecting consecutive tanks, and α_{12}, α_{23} are flow coefficients.

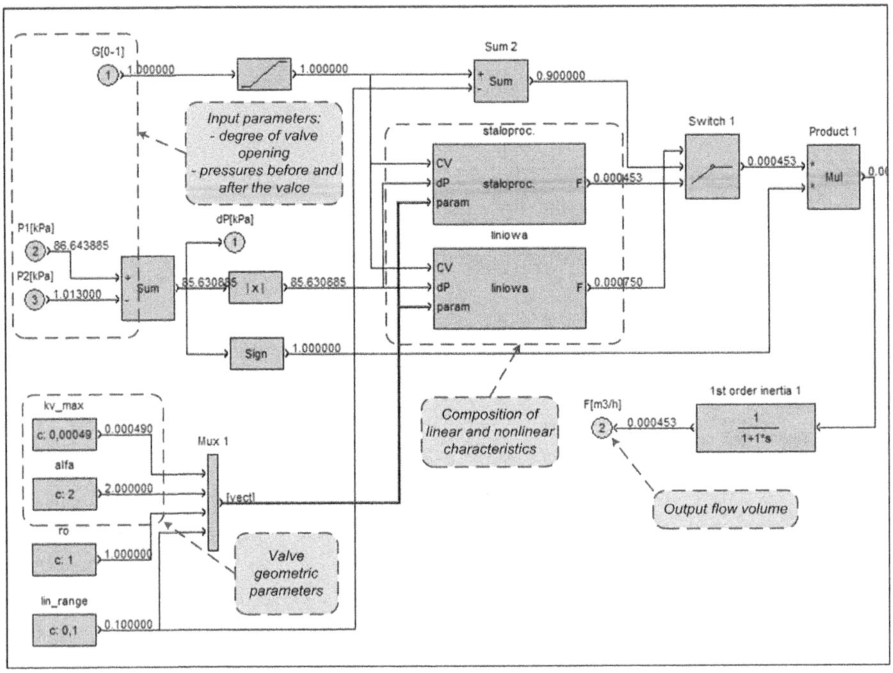

Fig. 7.9 Block diagram of the control valve model realized in the *PExSim* module. The "kv_max", "alfa", "ro" and "lin_range" blocks generate the parameters used to reconstruct the $K_v(S_v)$ function in (7.1).

Geometric parameters of the tanks were defined based on the dimensions of tanks in a real, laboratory set-up. Values of the $\alpha_{xy}S_{xy}$ products, determining the water flow between the tanks, were assigned experimentally in such a way that the time constants for the simulator were close to those achieved in a real laboratory set-up. A similar procedure was used for calculating coefficients connected with leakages (time constants of tank emptying).

A block diagram of the modeled three tanks in the *PExSim* calculation module is presented in Fig. 7.10.

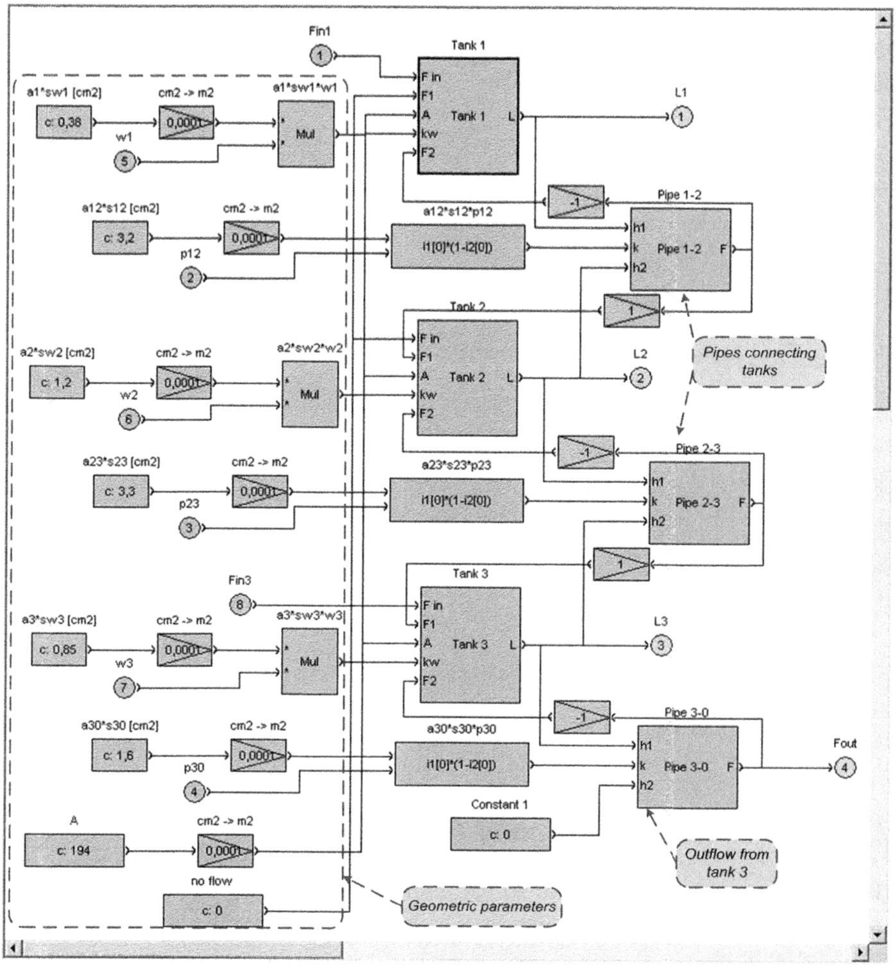

Fig. 7.10 Block diagram of a three parallel tanks model realized in the *PExSim* module

Possibility of fault simulation. The elaborated simulator provides the capability of simulating all faults considered for the TT system. Fault simulation is carried out by

- functional blocks that modify signal values (additive or multiplicative changes) for the faults of measurement paths,
- physical modeling (e.g., additional gravity water outflow from the tank) for the faults of process components and actuators.

An example of physical fault modeling is presented in Fig. 7.11 The leakage from Tank 1 (fault f_9) is calculated according to the following formula for a free gravity outflow:

$$F_{10} = k_w \sqrt{2gL_1}, \tag{7.5}$$

where F_{10} denotes the volume of the gravity outflow and k_w denotes a constant coefficient dependent on the resistance and cross-section of the leakage (hole).

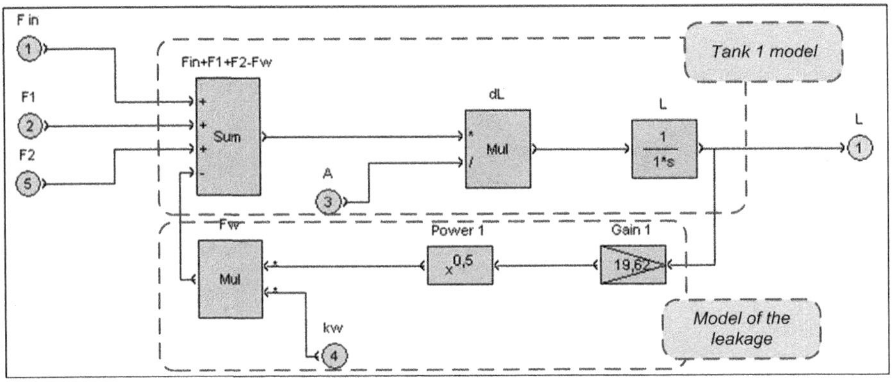

Fig. 7.11 Example of physical fault modeling: leakage from Tank 1

7.4.2 *Self-tuning: Selection of PID Settings*

Blocks from the *Self-tuning and Adaptation* library and the three tank plant described above are used here to demonstrate the operation of self-tuning loops while selecting settings.

7.4.2.1 Control Loop Diagram

The controller governs the water level in the third tank by an inflow to the first tank. The main path of control involving a general PIDSELF controller and accompanying blocks is shown in Fig. 7.12. "SP generator" provides the set-point *SP* (internal or external), the constants "CV Man", "Man/Auto" and "Tune" set manual control and operating modes. The water level expressed in meters is the process variable *PV* (up to 0.5 m), and the opening of the inflow valve in percentage points is the controlled variable *CV*. "Plant monitor" displays *SP*, *PV* and *CV* (m, %).

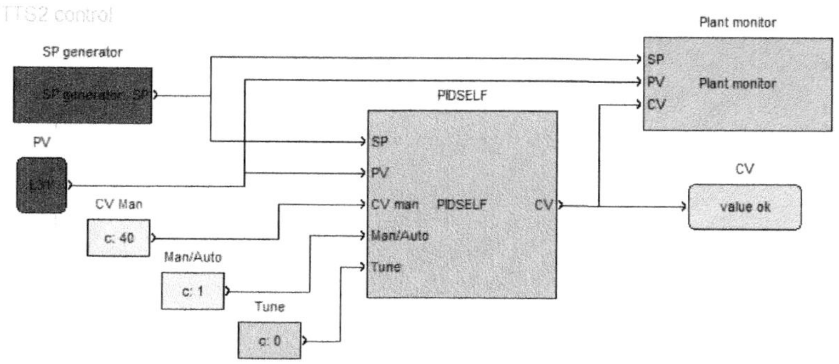

Fig. 7.12 Main path of control with the PIDSELF controller

PIDSELF. The internal structure of "PIDSELF" (subpath) is shown in Fig. 7.13. "SELF_step" executes tuning using the step response. PID is a standard controller with P, I and D components driven by the control error $SP - PV$. "SELF_step" and PID perform calculations on numbers in a normalized interval $< 0.0, 1.0 >$. Since SP and PV belong to the interval $< 0.0, 0.5 >$ m, the gains of "Gain SP" and "Filter PV" are 2. Likewise, because the plant requires CV in $< 0.0, 100 >$ %, we have 0.01 in "Gain CVm" and 100 in "Gain CV". The filter time constant of 3 seconds is about 20 times smaller than the time constant of the plant (see Table 7.3 below). The status STA indicates the current state of the "SELF_step" block (*Ready*, *Busy*, *Success*, an error code). The diagram also involves two display blocks, "CV, SP, PV, CVm" for the observation of signals (normalized), and "Kp, Ti, Td" to read out the settings.

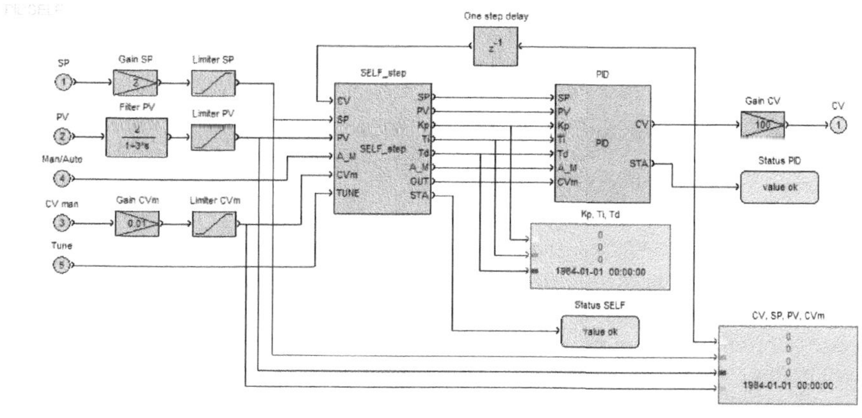

Fig. 7.13 Subpath of PIDSELF for step response tuning

Plant responses. To get some feeling of the plant dynamics, three step responses recorded for "CV man" = 40, 70 and 100% are shown in Fig. 7.14 ($ManAuto = 0$, the *Manual* mode). The responses exhibit a single dominating time constant and small delay. Hence $k_o e^{-\tau s}/(Ts+1)$ may be expected as an appropriate model. It is clear by looking at settling times that the plant is non-linear.

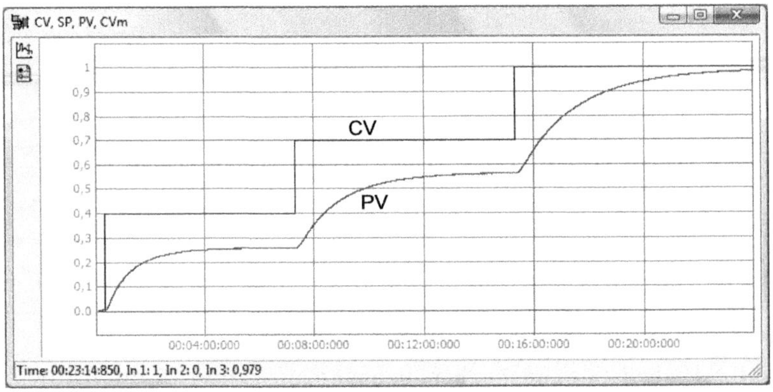

Fig. 7.14 Step responses of the TT plant

7.4.2.2 Step Response Self-tuning

Due to non-linearity, self-tuning will be executed at a few operating points. Besides, to avoid excessive sensitivity of the controller, the settling time of overdamped responses, and not overshoot, will be chosen as the tuning specification (in "SELF_step" properties). Recall that, before self-tuning is on, the plant must remain in a steady-state.

Operation. Self-tuning begins when "Tune" is switched from *off* (value 0) to *on* (1). "SELF_step" sets PID into *Manual* ($A\_M = 0$), changes CVm from the current value by the CV step, and replaces *Ready* (0) as the status *STA* by *Busy* (1). After some time, when the process variable *PV* settles down at the new level, "SELF_step" announces *Success* (2) at *STA*, presents calculated settings at the outputs Kp, Ti, Td and places the PID controller in *Automatic* ($A\_M = 1$). The user may now change the set-point *SP* to verify the closed loop response. By setting *SAVE* (1) in "SELF_step" properties, the settings are stored permanently. Self-tuning is terminated by switching "Tune" back to *off*. The controller returns to the pre-tuning mode (*Manual* or *Automatic*), *STA* becomes *Ready* again.

Switching "Tune" to *off* before *Success* stops the procedure, leaving the settings unchanged. The same happens if for some reason "SELF_step" fails and outputs an error code at the *STA* output.

Results. The following data are selected for self-tuning demonstration:

- operating points at $SP = 0.1, 0.3, 0.5, 0.7, 0.9$,
- control steps of magnitude 0.2 (increase/decrease),
- the noise level 0.002 (0.2%),
- a multiplier $\alpha = 1$ for settling time specification $t_s = \alpha T$ (6.83).

$t_s = T$ means that the closed loop response will settle down four times faster than the plant itself. Table 7.3 presents plant parameters k_o, T, τ and calculated settings K_p, T_i for the five operating points. As seen, "SELF_step" chooses PI control ($T_d = 0$) due to the single time constant. The ratio τ/T is in $< 0.11, 0.22 >$, so the plant may be considered "easy".

Table 7.3 Plant model parameters and controller settings

SP	CV step	k_o	T	τ	K_p	T_i
0.1	+0.2	0.46	49.5	11.2	3.11	49.5
0.3	+0.2	1.02	86.2	12.8	2.12	86.2
0.5	–0.2	1.01	77.9	12.8	1.94	77.9
0.7	–0.2	1.29	95.2	12.8	1.85	95.2
0.9	–0.2	1.43	112.8	12.8	1.99	113

Single settings. To verify how the control loop behaves when a single set of settings is applied in the whole range, the set-point SP is changed from 0.0 to 0.9 in three equal steps, while keeping the settings at constant values determined for $SP = 0.5$, i.e., as $K_p = 1.94$, $T_i = 77.9$ (Table 7.3). Responses are shown in Fig. 7.15. They do not differ much (the upper limit of CV is reached in the third case), so a single set of settings suffices for this plant.

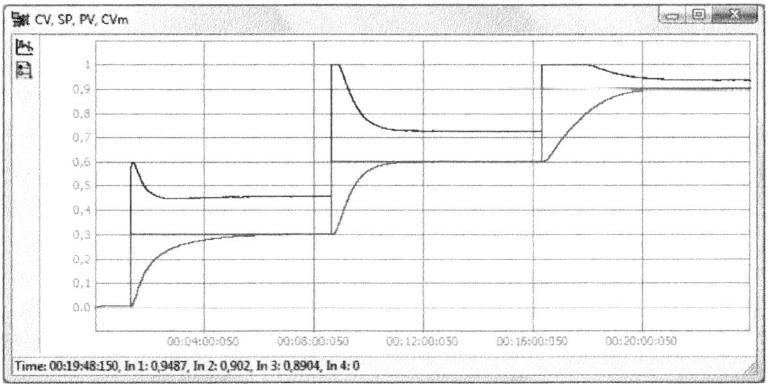

Fig. 7.15 Closed loop responses for a single set of PID settings

7.4.2.3 Relay Self-tuning

A block diagram of Fig. 7.13 should now involve "SELF_relay" instead of "SELF_step" and a dedicated PIDTC controller instead of standard PID. Due to the internal filter and a non-standard structure (Fig. 6.27), PIDTC provides none or small overshoot. "SELF_relay" does not have *CV* input, since relay tuning is executed in a closed loop. Besides, the plant does not have to be in a steady-state.

Sample run. Tuning transients and resulting responses shown in Fig. 7.16 are obtained for the following steps:

1. the initial *Manual* mode of PIDTC ($A\_M = 0$) with $CVm = 0.0$;
2. $SP = 0.5$ as a relay set-point ($0.25\,m$ in the third tank);
3. "Tune" switched from *off* to *on*, relay control begins;
4. waiting for *Success* at the *STA* output;
5. reading out Kp, Ti, Td and *SAVE* in "SELF_relay";
6. the mode changed to *Automatic* ($A\_M = 1$), "Tune" switched back to *off*;
7. the verification of closed loop responses by decreasing *SP* to 0.4 and increasing to 0.6 later.

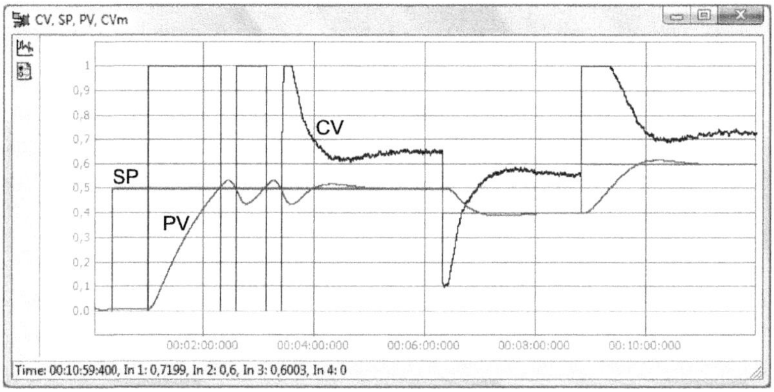

Fig. 7.16 Relay tuning transients and closed loop responses

As seen, "SELF_relay" is able to complete self-tuning after five switchings. The oscillation period T_{cr} is 57 seconds, relay control exhibits slight asymmetry with the filling ratio $t_{on}/T_{cr} = 0.64$ (Fig. 6.28). Settings of PIDTC are $K_p = 4.3$, $T_i = 15.8$, $T_d = 3.86$. Note that they differ substantially from Table 7.3, despite the fact that responses look reasonably similar (5 % overshoot in Fig. 7.16). The difference should not be surprising, however, since step response tuning and relay tuning use different information on plant properties. The settling time is now about 50 seconds.

Disturbance suppression. The reaction of the loop to a temporary leak of water from one of the tanks is shown in Fig. 7.17. The leak is fairly large, since a steady-state increase of *CV* by 40% from the last value is needed to compensate for it. Nevertheless, the maximum drop of the water level in the third tank does not exceed 10% (from the last value).

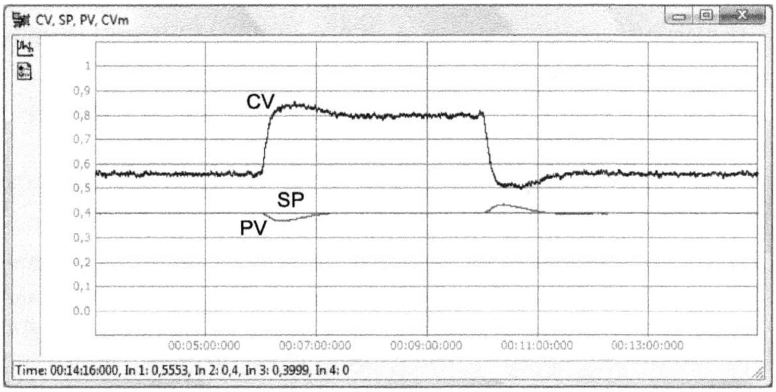

Fig. 7.17 Response to temporary step disturbance

7.4.3 *Reconstructing Process Variables with TSK Models*

This section presents the use of TSK models to build partial models reconstructing selected process variables. These models were identified using archival process data. For this purpose, the *MITforRD* module was used. The main window of the application is presented in Fig. 7.18.

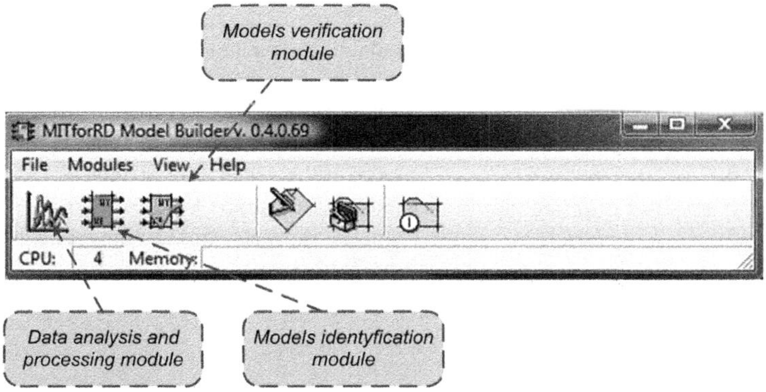

Fig. 7.18 *MITforRD* module main window

The identification of the models was conducted using the following algorithm parameters:

- the model reference output: the type of simulation in the which the output value is estimated on the basis of the model outputs from previous steps of simulation. The absolute error index was optimized;
- the number of membership functions for each fuzzyfied variable could change from 1 to 4;
- all fuzzyfication parameters, determining the position and shape of membership functions, were optimized;
- the structure of the successor function is not assumed to be known:
  - the order of the equation was limited to 2,
  - the delay of the input signal could change from 0 to 3, the output signal from 1 to 3 (in the case of the MRO);
- the following genetic operators were used: the initialization operator (numerical parameters describing the structure of the conditions and the conclusions are randomly selected from the range of the variation of the specified parameter, making the selection according to homogeneous distribution), the cross operator (crossing conducted using the information contained in the encoded genome) and the mutation operator (carrying out random disturbances of randomly selected parameters).

The process of model building using the *MITforRD* module consists of four stages during which the user is supported by wizard windows. The identification process is shown in the example model of the level in Tank 1 (no. 3 in Table 7.4). The time series of the modeled signal is shown in Fig. 7.19.

Table 7.4 List of three tank system partial models

No.	Base partial model	Description
1	$\hat{G} = f_1(CV)$	Model of the pneumatic actuator
2	$\hat{F_1} = f_2(G)$	Model of the control valve
3	$\hat{L_1} = f_3(F_1, L_{21})$	Model of the water level in Tank 1
4	$\hat{L_2} = f_4(L_{11}, L_{31})$	Model of the water level in Tank 2
5	$\hat{L_3} = f_5(L_{21})$	Model of the water level in Tank 3
6	$d\hat{P_1} = f_6(G)$	Model of the pump (+ outlet pipe)
7	$\hat{F_2} = f_7(L_3)$	Model of the water flow on the outflow from Tank 3

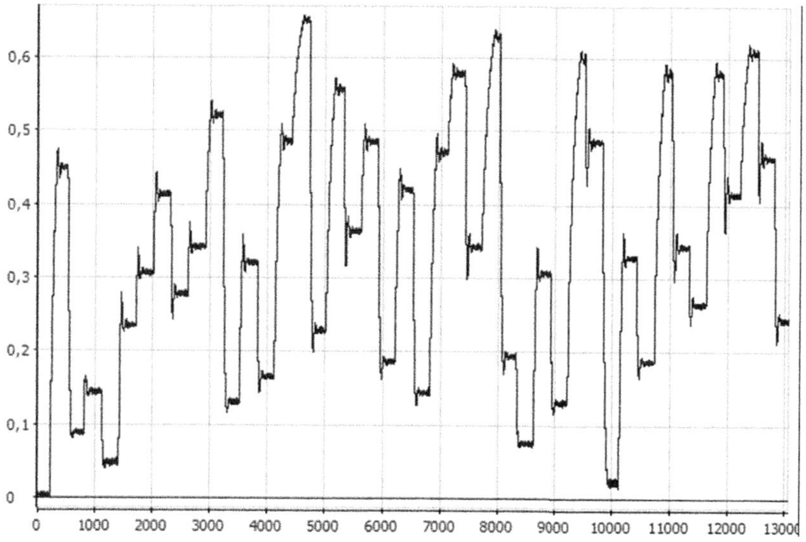

Fig. 7.19 Example time series of the modeled L_{11} signal (water level in Tank 1)

Stage I. Defining the modeled signal (model output) is the first phase of model identification. In our case, the L_1 signal was selected (water level in Tank 1).

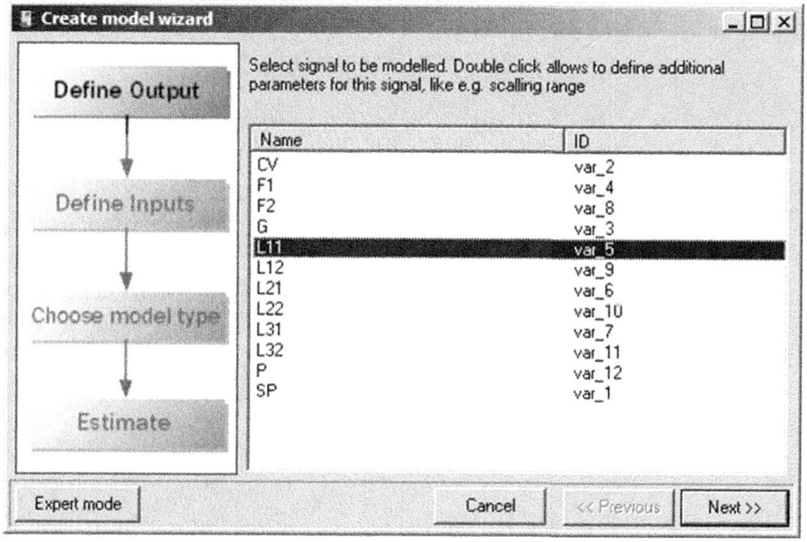

Fig. 7.20 Modeled signal selection window

Stage II. Then the input signals were selected. In the presented example, the model inputs (Fig. 7.21) are the water inflow to Tank 1 (F_1), the water level in Tank 2 (L_2) and the water level in Tank 1 (L_1, *MRO mode*).

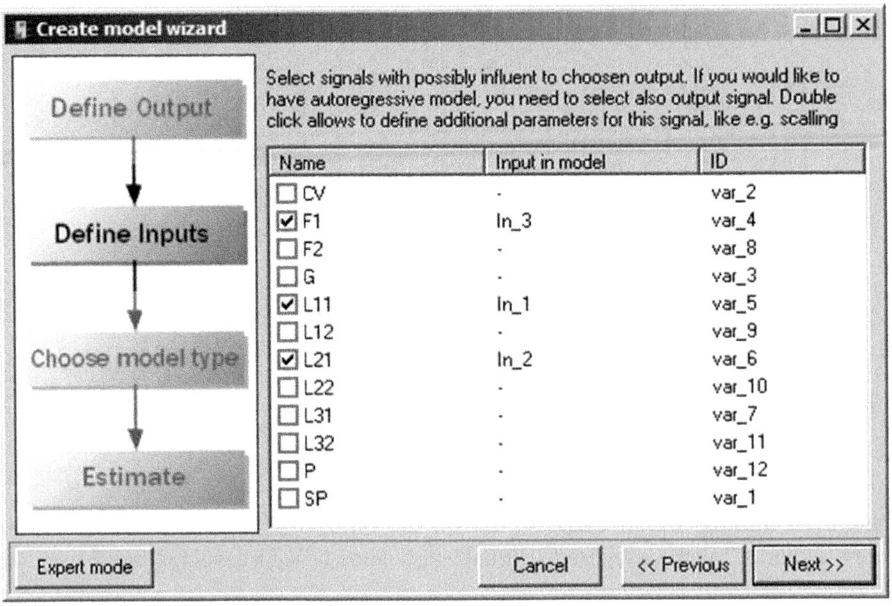

Fig. 7.21 Input signals selection window

Stage III. The next step is the definition of the model type. The list of available model types is dependent on the number of available plug-ins implementing the various modeling methods (classical linear models, fuzzy, neural, etc.) Using the *Configure* button, it is possible to introduce additional parameters specific to the particular algorithm. In our case, a TSK model was selected. The parameters of the algorithm were set in accordance with the above-described assumptions.

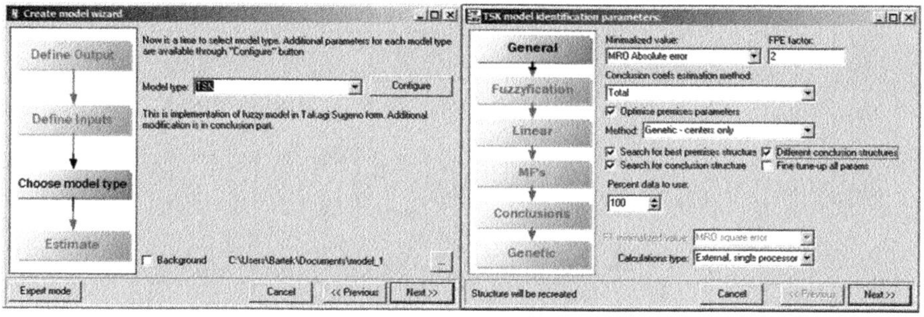

Fig. 7.22 Windows for model definition and defining algorithm parameters

Stage IV. The final stage of the model identification process is the estimation of model parameters based on the analysis of process data. A special dialogue window is used to display the progress of the identification process, allowing the control of its course (Fig. 7.23).

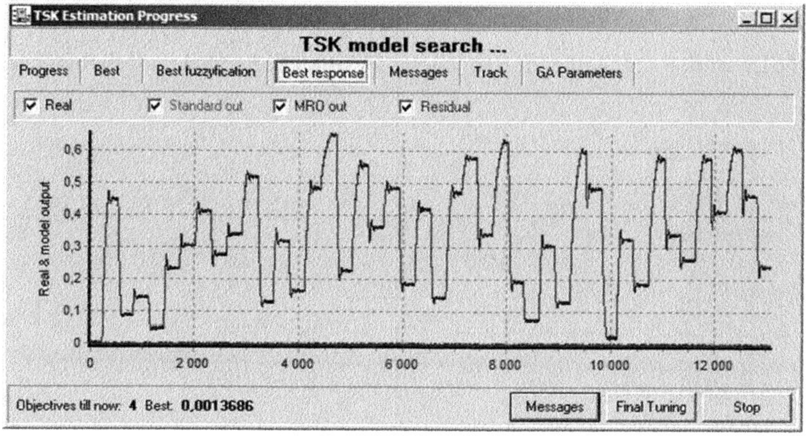

Fig. 7.23 Parameter estimation window

As a result of the identification process, a TSK model, which reproduces the water level in Tank 1 with specific input signals, was obtained. The model response, obtained during the identification process, is shown in Fig. 7.24.

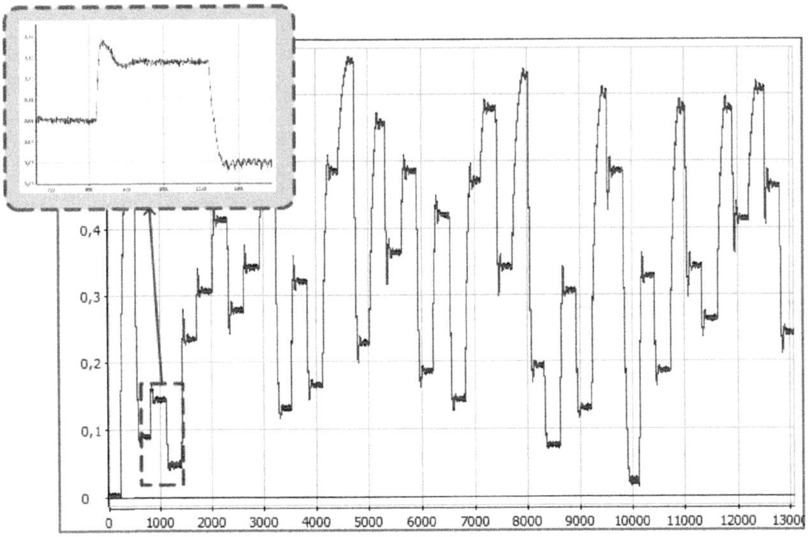

Fig. 7.24 Simulation results of the TSK model of the water level in Tank 1 (real and modeled output signal)

The division into membership functions for fuzzyfied signals is shown in Fig. 7.25.

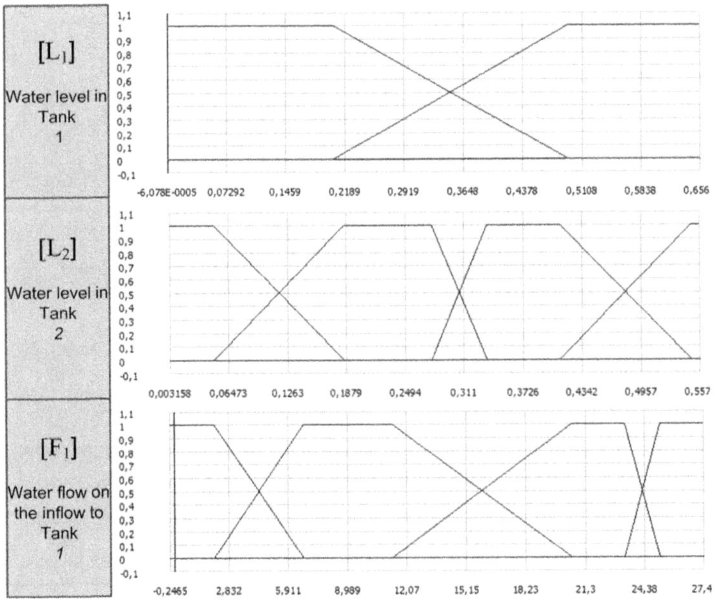

Fig. 7.25 Division into partitions for fuzzyfied signals with trapezoidal membership functions

The identification process was conducted analogously for the other partial models presented in Table 7.4.

7.4.4 *Process Modeling with Neural Networks*

In this section, neural networks of the dynamic type are used to model the water level in the three tank laboratory system. The neural models considered realize the dynamics using internal feedback, but both approaches provide one global model of a system, unlike TSK methodology, which uses local models.

7.4.4.1 Locally Recurrent Networks

Here, the locally recurrent network is applied to build a model of the three tank system. Analyzing knowledge about the system and measurably available data, the following signals were selected as inputs: the control value of the valve on the inlet of Tank 1—CV, the flow on the inlet of Tank 1—F_1, and the pressure before the control valve—P. The output was the water level in Tank 3—L_{32}. Generally, the three tank system model may be represented in the following form:

$$L = f_{NN}(CV, F_1, P), \tag{7.6}$$

where f_{NN} is non-linear mapping realized by the neural network. In the next step, it is necessary to choose the modeling/identification method. In the case considered, it is a locally recurrent network realized in the form of the *LRGF* plug-in. The selection of process variables and the model type is carried out analogously to the procedure presented in Section 7.4.3, Stages I–III.

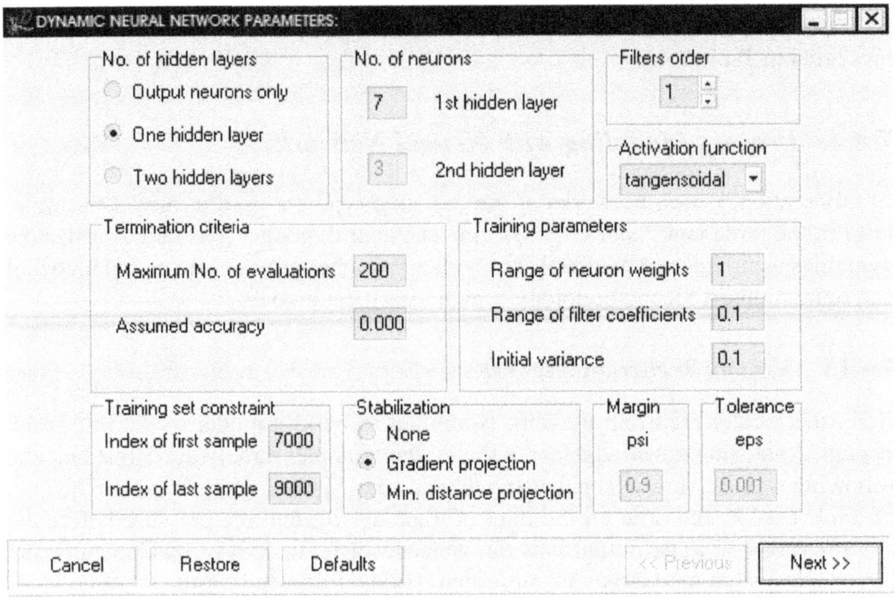

Fig. 7.26 Network parameter selection dialogue window

After that, one can configure the structure of the neural model. This process is shown in Fig. 7.26. The structure of the neural model was selected using the trial and error method, taking into account the generalization ability of the model. The best performing model has one hidden layer with seven neurons with the IIR filter of the 1-st order and a hyperbolic tangent activation function. The output layer consists of one linear neuron. The model has three inputs according to (7.6). The neural network was trained using the ARS algorithm with the initial variance $\sigma_0 = 0,1$. Before the training, input and output data were preliminarily preprocessed using linear scaling to the range $< 0,1 >$. To perform this, the minimum and maximum values of the process variables listed in Table 7.1 were used. The training was carried out off-line for 200 steps with the training set including 2001 samples. The stability of the model was guaranteed by the gradient projection method. The training process is shown in Fig. 7.27. As one can see, the *MITforRD* component prepares training samples (Fig. 7.27(a)), informs about the progress of the training (Fig. 7.27(b)) and displays model configuration settings (Fig. 7.27(c)).

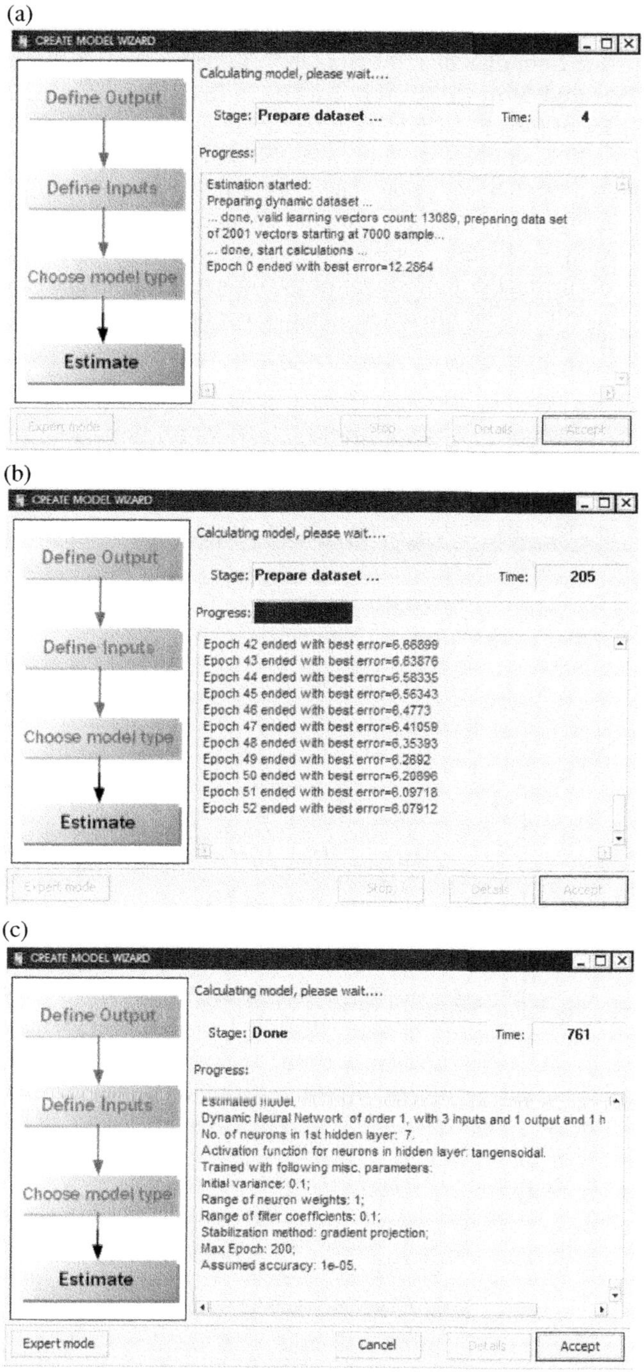

Fig. 7.27 Training of the neural model: preparing the data (a), progress of the training (b), configuration of the model (c)

Performance of the model in the fault-free state. The quality of training expressed in the form of the Sum of Squared Errors (SSE) is equal to $0,57$, and in the form of the Mean Squared Error (MSE) it is $2,8 \cdot 10^{-4}$. In order to investigate the generalization abilities of the model, it was tested using a very long sequence of testing data including 13089 samples. In this case, SSE=11,67 and MSE=$8,9 \cdot 10^{-4}$. The achieved results are relatively good. The testing results (for about 7000 samples) are presented in Fig. 7.28. The model mimics changes of the reference signal pretty well. The residual, oscillating around zero, has a relatively small value, excluding narrow spikes occurring during changes of the reference value. Data were collected in closed loop control. The reference signal was composed of random steps, and each step lasted 300 seconds.

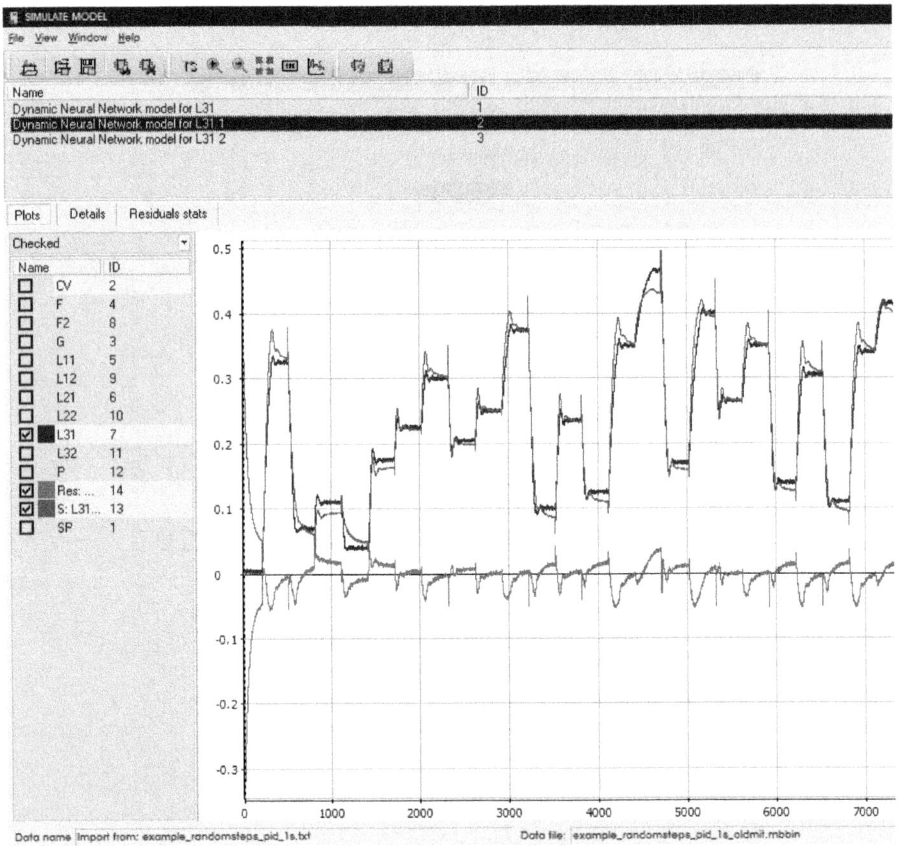

Fig. 7.28 Neural model testing

Performance of the model in the case of faults. The neural model was designed for normal operating conditions. The question is how it will behave in the case of faults. The *DiaSter* system makes it possible to simulate a number of faulty scenarios. The specification of faults is presented in Table 7.2.

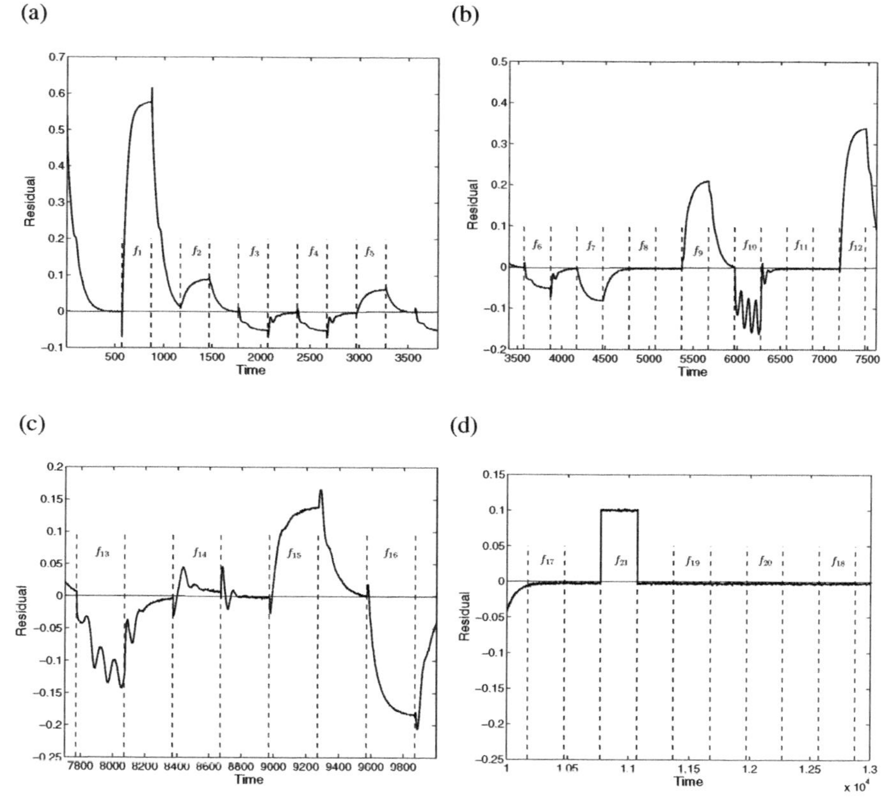

Fig. 7.29 Fault detection: faults f_1–f_5 (a), faults f_6–f_{12} (b), faults f_{13}–f_{16} (c), faults f_{17}–f_{21} (d)

Figure 7.29 shows the residual in the case of all simulated faults. As one can see, the residual is insensitive to the faults f_8 (sensor fault L_{11}), f_{11} (sensor fault L_{21}), f_{17} (sensor fault F_2), f_{18} (sensor fault L_{12}), f_{19} (sensor fault L_{22}), f_{20} (sensor fault L_{32}). It is impossible to detect these faults because the model does not use signals connected to them. In order to detect these faults, a more complex model representing water levels in all tanks should be designed.

7.4.4.2 Dynamic GMDH Neural Networks

The GMDH neural network with dynamic neurons (described in Chapter 3) is used to design the model of a three tank system. The selection of input and output signals is performed in the same way as in Section 7.4.2. In a general way, the model of a three tank system can be described by (7.6), where f_{NN} stands for a non-linear function represented by the GMDH neural network. The next step boils down to the selection of the identification/modeling method. In this case, the GMDH neural network, implemented as a plug-in of the *MITforRD* system, is used. The subsequent step consist in the selection of the appropriate configuration of the neural model (Fig. 7.30).

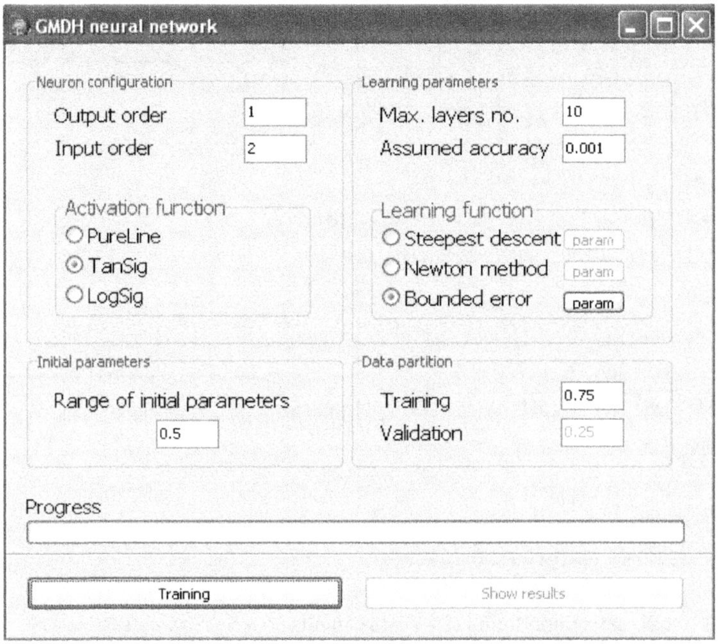

Fig. 7.30 Parameter selection panel of the GMDH model

In the presented example, the algorithm is stopped when the identification error achieves the level of 10e-3. Additionally, the structure of a neural network is constrained to five layers, while the maximum number of neurons in a layer is 100. Similarly as in the case of a locally recurrent neural network, it is assumed that the neuron activation function is tanh(·). Finally, the dynamics order of the input and output of a neuron is 1. The next very important step is concerned with the selection of the settings of the parameter estimation method. In the case of the bounded-error estimation technique described in Chapter 3, the maximum bounds of the error describing the difference between the model and the system should be provided. The

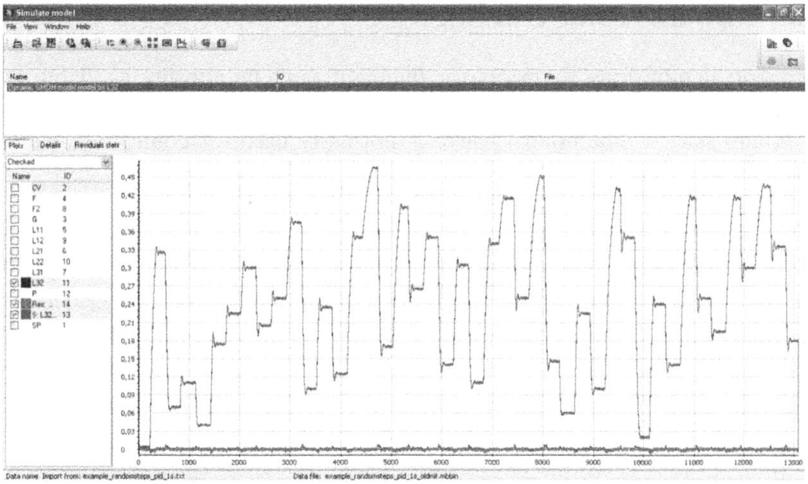

Fig. 7.31 Validation of the GMDH model

last parameter describes the proportion between the training and validation data sets. It should also be pointed out that before learning the data were pre-processed in a similar way as in the case of locally recurrent neural networks.

Performance of the model in the fault-free case. As a result of the identification procedure, the training was finalized after two iterations, which means that the GMDH neural network consists of two layers. The validation results, being a measure of the generalization abilities of neural networks, are presented in Fig. 7.31. As can be observed, the model approximates the system with very high accuracy, which confirms its high performance. The training data were collected for a three tank system working with the PID controller. The reference signal was composed of randomly selected values, where each one was held for 300 seconds. Comparing the results presented in Figs. 7.28 and 7.31, it can be observed that the GMDH model has better generalization abilities than the one designed with the locally recurrent neural network, which is clearly confirmed by the smaller amplitude of the residual for the fault-free case. On the other hand, the neural models being considered were designed with different criteria and optimization techniques, and hence their quantitative comparison cannot be performed.

Performance of the model in the faulty case. The GMDH neural model was designed with fault-free data. Thus, a natural question arises: What is the performance of the GMDH neural model in the case of faults? The set of faults, along with a short description, is given in Table 7.2. Apart from the fact that the GMDH neural model has higher performance than the locally recurrent neural network, the fault detection results were exactly the same for both networks. Indeed, the general structures of the networks are the same, and hence all the limitations regarding fault detection with a locally recurrent neural network are exactly the same for the GMDH neural network.

7.4.5 Incipient Fault Tracking

In order to demonstrate the tools for incipient fault monitoring available in the *DiaSter* system, the algorithm for monitoring control valve sedimentation was configured and started. The control valve is normally used for controlling the water level in Tank 3.

The difference between the observed value of the flow and the flow retrieved from the model was used to evaluate the degree of valve sedimentation. The TSK model was previously identified and then used to reconstruct the value of the flow. The model was achieved with the use of process data and the tools described in Section 7.4.3. The data for identification were collected during normal process operation, when the control valve was in a fault-free state (see Fig. 7.32). The evaluation of the valve state was performed by assessing residual changes.

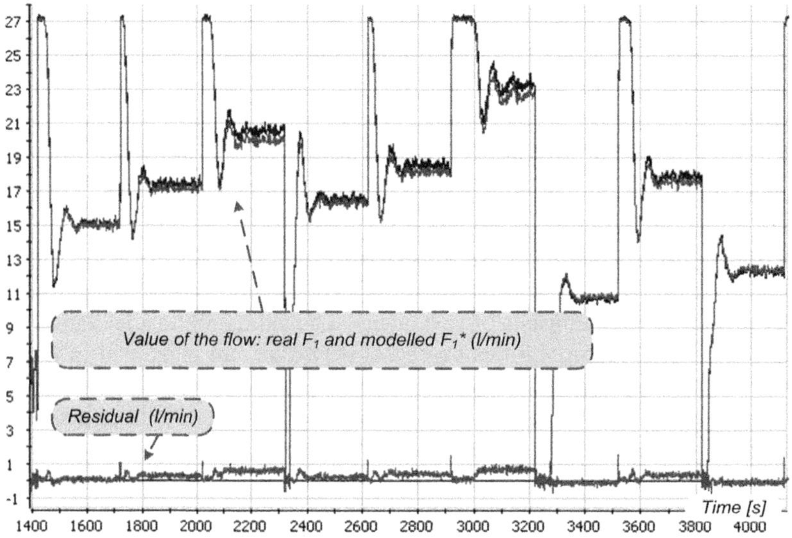

Fig. 7.32 Example of reconstructing the water flow through the control valve for the fault-free state

Valve sedimentation was simulated as the process of a slow change of the valve cross-section (percentage change of the current cross-section dependent on the current valve position). After the fault simulation was started (in the first stage the valve was completely efficient), the cross-section was decreased with the speed of 1 %/min.

The task of fault detection (including the determination of the moment of its initiation) and the monitoring of its development was realized with the use of the function blocks of the incipient fault monitoring library *slowFDI* of the *PExSim* calculation module. This task is conducted in two steps:

- **fault detection and monitoring.** The value of the parameter used to evaluate the state of the monitored component (in this case, the value of the calculated residual) is evaluated in an on-line mode. Upon detecting significant variations, the fault detection signal is generated and the moment when the fault started is evaluated (in this case, the moment when fault simulation was started);
- **fault development tracking.** The detection signal triggers the procedure of fault monitoring (development tracking) and estimating the time left to reach the critical component state. The critical state is specified by determining the limiting (threshold) value of the evaluated parameter (signal).

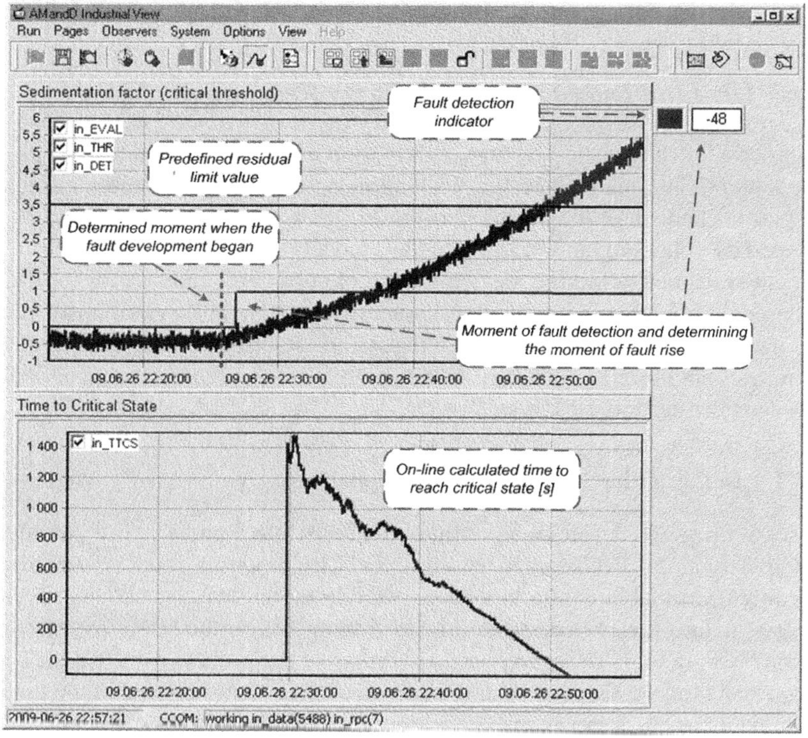

Fig. 7.33 Operator interface used to visualize the control valve sedimentation process. Time series inform about the changes of the value of the monitored parameter (in this case, the difference between the measured and the estimated flow through the control valve) and the calculated time to reach the critical component state.

In the presented example, the threshold value of the evaluated parameter was set to 3.5 l/min. The algorithm of fault detection and searching for the moment of its initiation was analyzing the data in a 60 s time window. Upon fault detection (the detection signal is pointed out on the top graph in Fig. 7.33), the algorithm determined that the fault started 48 s earlier. The detection signal was generated and the procedure of fault development tracking was started. It is important to note that the detection of the trend in the observed signal occurred very quickly, when the trend was practically impossible to be noticed by a human observer.

The automatic determination of the time left to reach the critical state was started just after an appropriate number of data samples from the process with an existing fault were collected (the width of the time window was set to 180 s). The first evaluated value of the predicted time was equal to about 1500 s. This value was updated in consecutive calculation steps, when new data from the process were achieved and processed. Analyzing Fig. 7.33 (bottom graph), one can observe that, when the fault was developing and the time was passing, the predicted time to the critical state gradually decreased.

7.4.6 On-Line Diagnostics with Fuzzy Reasoning

The process of diagnostic system configuration can be divided into several, contractual steps. The main stages correspond to designing detection algorithms and a proper mechanism of diagnostic reasoning. Others are related to additional tasks such as data collecting and pre-processing, modeling, system tuning or tests. The consecutive sections describe the application of current diagnostics for the TT system based on partial models, fuzzy residual evaluation and fuzzy diagnostic reasoning. Firstly, individual implementation phases are describes. Then, the examples of system tests are described. The presented application takes into account all kinds of faults considered for the TT system.

7.4.6.1 Defining the Set of Detection Algorithms

The set of proposed detection algorithms is mainly based on the set of partial models reconstructing selected process variables. The proposed models were defined according to available measurements as well as the knowledge about the process and its structure (described in the information model by the set of assets and the relation “belong to"). They cover as small parts of the process as possible. The set of proposed models, together with the corresponding assets defined in the platform information model, is shown in Fig. 7.34.

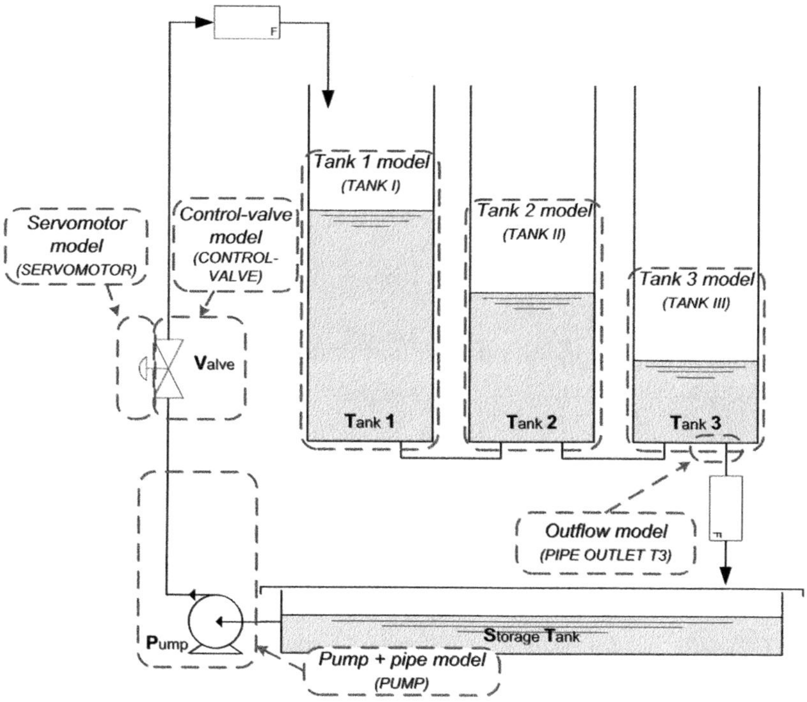

Fig. 7.34 Separated partial models for the TT system

The first five models strictly correspond to process components: the servomotor, the control valve and three tanks. Two other models reconstruct the outflow from Tank 3 (kind of output pipe model) and the pressure before the control valve (kind of pump + connecting pipe model). The second model is based on the relation between the pressure produced by the pump and the position of the valve, e.g., the resulting water flow.

The above-mentioned models are used to generate the first five residuals. Three additional residuals are based on the existing hardware redundancy of the water level sensors in consecutive tanks. No other additional heuristic tests (detection algorithms) that would carry extra knowledge about the relations between process variables were used.

The set of proposed residuals (together with the algorithm of their calculation) is presented in Table 7.5. The set was defined taking into account the achieved and the desired fault isolability. It is also possible to specify other residuals (based on other models and dependencies) that would allow achieving similar (high) fault isolability.

It is worth stressing that the usability of models for fault detection purposes was one of the most important features considered during the selection of the model structures. Sometimes simple model structures, e.g., partial (local) models vs. global ones, permit defining the diagnostic relation more easily and more certainly. As a

Table 7.5 Detection algorithms defined for the TT system

Residual	Description		Evaluation threshold
r-TTS2-01-m	$= G - G^*$ $G^* = f(CV)$	Servomotor (+ positioner) model	$\pm 5(2)$
r-TTS2-02-m	$= F_1 - F_1^*$ $F_1^* = f(G)$	Control valve model	$\pm 1.5(1)$
r-TTS2-03-m	$= L_1 - L_1^*$ $L_1^* = f(F_1, L_2)$	Tank 1 model	$\pm 1(0.5)$
r-TTS2-04-m	$= L_2 - L_2^*$ $L_2^* = f(L_1, L_3)$	Tank 2 model	$\pm 1(0.5)$
r-TTS2-05-m	$= L_3 - L_3^*$ $L_3^* = f(L_2)$	Tank 3 model	$\pm 1(0.5)$
r-TTS2-07-m	$= F_2 - F_2^*$ $F_2^* = f(L_3)$	Outflow model	$\pm 1.5(1)$
r-TTS2-08-m	$= P_1 - P_1^*$ $P_1^* = f(G)$	Pump model	$\pm 3(1)$
r-TTS2-01-r	$= L_{11} - L_{12}$	Utilization of hardware redundancy of Tank 1 level measurement	$\pm 0.8(0.2)$
r-TTS2-02-r	$= L_{21} - L_{22}$	Utilization of hardware redundancy of Tank 2 level measurement	$\pm 0.8(0.2)$
r-TTS2-03-r	$= L_{31} - L_{32}$	of Tank II level measurement of Tank 3 level measurement	$\pm 0.8(0.2)$

consequence, the operation of the diagnostic reasoning algorithm is more robust, although perhaps a little bit less accurate. Such a situation occurs in the case of the residual *r-TTS2-02-m*, where flow F_1 is reconstructed based on signal G without taking into account signal P_1. In this case, the pressure difference can be substituted with the pressure before the valve because that after the valve is constant (equals atmospheric pressure).

7.4.6.2 Residual Evaluation and Diagnostic Relation Determination

In the presented example, three-valued residual evaluation was used. Distinguishing positive and negative residual values allowed increasing fault isolability. This results from the fact that there are several faults which cause different (opposite with respect to their sign) fault effects, e.g., leakages and pipe clogging.

The diagnostic relation was defined based on expert knowledge and the structure of diagnostic tests. The achieved diagnostic matrix is shown in Fig. 7.35. One can notice that complete fault detectability and almost complete isolability were achieved. There exist only two groups of unconditionally unisolable faults: (f_3, f_4) and (f_{10}, f_{12}).

r_{FS}	f_1	f_2	f_3	f_4	f_5	f_6		f_7	$\downarrow f_8 \uparrow$		f_9	f_{10}	$\downarrow f_{11} \uparrow$		f_{12}	f_{13}	$\downarrow f_{14} \uparrow$		f_{15}	f_{16}	f_{17}	f_{18}	f_{19}	f_{20}	f_{21}
$S_{1\text{-}m}$		-	±	±	±																				
$S_{2\text{-}m}$	-				±	-	+	±																	
$S_{3\text{-}m}$								±	-	+	-	+	+	-	+										
$S_{4\text{-}m}$									+	-		-	-	+	-	+	+	-							
$S_{5\text{-}m}$													+	-		-	-	+	+	+					
$S_{7\text{-}m}$																	+	-		-	±				
$S_{8\text{-}m}$	-	-			±	+	-																		±
$S_{1\text{-}r}$									-	+												±			
$S_{2\text{-}r}$													-	+									±		
$S_{3\text{-}r}$																	-	+						±	
			❶	❶								❷			❷										

where: ❶, ❷ - groups of unconditionally unisolable faults

Fig. 7.35 Diagnostic matrix (notation of diagnostic relation) for the TT system. With respect to simplifications only the pattern directions of residual changes are depicted in the table. Particular rows correspond to pattern (expected) residual values in the case of particular fault existence. The sign "±" denotes that the direction of the residual change is unknown (or unimportant) but the change would occur. Two different columns (fault signatures) attributed to one fault (in the case of sensor faults) denote two different, possible combinations of residuals that can be observed according to the type of measurement disturbance (lower or higher value).

The faults (f_3, f_4) cannot be distinguished without installing additional sensors. Possible isolation of the faults (f_{10}, f_{12}) could be achieved by applying additional, more complex models or reasoning techniques, e.g., directional residuals. However, such solutions were not considered.

7.4.6.3 Data Collecting and Model Identification

Data recorded during normal process operation in a fault-free state were used for the model identification purpose. During the experiment, the step changes of *SP* covering the whole operation range were used. The data were collected with a 1s sampling time.

For model-based residuals (1...5), TSK models, described in Section 7.4.2, were used. The achieved accuracy was about 1–1.5% of the signal range. Only the model reconstructing valve position (signal *G*) was less precise. In this case, the accuracy was about 5%. The lower accuracy of this model results mainly from the assumed sampling time, which is too long with respect to the valve dynamics. An example result of the reconstruction of the water level in Tank 3 is shown in Fig. 7.36.

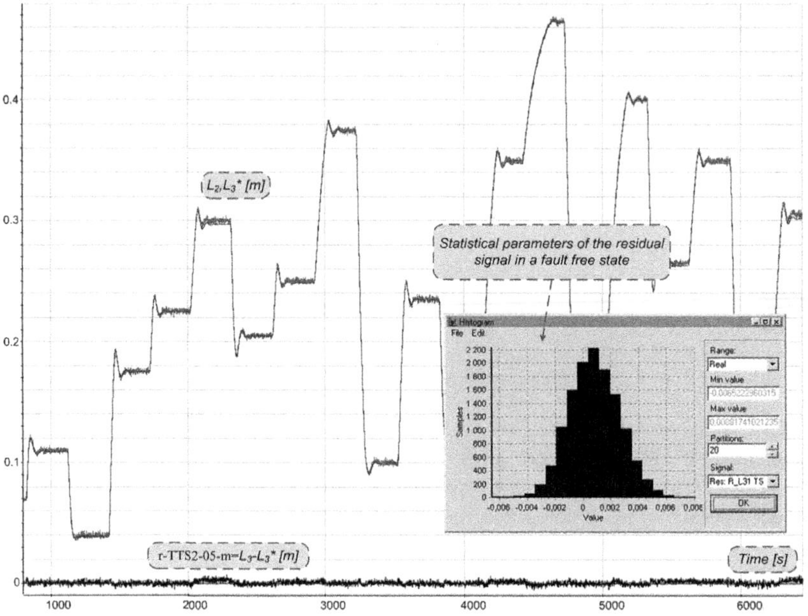

Fig. 7.36 Example of the modeling quality for the model reconstructing the level in Tank 3

7.4.6.4 Setting the Parameters of Fuzzy Residual Evaluation

Trapezoid membership functions were used for fuzzy residual evaluation. The function parameters were determined based on the evaluation of statistical parameters of residual time series for normal operation in a fault-free state. In fact, only the parameters of a fuzzy set Z describing normal residual values (in a fault-free state) had to be calculated. The parameters of two other sets, N and P, corresponding to positive and negative residual values, were selected automatically according to the rule stating that the sum of all membership functions should equal 1.

The boundaries of the residual values in a fault-free state together with the widths of the transition regions between fuzzy set Z and sets P and N, are given in Table 7.5. Figure 7.37 shows the window of the fuzzy residual parameters set-up.

Additionally, simple filtering in a time window of three samples width was used for all the residuals. Such an operation is conducted very often because it introduces small delay in detection and significantly decreases the possibility of generating false alarms arising from residual value variations.

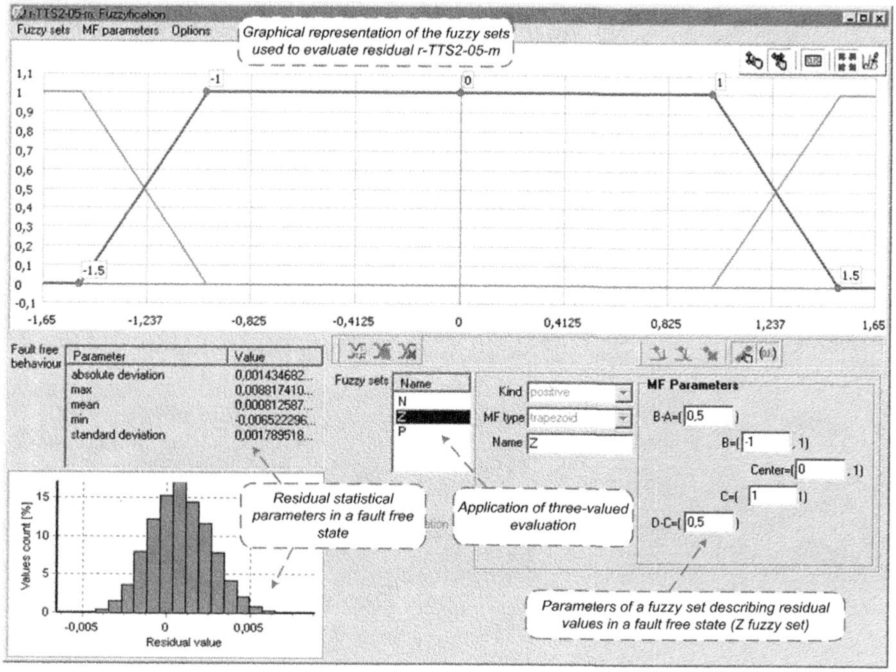

Fig. 7.37 Example of selecting fuzzy residual evaluation parameters for the $r-TTS2-05-m$ residual

7.4.6.5 Setting the Parameters of the Reasoning Algorithm

Proper diagnostic reasoning is conducted with the use of the *iFuzzyFDI* package. The algorithm implemented in this package allows conducting fuzzy reasoning in a hierarchical structure. This algorithm has several parameters that control its behavior. They had to be selected during the configuration stage. Finally, the following diagnostic reasoning parameters were set up:

- only *one reasoning subsystem* containing all faults and residuals was used. There was no need to perform decomposition in the case of such a simple process as the *TT* system;
- the option of *dynamic decomposition* was switched on. In this mode, the first observed symptom causes automatic determination of the subset of possible faults and useful residuals (in reasoning). Further reasoning is conducted only for a subsystem created in such a way;
- the fault isolation mode was set to *a safe mode that takes into account only observed symptoms*. This mode is useful in the case when the symptom dynamics are not taken into account. This provides protection against false diagnoses in the symptoms formulation phase;

- when detecting contradictory diagnoses elaborated at the basic level (assuming single fault scenarios), the system was set to switch automatically to *a special mode dedicated to analyze multiple faults scenarios.*

7.4.6.6 Configuration of the Operator Interface

In the final step, a specialized graphical interface for the operator of the diagnostic system was configured. The interface was realized in the *InView* visualization module of the *DiaSter* platform. It consists of several screens used to display diagnoses as well as to evaluate them. The main screen is presented in Fig. 7.38.

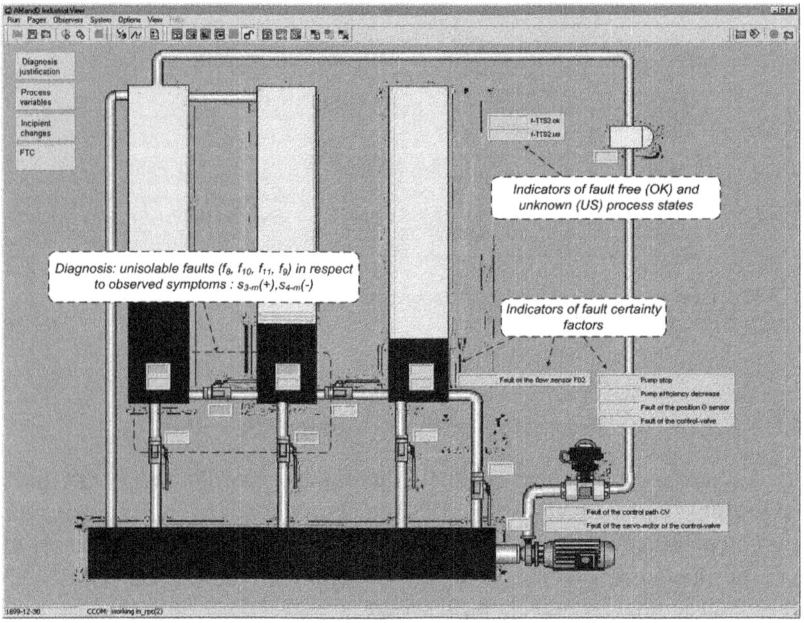

Fig. 7.38 Visualization of the process state in the *InView* module. Example of pointing out unisolable faults (f_8, f_{10}, f_{11}, f_9): case when the fault f_{10} was introduced. Additional faults (f_8, f_{11}, f_9) are unisolable according to the selected isolation mode that takes into account only observed symptoms.

It contains a graphical representation of the process structure (similar to the one used during normal process control) and special indicators displaying the fault certainty factors. These indicators are visualized in the form of the bar graphs placed on the standard process mimics near to the corresponding components and measurements. Additionally, the certainty factor of fault-free and unknown process states are displayed. The colour of a graph bar is connected with the value of the fault certainty factor—faults with high certainty factors are marked with red while those with smaller factor are displayed as yellow.

7.4.6.7 Examples of System Tests

The utilization of the process simulator allowed investigating several diagnostic system tests. Below, examples of results for two fault scenarios are presented:

- a single fault of the process component—partial clogging of the pipe between Tanks 1 and 2 (f_{10}),
- multiple fault scenario—simultaneous simulation of the faults of the measuring path L_{21} (f_{11}) and P_1 (f_{21}).

Fault of the process component f_{10}. In this scenario, partial clogging (50% decrease of the pipe cross-section) of the pipe connecting Tanks 1 and 2 was simulated. The time series of the residuals and the elaborated diagnosis are presented in Fig. 7.39.

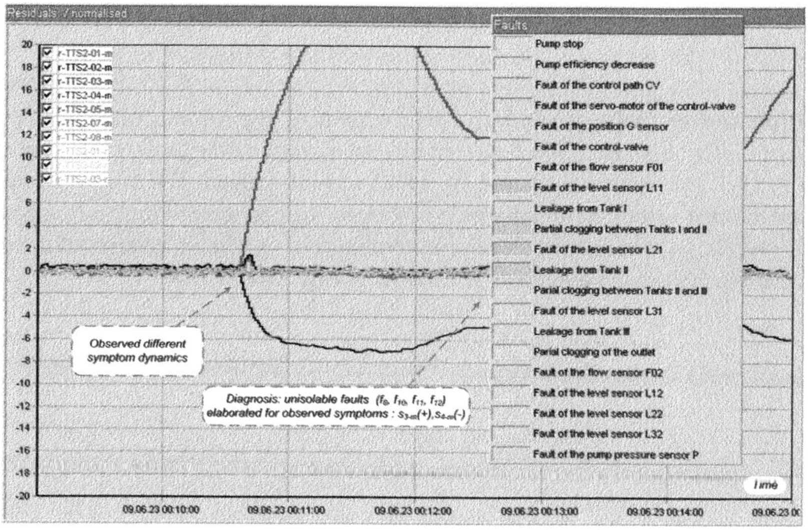

Fig. 7.39 Time series of residuals and the elaborated diagnosis in the case of fault f_{10} simulation. The presented diagnosis was elaborated for the moment 00 : 15.

The preliminary diagnosis elaborated in the first stage, after the first symptom occurred, was very imprecise. However, the final diagnosis was elaborated already after about 5 s—the faults f_8, f_{10}, f_{11} and f_{12} were pointed out as possible ones. They are unisolable, in this case, due to the selected isolation mode, which takes into account only observed symptoms. In the case of the fault f_8 or f_{11}, the occurrence of additional symptoms of residuals *s5-m* or *s2-r* would increase isolability and the elaborated diagnosis would be more precise (Fig. 7.40).

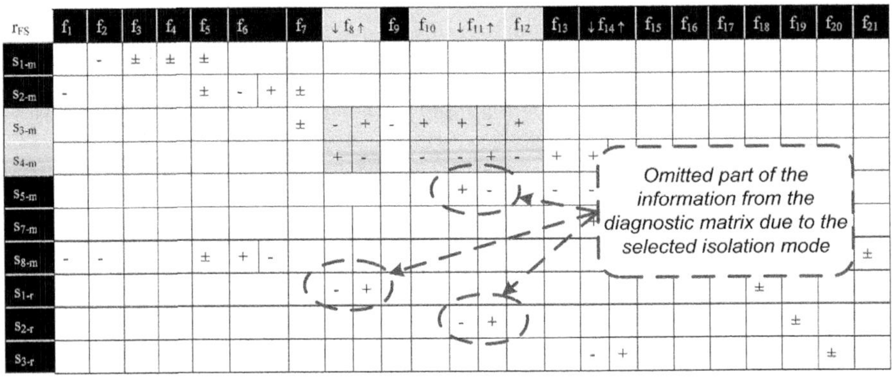

Fig. 7.40 Simulation of the fault f_{10}: problem of variable fault isolability in the case of the utilization of the isolation mode based only on observed symptoms. When the fault f_{11} is simulated, the additional symptoms (*s5-m*, *s2-r*) make unique isolation of that fault possible

Multiple sensor faults f_{11} and f_{21}. The reasoning algorithm applied also has the ability to infer about multiple faults. In this scenario, first the fault of the measuring path P_1 ($f_2$1, multiplicative change of -5%) was introduced. The diagnostic system started the first isolation thread which produced a diagnosis pointing out unisolable faults (f_5, f_6, f_{21}) for the observed symptoms *s8-m*$(-)$. The time series of the residuals and the elaborated diagnosis are shown in Fig. 7.41.

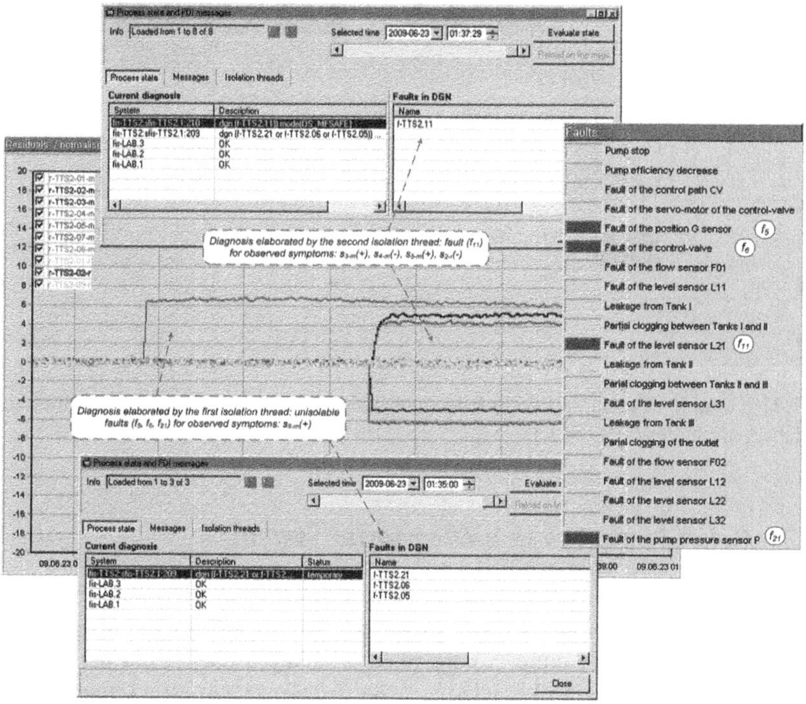

Fig. 7.41 Time series of residuals and the elaborated diagnosis in the case of simultaneous simulation of the faults f_{21} and f_{10}

While the first fault was still present, the second one was introduced—the fault of the measurement path L_{21} (f_{11}, additive fault of the value $-0.05\,m$). The reasoning system started the second isolation thread. For the second thread, disjoint subsets of possible faults and useful residuals were selected. This unequivocally pointed out the fault f_{11}. The idea of creating two independent isolation threads (with disjoint subdiagnostic relations) is illustrated in Fig. 7.42.

The final diagnosis elaborated by the diagnostic system took the following conjunction-alternative form: $DGN = f_{11} \vee f_5 \vee f_6 \vee f_{21}$ (the user interface visualizes the diagnosis in a simple form; the diagnostic messages explain its conjunction-alternative form).

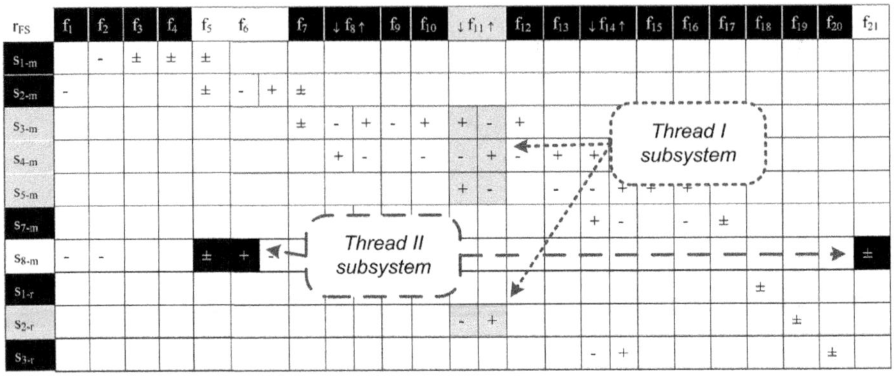

Fig. 7.42 Example of creating two independent isolation threads in the case of simultaneous simulation of the faults f_{21} and f_{11}

7.4.7 *Belief Networks in a Diagnostic System*

DiaSter allows employing diagnostic models which are the basis for reasoning on a technical state of different objects or processes. An example is the *belief-network-based model*, where reasoning is performed with the use of belief networks. The model allows incorporating knowledge from different sources: passive and active diagnostic experiments, experts who formulate subjective opinions, general domain knowledge about the observed objects or processes considered. The implementation of this model was contained in a plug-in of the *MITforRD* module.

This section presents an example of the identification of the diagnostic belief-network-based model for the diagnosis of the TT system. A basic introduction to the subsequent identification phases is also provided. Detailed explanations concerning the theoretical basis of BNBMs were presented in Chapter 5.

7.4.7.1 Identification of the BNBM

The BNBM is a special diagnostic classifier which assigns learning examples represented by values of process variables to the classes of the technical state. In the example considered, the structure of the BNBM consisted of

- the first stage (OCCs), which was represented by the set of parallel connected one-class classifiers. This stage allows preprocessing the input data and mapping them into the space of additional variables. They perform the function of instrumental variables which allow obtaining the interpretability of the parameters of the model and incorporating additional knowledge;
- the second stage, which contained the procedure of adjustment calculus (equalization and balance) of additional variables according to the set of user-defined constraint equations. Adjusted variables are the outcome of this stage;

- the third stage, which was represented by the belief network. In BNBMs, belief networks deal with the interpretation of adjusted (or non-adjusted) additional variables and drawing conclusions about the technical state of the object.

The general structure of a BNBM can comprise two or three stages. One can use the BNBM as a composition of the first and the third stage only. When additional domain knowledge about an object is given and constraint equations can be defined, all three stages should be used. In this section, all stages are used intentionally in order to present the main capabilities of the BNBM.

A typical process of the identification of the BNBM consists in

- setting up the initial version, which is conducted in the example considered;
- tuning model parameters, which is intentionally omitted in this section.

The determination of the initial version requires two steps: the assumption of the structure of the model and the identification of the model parameters. Both steps had to be preceded by the preparation of learning examples.

In contrast to dynamical or regressive models, classifiers, like the BNBM, assign learning examples represented by values of process variables to the classes of the technical state. In the example considered, an individual class of technical states was represented by a binary individual output variable (e.g., f_{13} presented in Fig. 7.43). They corresponded to 1 if the technical state (fault) considered occurred and 0 otherwise.

The binary values of output variables (presented in Fig. 7.43) which equal 1 were established on the basis of known time intervals for the learning data set when individual faults occur. Because the intervals for individual faults comprised also unsteady states, the learning examples which correspond to these states were excluded additionally on the basis of the known time constant of the TT system. Finally, the set of these variables formed the output data set of the BNBM.

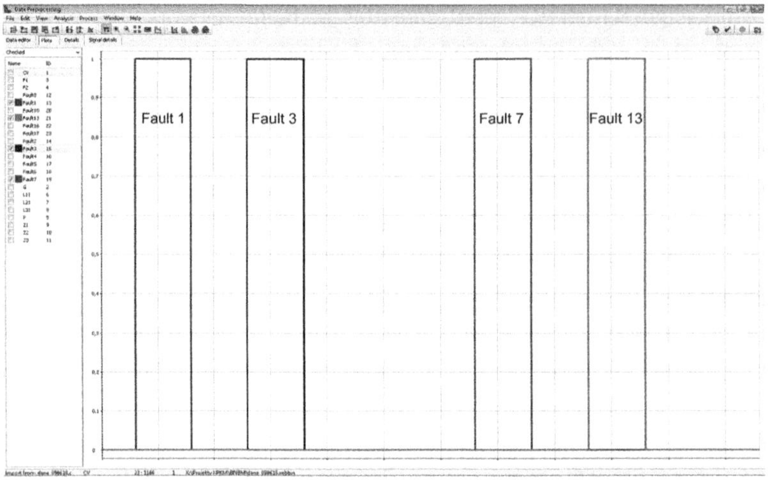

Fig. 7.43 Example output binary signals for different faults

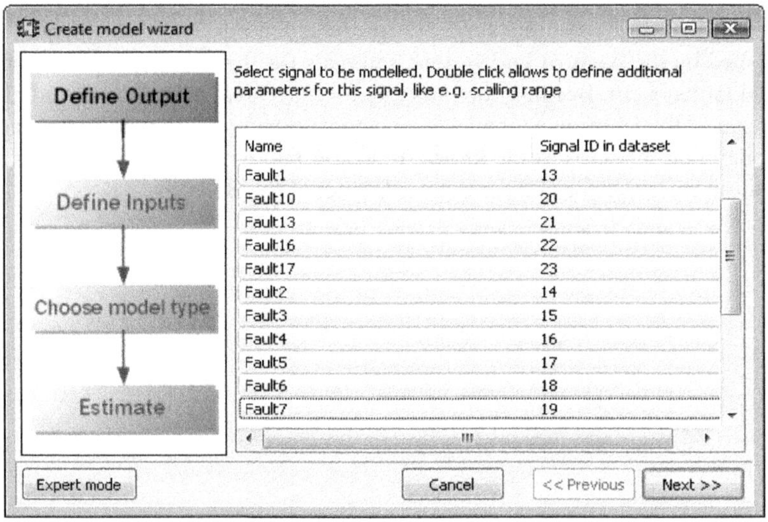

Fig. 7.44 Selection of output binary signals which represents the technical states considered

After the preparation of output variables, the set of inputs was indicated. To diagnose the TT system, the following monitored variables were selected:

$$\underline{x}_D = \{F_1, F_2, G, CV, P\}. \tag{7.7}$$

In the example considered, the set of input variables was established on the basis of subjective evaluation of the usefulness of process variables followed by the observation of their changes for different technical states. This approach is possible if the diagnosis of a simple object is considered. In more complex cases, for a large set of process variables, special methods of selection or extraction of process variables should be used.

Second stage of the BNBM. This stage allows incorporating general domain knowledge about an observed diagnostic object. This knowledge can be represented by physical laws expressed in the form of constraint equations, which often shows the relations between non-monitored additional variables and monitored process variables or known constants. The second stage is used to adjust additional variables according to the constraint equations applied.

The process of BNBM construction with the application of adjustment blocks in the second stage allows presenting additional capabilities of the model. To show it, an assumption associated with conscious resignation from the faults concerned with leakages was made intentionally. The subtraction of faults concerned with leakages allowed incorporating general domain knowledge about the TT system.

One can state, after omitting leakages, according to the mass balance equations, that the amount of liquid which inflows and outflows the TT system and the change in the storage of liquid for a limited increment of time must be in balance and can be expressed in the following form:

$$F_1 - F_2 = Z_1 + Z_2 + Z_3, \tag{7.8}$$

where

$$Z_1 = S_1 \frac{dL_{11}}{dt}, Z_2 = S_2 \frac{dL_{21}}{dt} \text{ and } Z_3 = S_3 \frac{dL_{31}}{dt}$$

are additional variables which describe changes of the liquid amount in Tank 1, Tank 2 and Tank 3, where S_1, S_2, S_3 are cross-sections of the tanks, and L_{11}, L_{21}, L_{31} are the monitored liquid levels in containers. The additional variables Z_1, Z_2, Z_3 resulted from numerical differentiation of the monitored variables L_{11}, L_{21}, L_{31}. In the example considered, the variables Z_1, Z_2, Z_3 were treated as directly non-monitored, inaccessible variables in a real monitoring process whose values will be reconstructed by the set of one-class classifiers during real time use of the diagnostic model.

For the purpose of incorporating the constraint equation in the BNBM, the following set of additional variables was defined:

$$\underline{z} = \{Z_1, Z_2, Z_3\}. \tag{7.9}$$

The *BNBM* plug-in allows representing the additional variables as discrete, in the form of the classes of values of additional variables. Every additional variable may be described by many classes of its values (intervals). In the example considered, the choice of the number of classes for individual variables was preceded by the observation of changes of values of process variables. The choice had finally significant influence on the efficiency of the diagnostic system built on the BNBM. For the set considered, (7.9), seven classes for Z_1, six for Z_2 and Z_3 were defined in the examined example (Fig. 7.45).

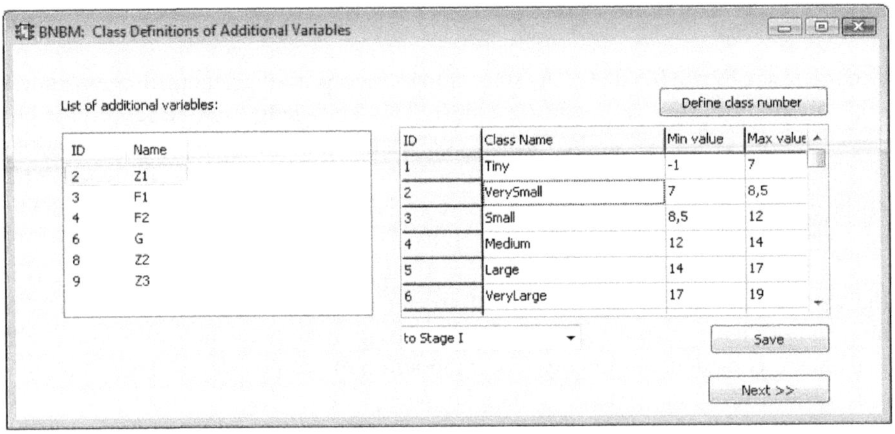

Fig. 7.45 Defining classes of additional values in the *MITforRD* platform. Six classes for variable Z_1 are shown as an example.

To form the constraint equation shown in (7.8), monitored variables F_1 and F_2 were added as continuous variables. Their difference expression is stated in the left-hand side of the equation (7.8). To obtain the equality of the equation considered, the vector of corrections c_{z1}, c_{z2}, c_{z3} for estimated additional variables $\hat{Z}_1, \hat{Z}_2, \hat{Z}_3$ was calculated by the *BNBM* plug-in. The equation (7.8) takes the form

$$F_1 - F_2 = (\hat{Z}_1 + c_{z1}) + (\hat{Z}_2 + c_{z2}) + (\hat{Z}_3 + c_{z3}), \tag{7.10}$$

where $\hat{Z}_1 + c_{z1}$, $\hat{Z}_2 + c_{z2}$, $\hat{Z}_3 + c_{z3}$ are reconstructed additional variables after adjustment.

The interface presented in Fig. 7.46 shows an example of a defined homogeneous constraint equation on the basis of different variable coefficients. Depending on employed values (1 ,-1, or 0) of coefficients, the variables are summed, subtracted or extracted from the equation.

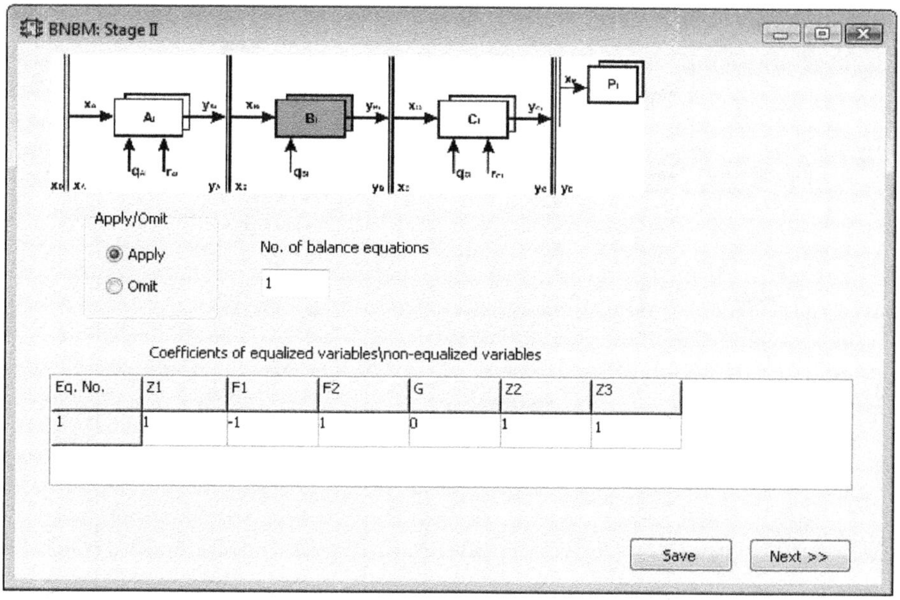

Fig. 7.46 Interface of the second stage. Coefficients for the homogeneous form of (7.8) are shown as an example

First stage of the BNBM: One-class classifiers. Generally, there is no universal classifier which allows obtaining sufficient efficiency of the classification task for every learning data set. This means that classifiers depend on the characteristics of the data to be classified and that classifier systems need to be especially evaluated.

For this purpose, various types of classifiers had been tested and selected before they were used in the first stage. This concerned their types as well as parameters, which were also initially determined during testing with the use of different validation methods. Good classification efficiency as well as sensitivity results were obtained for the nearest neighbor type of one-class classifiers. These classifiers presented good generalization properties during the reconstruction of additional variables and were finally used in the example considered.

The interface of the first stage in Fig. 7.47 presents individual one-class classifier parameters defined for an individual class of values of the additional variable. Another one is the fraction of rejected objects. This parameter deals with the fraction of examples which are not taken into account during the learning process. In nearest neighbor one-class classifiers, rejected objects are the outermost distance examples (objects), which often results from mistakes or non-extracted unsteady states and can be additionally omitted. The value of this parameter was adjusted iteratively to obtain sufficient efficiency of individual one-class classifiers.

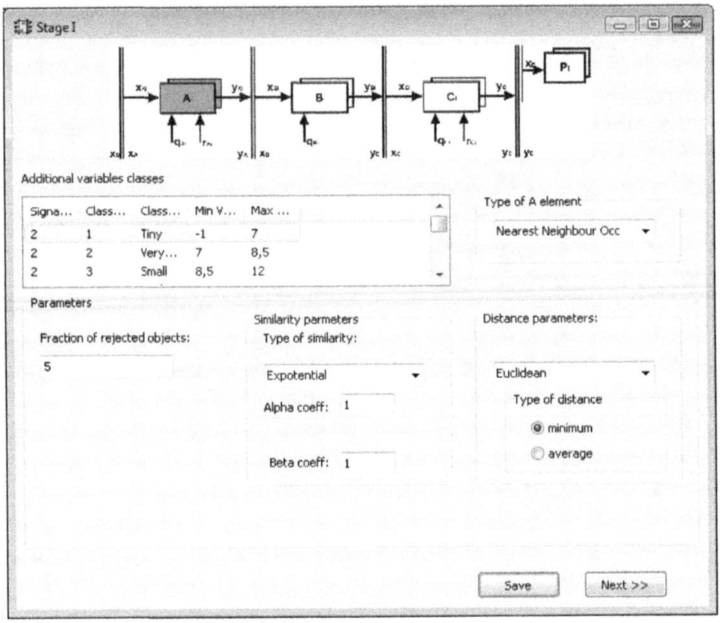

Fig. 7.47 First stage interface of the BNBM

Third stage of the BNBM: The belief network. The learning process of the belief network can be carried out in two ways. The first approach allows defining the conditional probabilities and prior probabilities by hand, based on one's own experience. In the second approach, examples in the form of additional (or reconstructed additional) and output variables are used to learn the parameters of the network. Both ways were used in the example considered.

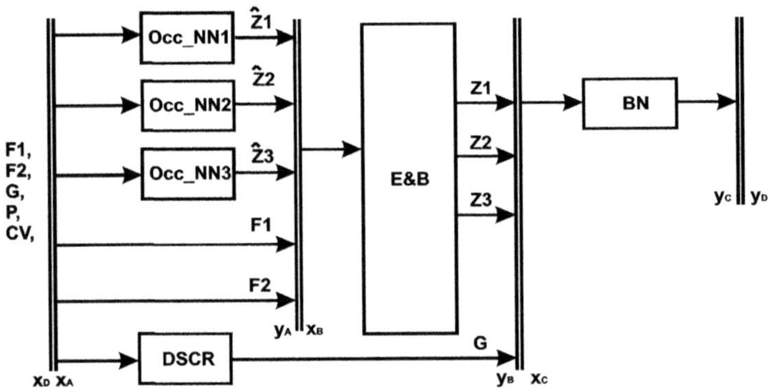

Fig. 7.48 Example structure of the BNBM for the TT system

The procedure of network learning stared from supervised learning on examples. The input set $\mathbf{x}_C$ (according to Fig. 7.48) of the third stage included

- adjusted additional values Z_1, Z_2, Z_3, which were the outcomes of the second stage;
- discrete values of variable G.

The output set of $\mathbf{y}_C$ contained all analyzed classes of technical states and was equivalent to the output set $\mathbf{y}_D$ of the BNBM.

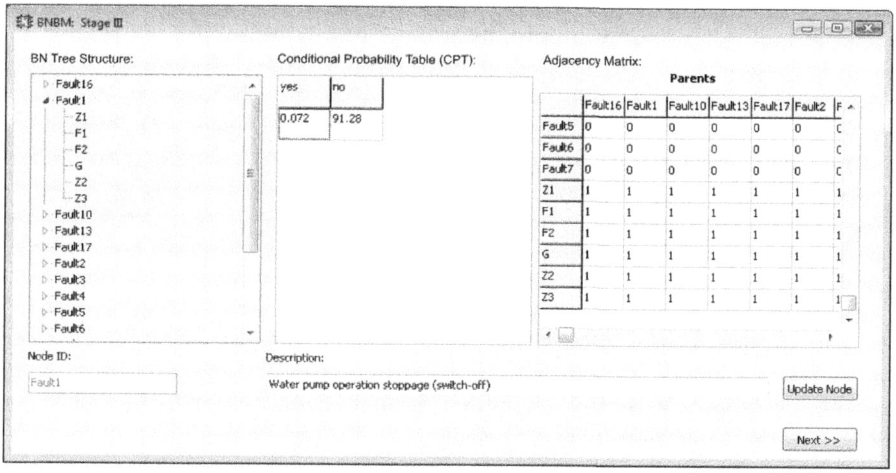

Fig. 7.49 Editor of belief network parameters: Stage III

In the *BNBM* plug-in, the Quick Medical Reference (QMR) structure of the belief network is a default structure. Two types of nodes can be distinguished in this structure: observation nodes and query nodes. The former represent diagnostic observation and the latter are answers of diagnostic queries. The observation nodes correspond to the input variables $\mathbf{x}_C$ and the query nodes to the technical state classes. In the example considered, all nodes were represented by discrete nodes. The calculation of Conditional Probability Tables (CPTs) was realized using the Expectation-Maximization (EM) algorithm for 600 of iterations and a $1,0e^{-}6$ learning error. The interface of Stage III is presented in Fig. 7.49.

The initial structure as well as the conditional probabilities obtained after learning on examples can be redefined for additional knowledge incorporation. If the additional expert knowledge is known, the change of the default structure and its parameters is especially required. To show it, the additional expert knowledge about the object considered was assumed intentionally in the analyzed example. The details are given in the next section. The redefined structure of the belief network is presented in Fig. 7.50.

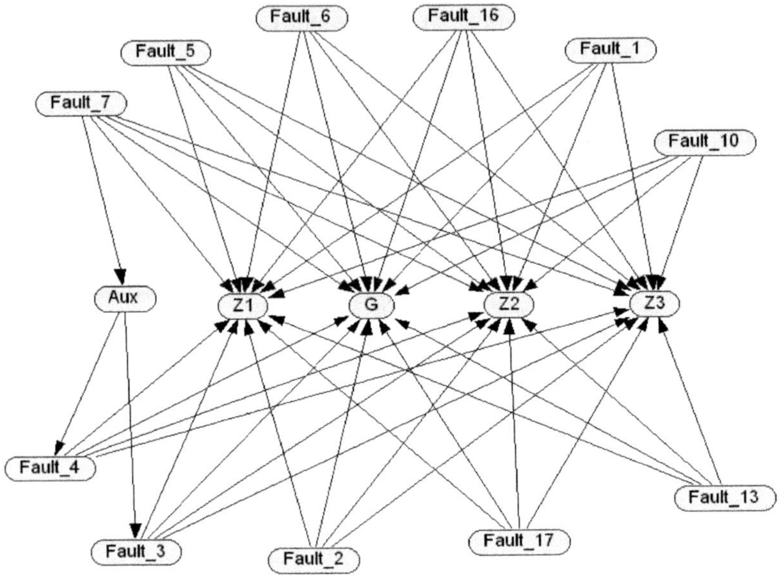

Fig. 7.50 Belief network structure of the BNBM for the TT system

The structure of the BNBM for the TT system and the input-output variables for individual stages are shown in Fig. 7.48.

7.4.7.2 Model Testing

The identified BNBM for the TT system was tested within the *MITforRD* framework. Individual faults were simulated, hence only the maximum likelihood value from the set of output states values of the belief network was treated as a conclusion of diagnostic BNBM operation. An example result of the recognition of f_1 (*Water pump operation stoppage*) is shown in Fig. 7.51.

The BNBM classifier was initially tested with the resubstitution method. Full recognition of the following faults was confirmed:

$$\mathbf{f} = \{f_1, f_2, f_5, f_6, f_7, f_{10}, f_{13}, f_{16}, f_{17}\}. \tag{7.11}$$

An indistinguishable pair of the faults f_3, f_4 was observed. In the BNBM, distinguishability can be improved using domain knowledge elicitation from experts. This knowledge can be expressed in the form of a subjective opinion about faults. In the example considered, information on the coexistence of the faults f_3, f_4 with other faults was employed. For instance, experts may state that the fault f_3 is more likely when faults (e.g., f_7) concerning measurement equipment appear simultaneously and unlikely for the fault f_4. This information can be incorporated in the network by direct association of nodes connected with measurement equipment with the nodes *Fault_*3 and *Fault_*4. However, if the number of measurement equipment faults is

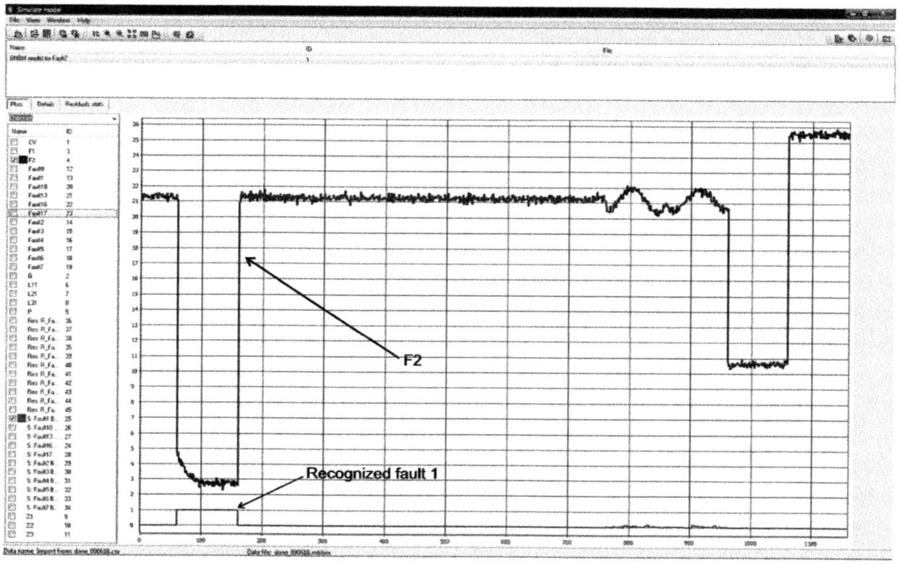

Fig. 7.51 Example results in recognizing f_1 during the verification of the BNBM with the *MIDforRD* module

large, then the specification of CPT numbers in the nodes *Fault_3* and *Fault_4* can be difficult. To avoid that, an additional node *Aux* was used between node measurement equipment faults (e.g. f_7) and the nodes *Fault_3* and *Fault_4*.

The new numbers of the CPT of the nodes *Aux*, *Fault_3* and *Fault_4* can be specified on the basis of individual expert opinions or, like in this example, on the aggregation of individual expert opinions by a knowledge engineer. The information that the fault f_3 is more likely when faults (e.g., f_7) concerning measurement equipment appear simultaneously and unlikely for the fault f_4 can be represented by conditional probabilities P for the nodes *Fault_3* and *Fault_4*. For example,

$$\begin{aligned} &P(Fault\_3 = Yes)|Aux = Yes) = 97\% \\ &P(Fault\_4 = No)|Aux = Yes) = 93\%. \end{aligned} \tag{7.12}$$

Analogously, one can state that the assertion *faults concerning measurement equipment occur* is more certain if more faults connected with this equipment appear simultaneously. In this case, the probabilities P for the node *Aux* can be specified. For example,

$$\begin{aligned} &P(Aux{=}Yes|Fault\_7 = Yes, Other\_faults = Yes) = 99\% \\ &P(Aux{=}Yes|Fault\_7 = Yes, Other\_faults = No) = 80\%. \end{aligned} \tag{7.13}$$

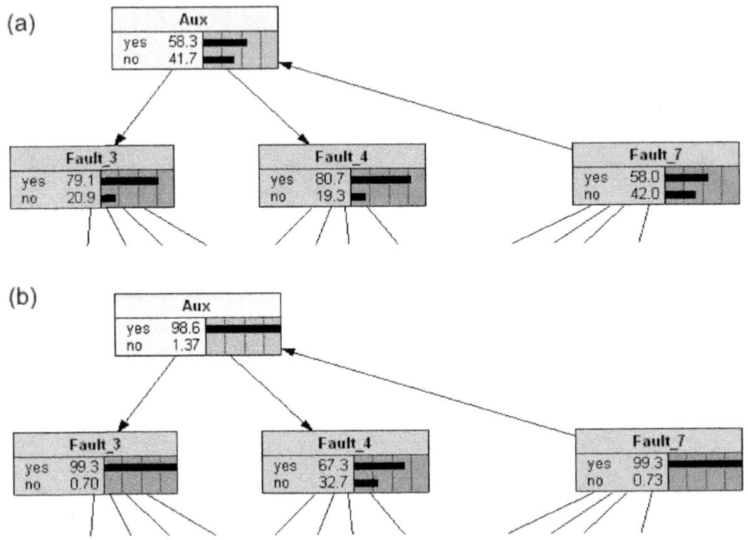

Fig. 7.52 Part of the belief network with an additional node *Aux* and query nodes with defined belief degrees as a result of an inference process: *Fault_3 = Fault f_3 has occurred. High belief degree*, *Fault_4 = Fault f_4 has occurred. High belief degree*, *Fault_7 = Fault f_7 has occurred. Low belief degree* (a); *Fault_3 = Fault f_3 has occurred. Very high belief degree*, *Fault_4 = Fault f_4 has occurred. Low belief degree*, *Fault_7 = Fault f_7 has occurred. Very high belief degree* (b).

The part of the belief network with the node *Aux* is presented in Fig. 7.52. Note that other associations between other measurement equipment faults which affect *Fault_3* and *Fault_4* are intentionally omitted in this picture. The examined nodes associated with variables represent the following statements:

Fault_3 Fault f_3 has occurred with states {Yes, No},
Fault_4 Fault f_4 has occurred with states {Yes, No},
Fault_7 Fault f_7 has occurred with states {Yes, No},
Aux Faults of measurement equipment have occurred with states {Yes, No}.

Figure 7.52(a) presents the case of the undistinguishable faults f_3 and f_4. The belief degrees of the variables *Fault_3* and *Fault_4* were defined as a result of the inference process. The belief degrees of the variable *Fault_3* (*Fault f_3 has occurred*) Yes = 79% and the variable F_4 (*Fault f_3 has occurred*) Yes = 80% are very high and show that these faults are very likely. The rest of the belief degrees of fault nodes are close to 50% and do not give any certain information to draw further conclusions.

Figure 7.52(b) presents an example showing that additional information about the variable *Fault_7* (*Fault f_7 has occurred*) Yes = 99% allows concluding that the fault f_3 is much more likely (*Fault_3* (*Fault f_3 has occurred*) Yes = 99%) than the fault f_4, where the belief degree is equal to Yes = 67% . The presented example shows that the use of additional knowledge can be helpful in obtaining the distinguishability of the faults f_3 and f_4.

An example process of BNBM construction was presented. It was shown how the BNBM allows integrating knowledge from different sources. Supervised learning on examples, the incorporation of general knowledge in the form of physical equations, and adding expert knowledge represented by subjective opinions were shown. Full distinguishability of the BNBM for the examined set of faults was achieved.

It is worth pointing out that the assumed structure of the BNBM for the object considered is not the only one. For instance, there are many equivalent belief network structures in the sense of the joint probability distribution which allow adding the same additional expert knowledge. Also, some associations between nodes may be unnecessary and could be removed with the use of structure learning algorithms. The BNBM may also require the application of the tuning procedure, especially for more complex diagnostic objects.

All the operations concerning the subsequent phases of BNBM identification were conducted in the *MITforRD* module. The application of an identified BNBM in on-line diagnostics can be done using the *PExSim* module.

7.4.8 Knowledge Discovery in Databases

The methods presented in Chapter 4 have been used for the data that had been prepared by means of the simulator described in Section 7.4.1. The *KDD* package includes four applications. Two of them—selecting process variables and building SVM models—require typical process data, such as those described in Section 7.2. Two other modules are used for modeling cyclic processes and thus require data of a special kind. Therefore, at the very beginning of this section we start with the description of a numerical experiment that has been run for preparing data suitable for modeling cyclic processes. Then we give examples of applications of programs that allow building models of dynamic processes.

7.4.8.1 Simulation Data for Cyclic Processes

Due to the nature of the two discussed methods of case-based reasoning (based on approximations models and on fuzzy models of processes), which are used within the *DiaSter* system to build knowledge bases for cyclic processes, it was necessary to prepare an adequate set of test data. For this purpose, it was necessary to use the TT system simulator realized in the *PExSim* module to allow an experiment to be carried out on the TT system simulator.

It was decided that the simulated process would be cyclic and built of three phases, with their nature differing substantially among the phases. All the phases have a fixed duration (Table 7.6). The whole cycle takes 1000 s. Phase 0 is used to determine the starting point of the process and defines the initial level in Tank 3.

Table 7.6 Times of process phase changes

Phase	Start time (s)
0	0
1	110
2	300
3	800

For each phase, a target level of the liquid in Tank 3 was applied using nine different configurations (Table 7.7). Figure 7.53 shows the model of the process in the *PExSim* module.

Table 7.7 Levels applied in Tank 3

Level L3	Phase 0	Phase 1	Phase 2	Phase 3
1	0	0,15	0,3	0,05
2	0,05	0,15	0,3	0,05
3	0,1	0,15	0,3	0,05
4	0	0,2	0,35	0,1
5	0,05	0,2	0,35	0,1
6	0,1	0,2	0,35	0,1
7	0	0,25	0,4	0,15
8	0,05	0,25	0,4	0,15
9	0,1	0,25	0,4	0,15

It was determined that single distortions, in the form of faults applied to the actuators and measuring elements, would only be introduced in the second phase. Such faults may be of a temporary or a permanent nature. The experiment was carried out for the two employed values of the fault levels of each component and for the faultless state. Table 7.8 shows a description of individual faults simulated during the experiment.

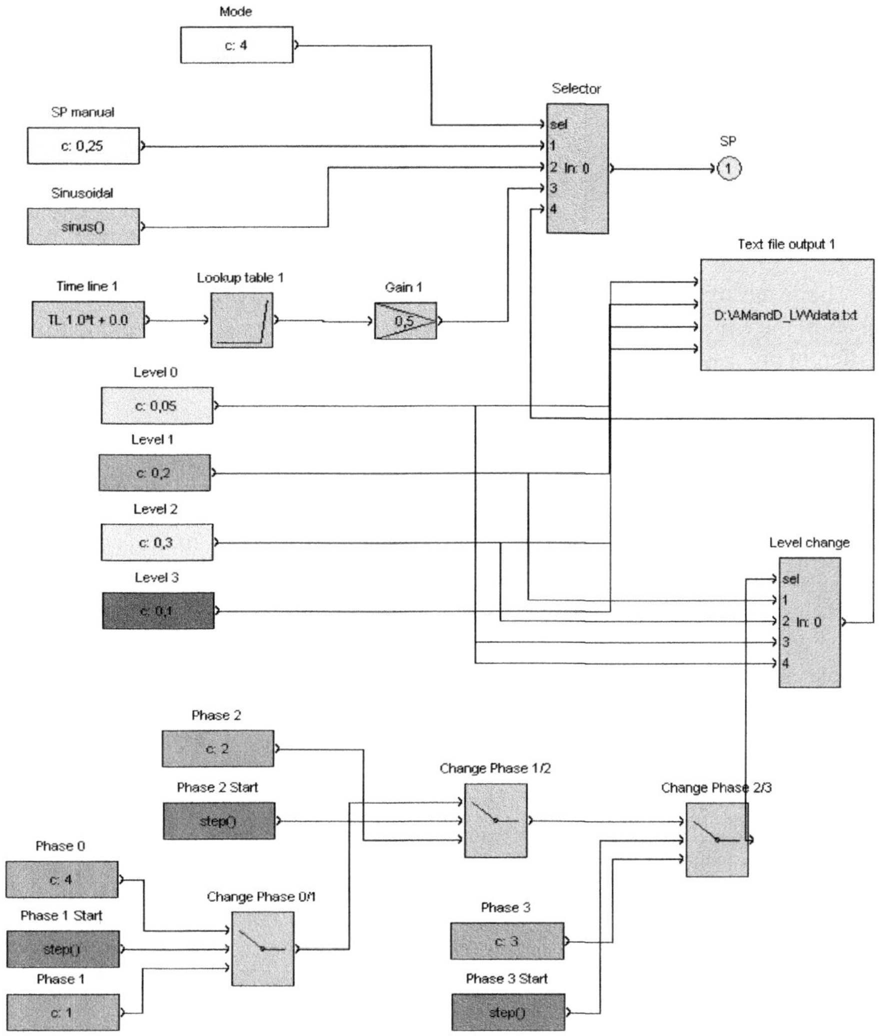

Fig. 7.53 Configuration of the simulator cycle

Table 7.8 Simulated faults

Fault	Index	Description
Damage of actuator	f_2	Efficiency drop in the pump
Leak	f_{15}	Leak from Tank 3
Block	f_{10}	Partial blocking of the channel between Tanks 1 and 2
Damage of holes	f_{14}	Damage of the measuring path of the level in Tank 3

The result of the experiment was simulation data obtained for 12 process variables of the simulator in the form of 81 cycles with the sample rate 1 s, representing different process states.

7.4.8.2 Selection of Input Variables

One of the stages of mathematical model identification on the basis of experimental data includes the stage of input selection. This stage may be, among others, automated by applying a method of feature selection developed within the domain of data mining. In this subsection, an example use of the *Feature Selection* plug-in for selecting model inputs is presented. The algorithm implemented in the plug-in is based on the approach known as filter feature selection. In particular, the presented plug-in is based on the algorithm of correlation-based feature selection with BF search.

The user interface of the plug-in includes three tab pages (Fig. 7.54). The user chooses the output variable in the *Output* tab page. The *Inputs* tab page facilitates the choice of input variables and the delays range for each variable. The third tab page, *Selection* (Fig. 7.54), allows defining parameters of the search process and displays the obtained results. Particularly, the user may not only set the direction of the search, but also determine the stop criterion.

(a)

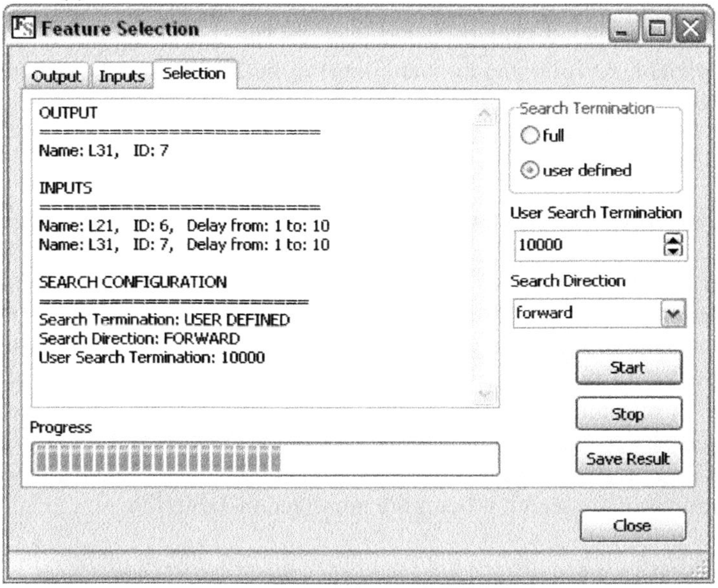

(b)

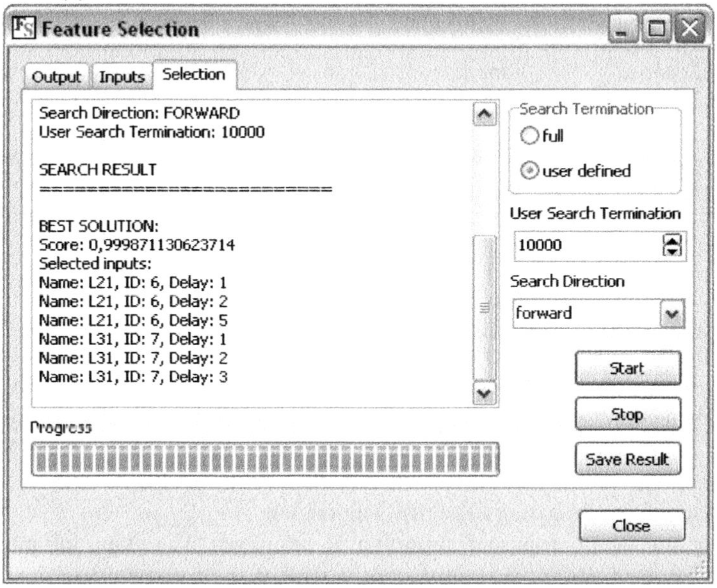

Fig. 7.54 User interface of the *Feature Selection* plug-in (*Selection* tab): calculation (a), obtained results (b)

The use of the plug-in was presented as an example of a task consisting in selecting input variables of the model (7.14). The purpose of the model is to emulate changes of the water levels in the 3-rd tank (L_{31}). The tank belongs to the TT system. The L_{21} variable representing the water level in the 2-nd tank is the second variable considered in the model:

$$\hat{L}_{31}(k) = f(L_{21}(k-1),\ldots,L_{21}(k-10),L_{31}(k-1),\ldots,L_{31}(k-10). \tag{7.14}$$

The essential data was acquired by means of the simulator. One of pre-defined simulation scenarios was applied. The chosen scenario includes programmed changes of the water level in the 3-rd tank (working point) and noise in measurement channels. During the simulation, no faults of the installation were considered. The system simulation took 220 minutes, whereas a sampling period was 1 second. As a result, time series of 12 process variables were obtained. The stage of data preparation consisted in selecting data samples with indices from 1000 to 4000. Due to a vast state space (approx. 2^{30}), the limited search method along with the stop criterion after 10000 iterations was applied in the experiment. The obtained results for various configurations of the search process are presented in Table 7.9.

Table 7.9 Example results for input selection of the model (7.14)

Search termination	Search direction	Selected inputs	Score
user defined	forward	$L_{21}(k-1)$, $L_{21}(k-2)$, $L_{21}(k-5)$, $L_{31}(k-1)$, $L_{31}(k-2)$, $L_{31}(k-3)$	0.9998711
user defined	backward	$L_{21}(k-1)$, $L_{21}(k-3)$, $L_{31}(k-1)$, $L_{31}(k-2)$, $L_{31}(k-3)$	0,9998712

7.4.8.3 Predicting Process Variables by SVM models

The method of support vector machines is one of the most frequently used methods of data modeling within the domain of data mining. Generally, it may be applied for tasks of data classification and approximation. In this subsection, an example use of the *SVM Regression* plug-in is presented. The plug-in is based on the SVM method and its aim is to identify autoregressive models. Particularly, the presented plug-in was developed on the basis of the SMO algorithm.

A user interface consists of a number of tab pages. The given tab pages allow determining the SVM model parameters as well as the parameters of the algorithm defining the model. In Fig. 7.55, main plug-in tab pages are presented. The *General* tab page (Fig. 7.55(a)) is used for defining the basic parameters of the SMO algorithm. Moreover, the *Inputs* and *Kernel* (Fig. 7.55(b)) tab pages facilitate precise determination of the features of the SVM model to be identified.

(a)

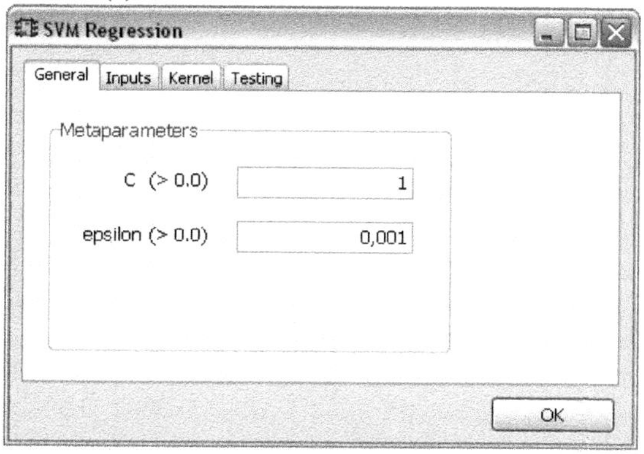

(b)

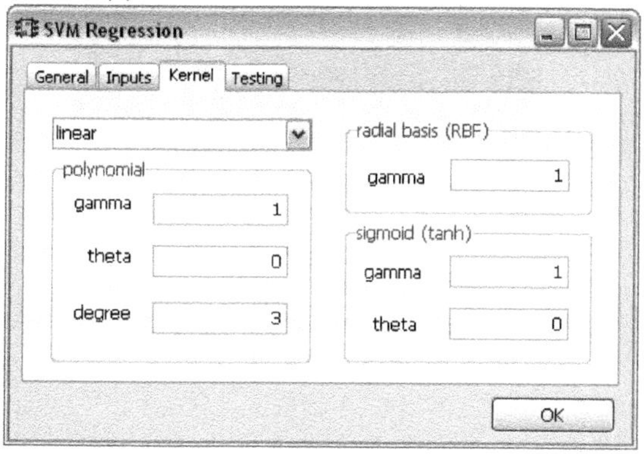

Fig. 7.55 User interface of the *SVM* plug-in: *General* tab (a), *Kernel* tab (b)

Similarly as in the case of input variables selection, the task of identifying the model (7.14) is presented as an example of plug-in use. The same set of data was applied. Variables identified in the selection process were introduced as the model inputs. Particularly, these included $L_{21}(k-1)$, $L_{21}(k-2)$, $L_{21}(k-3)$, $L_{21}(k-4)$, $L_{21}(k-5)$, $L_{21}(k-6)$, $L_{31}(k-1)$, $L_{31}(k-2)$, $L_{31}(k-3)$.

During the experiment, various configurations of the model and the learning algorithm were examined. Among others, models with linear and radial kernel functions were analyzed. In particular, a model consisting of 100 support vectors was obtained. Moreover, the following parameters of the learning algorithm were applied: $C = 1$ and $\varepsilon = 0.001$. Results of model verification, i.e., the model response and the residuum, are presented in Fig. 7.56. Furthermore, the mean square error as well as the Mean Absolute Error (MAE) were calculated:

- MSE: $e^2 = 2.24502e - 06$,
- MAE: $|e| = 0.00119368$.

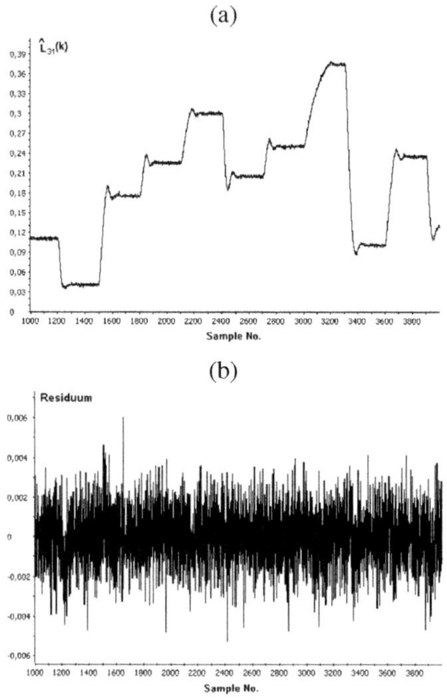

Fig. 7.56 Results of SVM-based model identification of the L_{31} variable: model response (a), residuum (b)

7.4.8.4 Approximation Models of Cyclic Processes

The *CBR* module allows building a process model of a cyclic nature in the form of a base of models describing similarities between applications of selected groups of process variables. During modeling, the *CBR* module uses archival data in the form of *MITforRD* module data files (*mbbin* files) and stores information about the model (tables of similarities), and the configuration of the individual module in a specialized *CBR* package database stored in the *DiaSter* platform configuration.

As a first step, it is necessary to prepare an appropriate archival database. Such a database must contain information regarding the cycles of the studied process in the form of appropriate indices of the cycles for the data stored in the archival database. During the preparation of the model, the user selects the desired variables from the set of process variables available in the *DiaSter* platform configuration and indicates the type of the model to be created.

After selecting the CBR model, the user interface window appears to configure additional information selected from the database (Fig. 7.57). It is now possible to enter information about the new cycle, or delete an existing cycle of the selected database for modeling. At this time it is also possible to configure the type for the selected process variables (*INPUT*, *OUTPUT*, *CONTROL*). This information is automatically stored in the database of the *CBR* module.

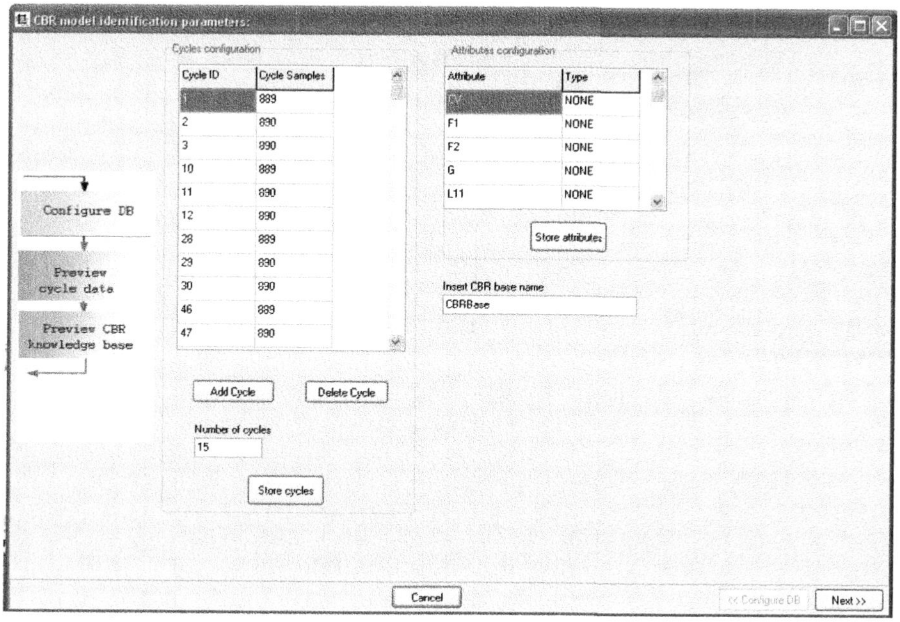

Fig. 7.57 *CBR* module windows: cycle configuration

In the next window of the interface (Fig. 7.58), the user has the ability to browse the archival data regarding the preparation of the modeling process. The user may view the approximated process cycles for different numbers of approximating points. Approximation is conducted in accordance with the described method of modeling cyclic processes.

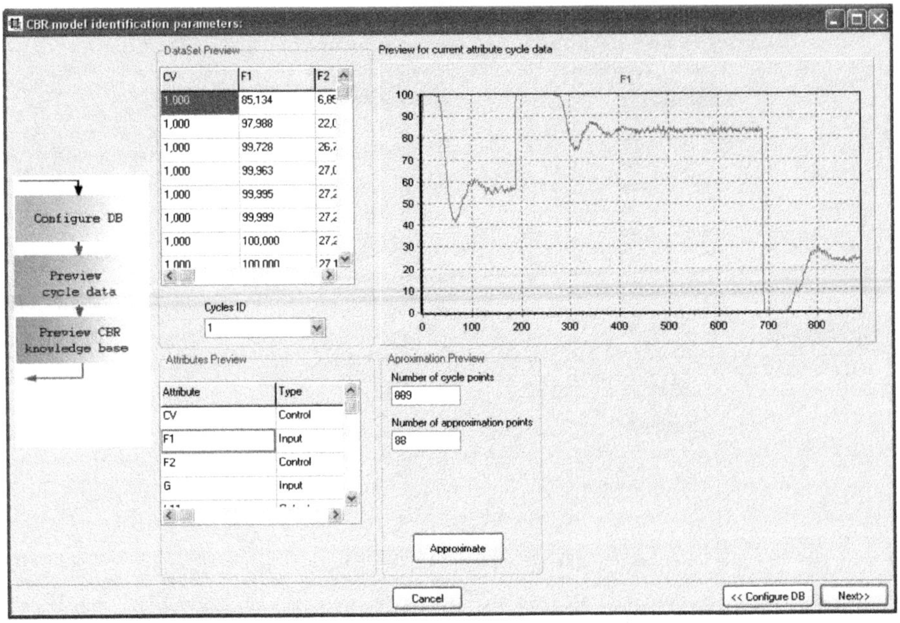

Fig. 7.58 *CBR* module windows: parameters of the method

After the configuration, the user can go to the model estimation window or open another window to analyze the model already created (Fig. 7.59). As a result of the *CBR* module, a knowledge base is created in the form of tables of similarities between archival realizations for the selected process variables. This table is then stored in the database.

During the search for a process similar to the one currently being conducted, the module returns the tables of values of similarities with respect to the process realization under consideration. The table contains values of similarities for process variables selected during the phase of identification. The similarities relate to the inputs, outputs and controls for the different process realizations. A complete table of similarities is returned, which allows sorting results by the user and provides him/her with the possibility to view selected realizations.

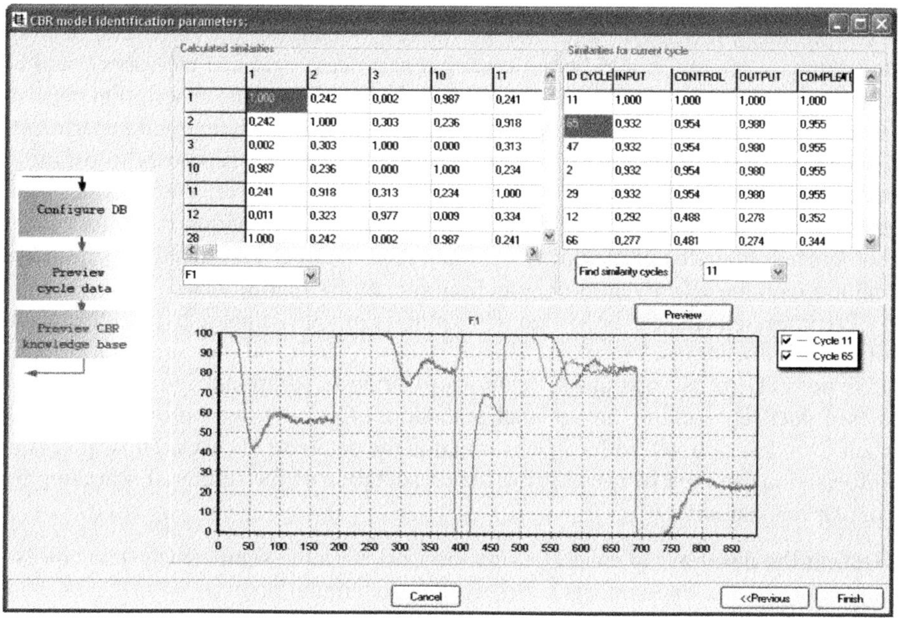

Fig. 7.59 *CBR* module windows: results of searching for a similar cycle

7.4.8.5 Fuzzy Models of Cyclic Processes

The *CBRFuzzy* module allows building a database of cases with fuzzy representation of every case. The database of cases can be used for searching for similar realizations using a special fuzzy similarity measure.

Building the database of cases. A database of cases stores historical control courses of a given process. Every case in the database consists of a description of process realization and the data representing control time series (Fig. 7.60).

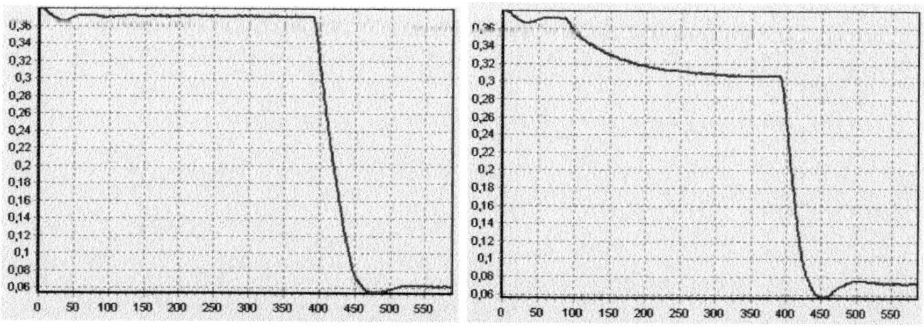

Fig. 7.60 Examples of cases representing different classes

The description of every case can be realized as a string describing the class of the example, e.g., process OK, a leak of the valve, a blockage of the valve, etc. Usually, the database of cases could be built using real process data and the cases could be described by experts conducting given a process. However, the description consists of analyses carried out using data obtained as a result of the numerical experiment.

Depending on the representation of the cases it might be necessary to set information about the representation of the parameters of a course. When using linear approximation, it is necessary to set the maximal approximation error, fuzzy sets borders and matrices of similarity of fuzzy sets. Parameter values have significant influence on the effectiveness of searching for similar realizations.

Preprocessed data can be loaded into a plug-in private database of cases by means of special procedures. Every case should be described by failure class membership. It is possible to use operators' opinions to provide additional information about stored data representing an archival realization. Operators can provide data with their own descriptions, both of them stored in the database of cases. Those pieces of information can then be used during the presentation of the results of searching for similar realizations.

Testing the database of cases. During the work with the simulated system consisting of three tanks, experiments with searching for similar realizations in the database of cases were conducted. The database contained realizations corresponding to the control of the valve. At a certain stage of the cycle, searching for courses similar to the current one was carried out. The moment of triggering the searching operation can be defined as fixed time of the realization of a certain stage of a given process. Every similar example found is described by the value of membership to the fault class. This information can be presented to the operator of the process.

The case much less similar looked like in Fig. 7.62. It is possible to limit the number of found realizations by setting the threshold value of similarity. The value of similarity is strongly determined by values of parameters of the module. The most similar case found can be used as the description of the current realization by the operators of the process. This description can be used by the operator as advice on how to deal with the current process. For a majority of cases in the database, the most similar case found belonged to the same class (Table 7.10).

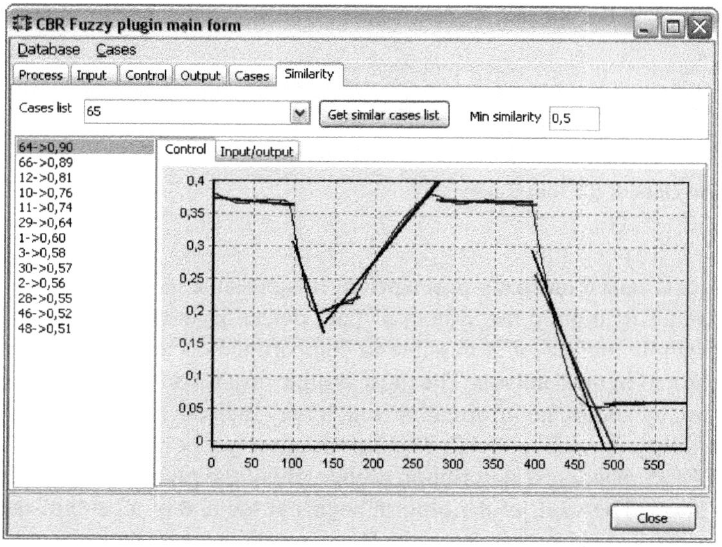

Fig. 7.61 Results of searching for similar courses: high similarity

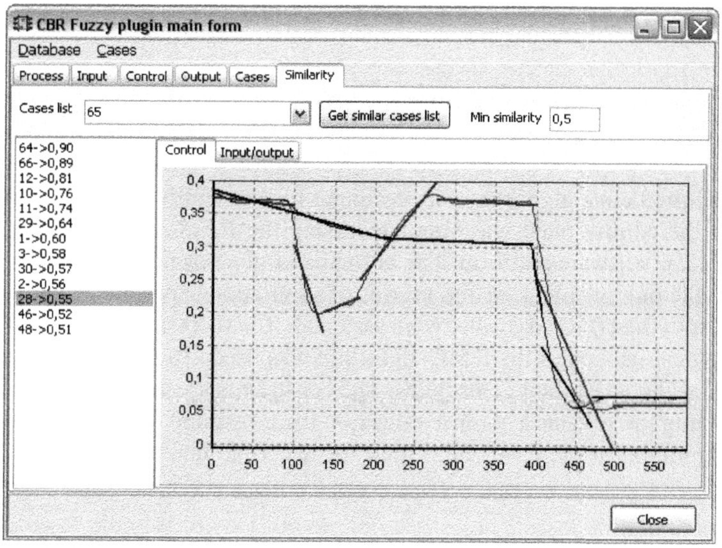

Fig. 7.62 Results of searching for similar courses: low similarity

Table 7.10 Performance of the database of cases and the similarity measure

Class	Number of cases	Performance index [%]
Number of all cases in the database	15	
Same class of the found case	12	80
Different class of the found case	3	20

Using models based on fuzzy description. It is possible to use a *DiaSter* solution founded on CBR to build the system of fault detection in the production process. For the example process, it is possible to compare the current control course with the cases stored in the database. The most similar case found should be described in the database by additional information which may indicate the state of the process, like "possibility of a valve fault". This information can be presented in a synoptic screen of *PExSim* as a text description of the state of the given process at the end of a certain stage. The work of the plug-in begins at the end of a certain stage of the process or is triggered by the operator. The control course of the current process is downloaded from the *DiaSter* system and processed to a form that can be compared with the cases stored in the knowledge base. Then searching for a similar realization in the database begins. Based on information regarding the most similar realization, a conclusion about the current process can be formulated. This conclusion can be presented in a synoptic screen of the *DiaSter* system.

7.4.9 Model Predictive Control with Constraints and Faulty Conditions

The DMC predictive algorithm was designed for the control plant. In order to do that, first the control plant was simulated using the *PExSim* module—an element of the *DiaSter* software platform. The normalized step response was obtained using the "Text file output" block. After a series of simulation experiments, the following parameters of the DMC controller were assumed: $\lambda = 0.0001, D = N = 100, N_u = 2$. The analytical version of the DMC algorithm was used. Thus, the control signal is generated using an explicit control law. The constraints are taken into consideration by projecting the obtained control values on the constraint set. Figure 7.63 shows the configuration of the DMC algorithm for the triple tank system in the *PExSim* module.

7.4.9.1 Control with Constraints

In Fig. 7.64, results obtained during control system simulation after a change of the set-point of the level in the third tank from 0 to 0.35 are shown. Two cases were considered: without and with constraints taken into consideration by the controller, using the mechanism of control projection. In the first case, the control signal generated by the controller is much higher than the upper allowed bound (100% of the valve opening). Because of that, relatively big overshoot was obtained. After using the control projection mechanism, the operation of the control system improved. The overshoot is now much smaller and the control signal is much shorter on the boundary.

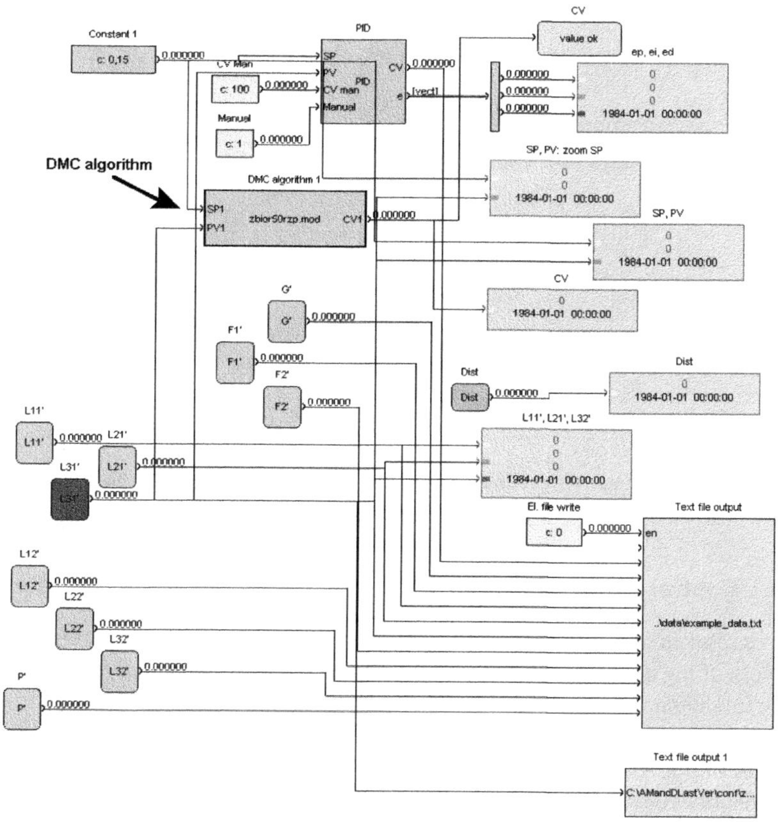

Fig. 7.63 DMC algorithm for the triple tank system simulated in the *PExSim* module

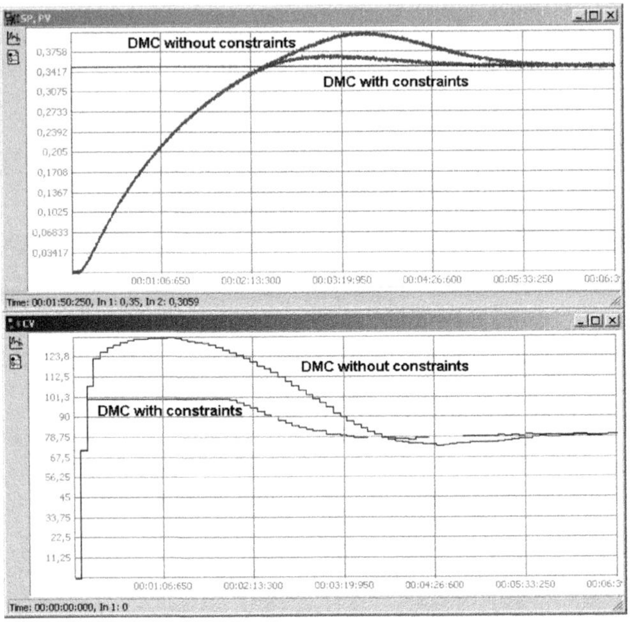

Fig. 7.64 Result obtained in the control system with the analytical DMC algorithm without (a) and with (b) constraints taken into consideration

7.4.9.2 Control in Faulty Conditions

The operation of the control system in faulty conditions is very interesting. The robustness of the control system with the predictive controller based on the nominal process model (no information about faults was used) was checked. Many faulty scenarios were tested. The most interesting results are presented in the subsequent figures. In Fig. 7.65, results of control system simulation in the case when a leakage from Tank 1 occurs (fault f_9) are shown. 50% of the maximal leakage was assumed. Despite such a huge disturbance, a satisfactory response was obtained. The control time is a little bit longer than in the case before fault occurrence, but there is no overshoot. By observing the control signal one can notice that the operating point changed (the control signal settles near 92% instead of 78%). Similar behavior of the control system was observed in the case of leakages from the other two tanks. However, in the case of huge leakages, it is impossible to achieve the assumed set-point value.

The next group of faulty situations which were simulated are jammings of the pipes leading from the tanks. In Fig. 7.66, results of control system simulation in the case when the pipe between Tanks 2 and 3 was jammed (fault f_{13}) are shown. 50% of the maximal flow rate was assumed. Despite the fault, the obtained responses are very close to those before fault occurrence. A similar result was obtained in the case of the jamming of the pipe between Tanks 1 and 2. These results are the evidence of considerable robustness of the controller.

Big differences in control system operation were observed after the jamming of the outlet from Tank 3 (fault f_{16}, Fig. 7.67). It is because Tank 3 can now be filled in much faster than in the case before fault occurrence. As a result, huge overshoot occurs. On the other hand, the control signal stays on the bound much shorter.

Summing up, in the case of most faulty situations, the predictive controller works unexpectedly well despite the simplicity of its structure and the constraint handling mechanism, as well as the fact that the controller was not reconfigured. The *PExSim* module, however, makes it possible to design a few controllers for different faulty scenarios and switching between them, depending on a particular fault.

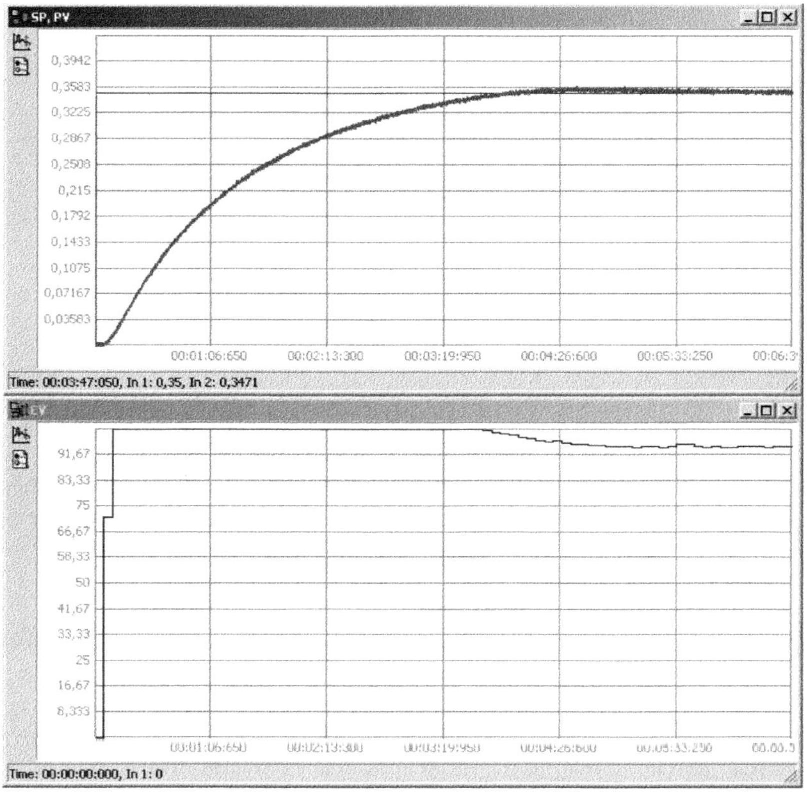

Fig. 7.65 Result obtained in the control system with the analytical DMC algorithm with constraints taken into consideration when a leakage from Tank 1 occurs (fault f_9)

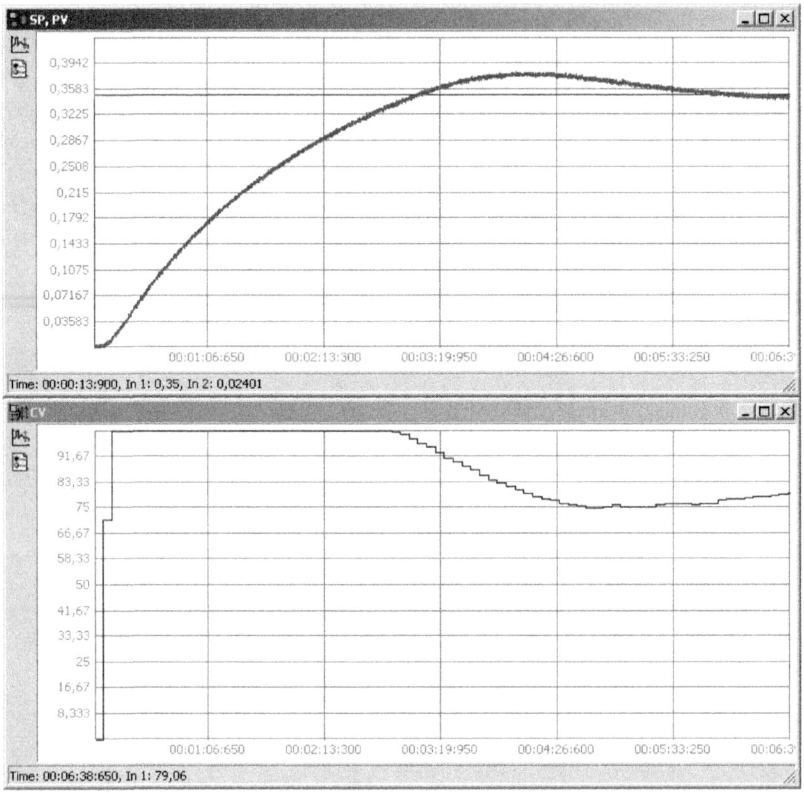

Fig. 7.66 Result obtained in the control system with the analytical DMC algorithm with constraints taken into consideration when the pipe between Tanks 2 and 3 was jammed (fault f_{13})

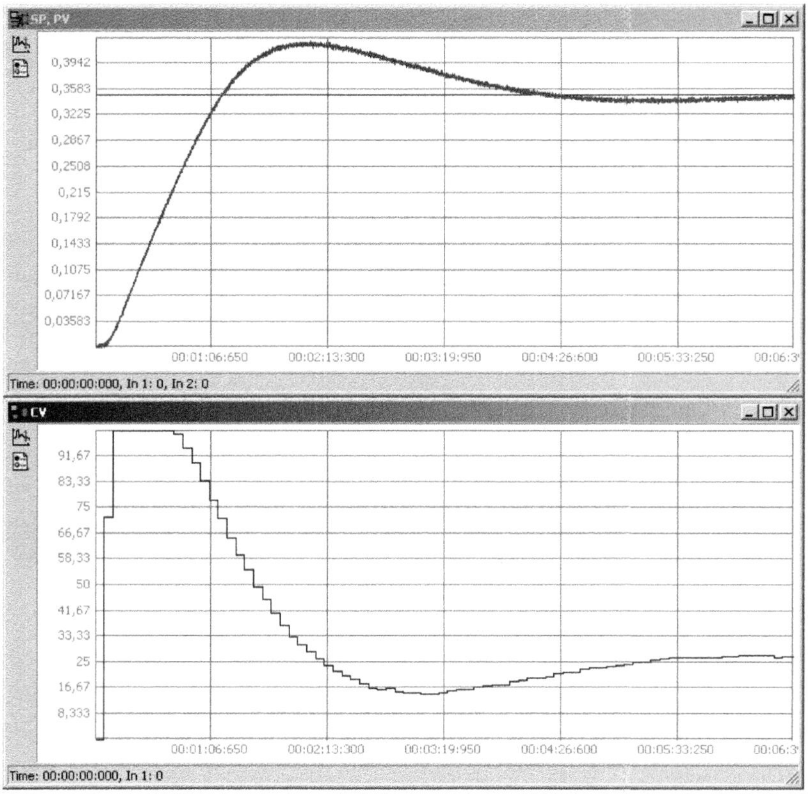

Fig. 7.67 Result obtained in the control system with the analytical DMC algorithm with constraints taken into consideration when the outlet from Tank 3 was jammed (fault f_{16})

References

Aamodt, A., Plaza, E.: Case-based reasoning: Foundational issues, methodological variations, and system approaches. AICom–Artificial Intelligence Communications 7(1), 39–59 (1994)

Abdessemed, F., Monacelli, E., Benmahammed, K.: A learning paradigm for motion control of mobile manipulators. International Journal of Applied Mathematics and Computer Science 16(4), 475–484 (2006)

Ackermann, J.: Robust Control. The Parameter Space Approach. Springer, Berlin (2002)

Aczel, A.D.: Complete Business Statistics, 5th edn. McGraw-Hill/Irwin, New York (2001)

DAMS, http://www.mscsoftware.com/products/adams.cfm.

Albertos, P., Sala, A.: Multivariable Control Systems. An Engineering Approach. Springer, Berlin (2002)

Allgöwer, F., Badgwell, T., Qin, J., Rawlings, J., Wright, S.: Nonlinear predictive control and moving horizon estimation—An introductory overview. In: Frank, P. (ed.) Advances in Control—Highlights of ECC 1999, ch. 12. Springer, London (1999)

Arbib, M.A. (ed.): The Metaphorical Brain, 2nd edn. Wiley, New York (1989)

Astolfi, A., Karagiannis, D., Ortega, R.: Nonlinear and Adaptive Control with Applications. Springer, Berlin (2008)

Åström, K.J., Hägglund, T.: PID Controllers, Theory, Design, and Tuning, Instrument Society of America, Research Triangle Park (1995)

Åström, K.J., Hägglund, T.: Automatic tuninig of simple regulators with specifications on phase and amplitude margins. Automatica 20(5), 645–651 (1984)

Åström, K.J., Hägglund, T., Hang, C.C., Ho, W.K.: Automatic tuninig adaptation for PID controllers—A survey. Control Engineering Practice 1(4), 699–714 (1993)

Åström, K., Wittemark, B.: Computer Controlled Systems. Prentice Hall, Upper Saddle River (1997)

Back, A.D., Tsoi, A.C.: FIR IIR synapses. A new neural network architecture for time series modelling. Neural Computation 3(3), 375–385 (1991)

Bańka, S.: Control of Multidimensional Dynamic System, Technical University of Szczecin Press, Szczecin (2007) (in Polish)

Bargiela, A., Pedrycz, W.: Granular Computing. An Introduction. Kluwer Academic Publishers, Boston (2003)

Bartyś, M., Patton, R., Syfert, M., Heras, S., Quevedo, J.: Introduction to the DAMADICS actuator FDI study. Control Engineering Practice 14(6), 577–596 (2006)

Basseville, M., Benveniste, A.: Detection of Abrupt Changes in Signal and Dynamic Systems. Springer, Berlin (1986)

Basseville, M., Nikiforov, I.V.: Detection of Abrupt Changes: Theory and Application. Prentice Hall, Englewood Cliffs (1993)
Batyrshin, I., Wagenknecht, M.: Towards a linguistic description of dependencies in data. International Journal of Applied Mathematics and Computer Science 12(3), 391–402 (2002)
Bertsekas, D.: Nonlinear Programming. Athena Scientific, Belmont (1995)
Blanke, M., Kinnaert, M., Lunze, J., Staroswiecki, M.: Diagnosis and Fault-Tolerant Control. Springer, New York (2003)
Blanke, M., Staroswiecki, M.: Structural design of systems with safe behaviour under single and multiple faults. In: Proceedings of the IFAC Symposium on Safeprocess 2006, Beijing, China, pp. 511–515 (2006)
Blevins, T., McMillan, G., Wojsznis, W., Brown, M.: Advanced Control Unleashed. ISA Society, Research Triangle Park (2003)
Bobál, V., Böhm, J., Fessl, J., Machácek, J.: Digital Self-tuning Controllers. Algorithms, Implementation and Applications. Springer, Berlin (2005)
Brdyś, M., Tatjewski, P.: Iterative Algorithms for Multilayer Optimizing Control. Imperial College Press, London (2005)
Calado, J., Korbicz, J., Patan, K., Patton, R., Sa da Costa, J.: Soft computing approaches to fault diagnosis for dynamic systems. European Journal of Control 7(2-3), 248–286 (2001)
Camacho, E., Bordons, C.: Model Predictive Control. Springer, London (2004)
Campolucci, P., Uncini, A., Piazza, F., Rao, B.D.: On-line learning algorithms for locally recurrent neural networks. IEEE Transactions on Neural Networks 10(2), 253–271 (1999)
Cempel, C.: Vibroacoustic Condition Monitoring. Ellis Horwood, Chichester (1991)
Charniak, E.: Bayesian networks without tears. AI Magazine 12(4), 50–63 (1991)
Chen, J., Patton, R.J.: Robust Model-Based Fault Diagnosis for Dynamic Systems. Kluwer Academic Publishers, Berlin (1999)
Chernous'ko, F., Ananievski, I.M., Reshmin, S.A.: Control of Nonlinear Dynamical Systems. Methods and Applications. Springer, Berlin (2008)
Cholewa, A.: Representation of Sequences of Events for the Needs of Concluding in Technical Diagnostics, PhD thesis, Silesian University of Technology, Department of Fundamentals of Machine Design (2004) (in Polish)
Cholewa, W.: Statement networks in expert systems for condition monitoring. In: Korbicz, J., Patan, K., Kowal, M. (eds.) Fault Diagnosis and Fault Tolerant Control, Akademicka Oficyna Wydawnicza EXIT, Warsaw, pp. 231–238 (2007)
Cholewa, W.: Mechanical analogy of statement networks. International Journal of Applied Mathematics and Computer Science 18(4), 477–486 (2008), doi:10.2478/v10006-008-0042-7
Cholewa, W., Urbanek, G.: Evolutionary search for relevant diagnostic features. In: Diez, P., Wiberg, E. (eds.) Adaptive Modeling Simulation, CIMME—International Center for Numerical Methods in Engineering, Barcelona, pp. 207–210 (2005)
Cholewa, W., White, M.: Inverse modelling in rotordynamics for identification of unbalance distribution. Machine Vibration 2(3), 157–167 (1993)
Cholewa, W., Wojtusik, J.: Identification of multilayer diagnostic models. In: Diez, P., Wiberg, E. (eds.) Adaptive Modeling Simulation, CIMME—International Center for Numerical Methods in Engineering, Barcelona, pp. 211–214 (2005)
Cholewa, W., Korbicz, J., Kościelny, J., Chrzanowski, P., Patan, K., Rogala, T., Syfert, M., Witczak, M.: Methods of diagnostics. In: Korbicz, J., Kościelny, J. (eds.) Modeling, Diagnostics and Supervised Advanced Control of Processes. Implementation in the DiaSter System, pp. 221–315. Wydawnictwa Naukowo-Techniczne WNT, Warsaw (2009)

Cholewa, W., Kiciński, J.: Technical Diagnostics. Inverted Diagnostic Models. Silesian University of Technology Press, Gliwice (1997) (in Polish)
Chiang, L.H., Russell, E.L., Braatz, R.D.: Fault Detection and Diagnosis in Industrial Systems. Springer, Berlin (2001)
Christofides, P.D., El-Farra, N.: Control of Nonlinear and Hybrid Process Systems. Designs for Uncertainty, Constraints Time-Delays. Springer, Berlin (2005)
Cichosz, P.: Learning Systems. Wydawnictwa Naukowo-Techniczne WNT, Warsaw (2000) (in Polish)
Cristianini, N., Shawe-Taylor, J.: An Introduction to Support Vector Machines and other Kernel-based Learning Methods. Cambridge University Press, New York (2000)
Clarke, D., Mohtadi, C., Tuffs, P.: Generalised predictive control—Parts I II. Automatica 23(2), 137–160 (1987)
Cutler, C., Ramaker, B.: Dynamic Matrix Control—A computer control algorithm. In: Proceedings of the Joint Automatic Control Conference, JACC, San Francisco, USA (1980)
Cybenko, G.: Approximation by superpositions of a sigmoidal function. Mathematics of Control, Signals, and Systems 2(4), 303–314 (1989)
Dawkins, R.: The Selfish Gene. Oxford University Press, New York (1976)
de Kleer, J., Williams, B.: Diagnosing mutliple faults. Artificial Intelligence 32(32), 97–130 (1987)
Deng, J., Becerra, V.M., Stobart, R.: Input constraints hling in an MPC/feedback linearization scheme. International Journal of Applied Mathematics and Computer Science 19(2), 219–232 (2009), doi:10.2478/v10006-009-0018-2
Ding, S.: Model-based Fault Diagnosis Techniques. Springer, Berlin (2008)
Doyle III, F., Ogunnaike, B., Pearson, R.: Nonlinear model-based control using second-order Volterra models. Automatica 31(5), 697–714 (1995)
Dreyfus, G.: Neural Networks. Methodology and Applications. Springer, Berlin (2005)
Duda, J.: Mathematical Models, Structures and Algorithms of Computer Supervisory Control. AGH University of Science and Technology Press, Cracow (2003) (in Polish)
Duda, R.O., Hart, P.E., Stork, D.G.: Pattern Classification. John Wiley & Sons, New York (2001)
Duzinkiewicz, K.: Set membership estimation of parameters and variables in dynamic networks by recursive algorithms with a moving measurement window. International Journal of Applied Mathematics and Computer Science 16(2), 209–217 (2006)
Eder, H.: MBPC benefits and key success factors. In: Proceedings of 5th European Control Conference, Karlsruhe, Germany. paper F1033-6, CD-ROM (1999)
Elman, J.L.: Finding structure in time. Cognitive Science 14(2), 179–211 (1990)
Fayyad, U.M., Irani, K.B.: Multi-interval discretization of continuous-valued attributes for classification learning. In: Proceedings of the 13th International Joint Conference on Uncertainty in Artificial Intelligence (IJCAI 1993), Chambery, France, pp. 1022–1029 (1993)
Findeisen, W.: Multilevel Control Systems. Państwowe Wydawnictwa Naukowe PWN, Warsaw (1974) (in Polish)
Flake, G.W., Lawrence, S.: Efficient SVM regression training with SMO. Machine Learning 46(1-3), 271–290 (2002)
Fletcher, R.: Practical Methods of Optimization. J. Wiley & Sons, Chichester (1987)
Frasconi, P., Gori, M., Soda, G.: Local feedback multilayered networks. Neural Computation 4(1), 120–130 (1992)
Frank, P.: Fault diagnosis in dynamic systems via state estimations methods. A survey. In: Tsafestas, S., Singh, M., Schmidt, G. (eds.) System Fault Diagnostics, Reliability and Related Knowledge Based Approaches, vol. 2, pp. 35–98. D. Reidle Publishing Company, Dordrecht (1987)

Frank, P.M., Köppen-Seliger, B.: New developments using AI in fault diagnosis. Engineering Applications of Artificial Intelligence 10(1), 3–14 (1997)
Frank, P., Marcu, T.: Diagnosis strategies and system: Principle, fuzzy and neural approaches. In: Intelligent Systems and Interfaces. Kluwer Academic Publishers, Norwell (2000)
Fuller, R.: Introduction to Neuro-Fuzzy Systems. Springer, Berlin (2000)
Garcia, C., Morshedi, A.: Quadratic programming solution of dynamic matrix control (QDMC). Chemical Engineering Communication 46(1-3), 73–87 (1986)
Gertler, J.: Fault Detection Diagnosis in Engineering Systems. Marcel Dekker, Inc., New York (1998)
Goldberg, D.E.: Genetic Algorithms in Search, Optimization and Machine Learning. Addison-Wesley Longman Publishing Co., Inc., Boston (1989)
Goodwin, G., Graebe, S., Salgado, M.: Control System Design. Prentice Hall, Upper Saddle River (2001)
Gori, M., Bengio, Y., Mori, R.D.: BPS: A learning algorithm for capturing the dynamic nature of speech. In: Proceedings of the International Joint Conference on Neural Networks, Washington, DC, USA, vol. II, pp. 417–423 (1989)
Grega, W.: Modules and Algorithms of Digital Control in Centralised and Distributed Systems. AGH University of Science and Technology Press, Cracow (2004) (in Polish)
Grimble, M.: Industrial Control System Design. John Wiley & Sons, Springer, Berlin (2001)
Gunnarsson, S.: On some asymptotic uncertainty bounds in recursive least squares identification. IEEE Transactions on Automatic Control 38(11), 1685–1688 (1993)
Gupta, M., Jin, L., Homma, N.: Static and Dynamic Neural Networks: From Fundamentals to Advanced Theory. John Wiley & Sons, Hoboken (2003)
Gupta, M.M., Rao, D.H.: Dynamic neural units with application to the control of unknown nonlinear systems. Journal of Intelligent and Fuzzy Systems 1(1), 73–92 (1993)
Hajiyev, C., Caliskan, F.: Diagnosis Reconfiguration in Flight Control Systems. Kluwer Academic Publishers, London (2003)
Hall, M.: Correlation-based Feature Selection for Machine Learning, PhD thesis, Waikato University, Department of Computer Science, Hamilton (1998)
Hart, W.E.: Adaptive global optimization with local search, PhD thesis, University of California, San Diego (1994)
Haykin, S.: Neural Networks. A Comprehensive Foundation, 2nd edn. Prentice-Hall, Englewood Cliffs (1999)
Hägglund, T., Åström, K.J.: Industrial adaptive controllers based on frequency response techniques. Automatica 27(4), 599–609 (1991)
Henrion, M., Breese, J.S., Horvitz, E.J.: Decision analysis and expert systems. AI Magazine 12(4), 64–91 (1991)
Hertz, J., Krogh, A., Palmer, R.G.: Introduction to the Theory of Neural Computation. Addison-Wesley Publishing Company, Inc., Reading (1991)
Héctor, B.P., Fabián, G.N.: Reconfigurable Distributed Control. Springer, London (2005)
Hornik, K., Stinchcombe, M., White, H.: Multilayer feedforward networks are universal approximators. Neural Networks 2(5), 359–366 (1989)
Hrycej, T.: Neurocontrol: Towards an Industrial Control Methodology. John Wiley & Sons, Hoboken (1997)
Huang, B., Kadali, R.: Dynamic Modeling, Predictive Control and Performance Monitoring. A Data-driven Subspace Approach. Springer, Berlin (2008)
Hunt, K.J., Sbarbaro, D., Zbikowski, R., Gathrop, P.J.: Neural networks for control systems—A survey. Automatica 28(6), 1083–1112 (1992)
Isermann, R.: Identifikation Dynamischer Systeme. Springer, Berlin (1988)

Isermann, R.: Fault Diagnosis Systems. An Introduction from Fault Detection to Fault Tolerance. Springer, New York (2006)

Isermann, R., Ballé, P.: Trends in the application of model-based fault detection and diagnosis of technical process. Control Engineering Practice 5(5), 709–719 (1997)

Isermann, R., Münchhof, M.: Identification of Dynamical Systems. An Introduction with Applications. Springer, Berlin (2009)

Isham, V.: An introduction to spatial point processes and markov rom fields. International Statistical Review 49(1), 21–43 (1981)

Ivakhnenko, A.G.: Polynominal theory of complex systems. IEEE Transactions on Systems, Man Cybernetics 1(4), 44–58 (1971)

Jang, J.-S.R., Sun, C.-T., Mizutani, E.: Neuro-Fuzzy and Soft Computing. Prentice Hall, Upper Saddle River (1997)

Janiszowski, K.: Modification of Tustin approximation. IEEE Transactions on Automatic Control 38(8), 1313–1317 (1993)

Janiszowski, K.: Identification of the Parametric Models in Examples. Akademicka Oficyna Wydawnicza EXIT, Warsaw (2002) (in Polish)

Janiszowski, K.: Adaptation, modelling of drives dynamic and controller design in servomechanism pneumatic systems. IEE Proceedings—Control Theory + Applications 151(2), 234–245 (2004)

Jensen, V.J.: Bayesian Networks and Decision Graphs. Springer, New York (2002)

Jordan, M.I., Jacobs, R.A.: Supervised learning systems with excess degrees of freedom. In: Touretzky, D.S. (ed.) Advances in Neural Information Processing Systems II (Denver 1989), pp. 324–331. Morgan Kaufmann, San Mateo (1990)

Kacprzyk, J.: Multistage Fuzzy Control: A Model-Based Approach to Fuzzy Control and Decision Making. John Wiley & Sons, Hoboken (1996)

Kassman, D.E., Badgwell, T.A., Hawkins, R.B.: Robust steady-state target calculation for model predictive control. AIChE Journal 46(5), 1007–1024 (2000)

Keogh, E., Chu, S., Hart, D., Pazzani, M.: Segmenting time series: A survey and novel approach. In: Last, M., Kel, A., Bunke, H. (eds.) Data Mining in Time Series Databases, pp. 1–22. World Scientific Publishing Company, Singapore (2003)

Kiciński, J.: Rotor Dynamics. Institute of Fluid-Fluid Machinery Publishers, Gdańsk (2006)

Kira, K., Rendell, L.: A practical approach to feature selection. In: Proceedings of the 9th International Workshop on Machine Learning (ML 1992), Aberdeen, UK, pp. 249–256 (1992)

Kohavi, R.: Wrappers for Permormance Enhancement and Oblivious Decision Graphs, PhD thesis, Stanford University, Department of Computer Science, Stanford (1995)

Korbicz, J.: Robust fault detection using analytical and soft computing methods. Bulletin of the Polish Academy of Sciences: Technical Sciences 54(1), 75–88 (2006)

Korbicz, J., Kościelny, J. (eds.): Modelling, Diagnostics and Process Control. Implementation in the DiaSter System. Wydawnictwa Naukowo-Techniczne, WNT, Warsaw (2009) (in Polish)

Korbicz, J., Kościelny, J., Kowalczuk, Z., Cholewa, W. (eds.): Fault Diagnosis. Models, Artificial Intelligence, Applications. Springer, Heidelberg (2004)

Korbicz, J., Mrugalski, M.: Confidence estimation of GMDH neural networks and its application in fault detection systems. International Journal of Systems Science 39(8), 783–800 (2008)

Korbicz, J., Witczak, M.: On fault detection under soft computing model uncertainty. In: Proceedings of the 17th World Congress of the International Federation of Automatic Control—IFAC 2008, Seoul, Korea, pp. 7901–7906 (2008) (CD-ROM)

Kościelny, J.: Fault isolation in industrial processes by dynamic table of states method. Automatica 31(5), 747–753 (1995)

Kościelny, J.: Diagnostics of processes in decentralized structures. Archives of Control Sciences 7(3-4), 181–202 (1998)

Kościelny, J.: Diagnosis of Automated Industrial Processes. Akademicka Oficyna Wydawnicza EXIT, Warsaw (2001) (in Polish)

Kościelny, J., Bartyś, M.: Multiple fault isolation in diagnostic of industrial processes. In: Proceedings of the European Control Conference, Cambridge, UK, vol. 4, pp. 1–6 (2003)

Kościelny, J., Bartyś, M., Rzepiejewski, P., Sá da Costa, J.: Actuator fault distinguishability study for the DAMADICS benchmark problem. Control Engineering Practice 14(6), 645–652 (2006)

Kościelny, J., Bartyś, M., Syfert, M.: The practical problems of fault isolation in large scale industrial systems. In: Proceedings of the 6th IFAC Symposium on Fault Detection, Supervision and Safety of Technical Processes, Beijing, China (2006) CD-ROM

Kościelny, J., Bartyś, M., Syfert, M.: Diagnostics of industrial processes in decentralised structures with application of fuzzy logic. In: Proceedings of the 17th IFAC World Congress, Seoul, Korea, pp. 6944–6949 (2008)

Kościelny, J.M., Syfert, M.: Fuzzy diagnostic reasoning that takes into account the uncertainty of the faults-symptoms relation. International Journal of Applied Mathematics and Computer Science 16(1), 27–35 (2006)

Kościelny, J.M., Syfert, M., Dziembowski, B.: Fault diagnosis with use of the knowledge about symptoms delays interval. In: Proceedings of the 17th IFAC World Congress, Seoul, Korea, pp. 7350–7355 (2008a) (CD-ROM)

Kościelny, J., Sędziak, D., Zakroczymski, K.: Fuzzy logic fault isolation in large scale systems. International Journal of Applied Mathematics and Computer Science 9(3), 637–652 (1999)

Kościelny, J., Syfert, M.: The issue of symptom based diagnostic reasoning. In: Recent Advances in Mechatronics, pp. 167–171. Springer, Berlin (2007)

Kościelny, J., Syfert, M., Dziembowski, B.: The issue of symptoms arising delay during diagnostic reasoning. In: Proceedings of the 5th Workshop on Advanced Control and Diagnosis, Grenoble, France (2007)

Krasnogor, N., Smith, J.: A tutorial for competent memetic algorithm: Model, taxonomy, and design issues. IEEE Transactions on Evolutionary Computation 9(5), 474–488 (2005)

Kraus, T.W., Myron, T.J.: Self-tuning PID controller using pattern recognition approach. Control Engineering 31(6), 106–111 (1984)

Kreisselmeier, G., Steinhauser, R.: Systematische auslegung von reglern durch optimierung eines vektoriellen gütektriterium. Automatiesirungstechnik 43(3), 1 (1979)

Lau, I.D., Zito, G.: Digital Control Systems. Design, Identification and Implementation. Springer, Berlin (2006)

Lauritzen, S.L.: Lectures on Contingency Tables. University of Aalborg Press, Aalborg (1982)

Leonhardt, S., Ayoubi, M.: Methods of fault diagnosis. Control Engineering Practice 5(5), 683–692 (1997)

Ligęza, A.: Logical Foundations for Rule-Based Systems. AGH University of Science and Technology Press, Cracow (2005)

Ligęza, A., Kościelny, J.: A new approach to multiple fault diagnosis. Combination of diagnostic matrices, graphs and rule based models. In: Korbicz, J., Patan, K., Kowal, M. (eds.) Fault Diagnosis and Fault Tolerant Control, pp. 219–230. Akademicka Oficyna Wydawnicza EXIT, Warsaw (2007)

Ljung, L.: System Identification—Theory for the User. Prentice Hall, Englewood Cliffs (1999)

Lyons, R.: Understing Digital Signal Processing. Addison Wesley Longman, Inc., Reading (1997)

Ławryńczuk, M.: A family of model predictive control algorithms with artificial neural networks. International Journal of Applied Mathematics and Computer Science 17(2), 217–232 (2007), doi:10.2478/v10006-007-0020-5

Ławryńczuk, M.: Modelling nonlinear predictive control of a yeast fermentation biochemical reactor using neural networks. Chemical Engineering Journal 145(2), 290–307 (2008)

Ławryńczuk, M., Marusak, P., Tatjewski, P.: Multilayer integrated structures for predictive control and economic optimisation. In: Proceedings of the 11th Symposium on Large Scale Systems Theory Applications, Gdańsk, Pol (2007) CD-ROM

Ławryńczuk, M., Marusak, P., Tatjewski, P.: Cooperation of model predictive control with steady-state economic optimisation. Control and Cybernetics 37(1), 133–158 (2008)

Ławryńczuk, M., Tatjewski, P.: Efficient predictive control integrated with economic optimisation based on neural models. In: Rutkowski, L., Tadeusiewicz, R., Zadeh, L.A., Zurada, J.M. (eds.) ICAISC 2008. LNCS (LNAI), vol. 5097, pp. 111–122. Springer, Heidelberg (2008)

Maciejowski, J.: Predictive Control. Prentice Hall, Upper Saddle River (2002)

Mahmoud, M., Jiang, J., Zhang, Y.M.: Active Fault Tolerant Control Systems: Stochastic Analysis and Synthesis. Lecture Notes in Control Information Sciences, vol. 287. Springer, Berlin (2003)

Marusak, P.: Efficient fuzzy predictive economic set-point optimizer. In: Rutkowski, L., Tadeusiewicz, R., Zadeh, L.A., Zurada, J.M. (eds.) ICAISC 2008. LNCS (LNAI), vol. 5097, pp. 273–284. Springer, Heidelberg (2008)

Marusak, P.: Advantages of an easy to design fuzzy predictive algorithm in control systems of nonlinear chemical reactors. Applied Soft Computing 9(3), 1111–1125 (2009)

Marusak, P., Tatjewski, P.: Effective dual-mode fuzzy DMC algorithms with on-line quadratic optimization and guaranteed stability. International Journal of Applied Mathematics and Computer Science 19(1), 127–141 (2009), doi:10.2478/v10006-009-0012-8

Matlab: System documentation, `http://www.mathworks.com`

Mayne, D., Rawlings, J., Rao, C., Scokaert, P.: Constrained model predictive control: stability and optimality. Automatica 36(6), 789–814 (2000)

Michalewicz, Z.: Genetic Algorithms + Data Structures = Evolution Programs, 3rd edn. Springer, London (1996)

Michalewicz, Z., Fogel, D.B.: How to Solve It: Modern Heuristics. Springer, Heidelberg (2004)

Milanese, M.: Set membership identification of nonlinear systems. Automatica 40(6), 957–975 (2004)

Milanese, M., Norton, J., Piet-Lahanier, H., Walter, E.: Bounding Approaches to System Identification. Plenum Press, New York (1996)

Moczulski, W.: Technical Diagnostics. Knowledge Gaining Methods. Silesian University of Technology Press, Gliwice (2002a)

Moczulski, W.: Problems of declarative and procedural knowledge acquisition for machinery diagnostics. Computer Assisted Mechanics and Engineering Sciences 9(1), 71–86 (2002b)

Moczulski, W., Żytkow, J.M.: Automated search for knowledge on machinery diagnostics. In: Kłopotek, M., Michalewicz, M., Raś, Z. (eds.) Proceedings of the Conference on Intelligent Information Systems, IIS 1997, pp. 194–203. Polish Academy of Sciences, Warsaw (1997)

Moscato, P.: Memetic algorithms: A short introduction. In: Corne, D., Dorigo, M., Glover, F. (eds.) New Ideas in Optimization, pp. 219–234. McGraw-Hill, Maidenhead (1999)
Morari, M., Lee, J.: Model predictive control: Past, present and future. Computer and Chemical Engineering 23(4/5), 667–682 (1999)
Mozer, M.C.: A focused backpropagation algorithm for temporal pattern recognition. Complex Systems 3(4), 349–381 (1989)
Mozer, M.C.: Neural net architectures for temporal sequence processing. In: Weigend, A.S., Gershenfeld, N.A. (eds.) Time Series Predictions: Forecasting the Future and Underst-ing the Past, pp. 243–264. Addison-Wesley Publishing Company, Inc., Reading (1994)
Mulawka, J.: Expert's Systems. Wydawnictwa Naukowo-Techniczne WNT, Warsaw (1996) (in Polish)
Mrugalski, M.: Neural Network Based Modelling of Non-linear Systems in Fault Detection Schemes, PhD thesis, University of Zielona Góra, Faculty of Electrical Engineering, Computer Science Telecommunications (2004) (in Polish)
Mrugalski, M., Korbicz, J.: Least mean square vs. outer bounding ellipsoid algorithm in confidence estimation of the GMDH neural networks. In: Beliczynski, B., Dzielinski, A., Iwanowski, M., Ribeiro, B. (eds.) ICANNGA 2007. LNCS, vol. 4432, pp. 19–26. Springer, Heidelberg (2007)
Mrugalski, M., Witczak, M., Korbicz, J.: Confidence estimation of the multi-layer perceptron and its application in fault detection systems. Engineering Applications of Artificial Intelligence 21(8), 895–906 (2008)
Narendra, K.S., Parthasarathy, K.: Identification and control of dynamical systems using neural networks. IEEE Transactions on Neural Networks 1(1), 12–18 (1990)
Natke, H., Cempel, C.: Model-Aided Diagnosis of Mechanical Systems. Springer, Berlin (1997)
Nelles, O.: Nonlinear System Identification. From Classical Approaches to Neural Network and Fuzzy Models. Springer, Berlin (2001)
Niederliński, A.: Computer Systems of the Industrial Control. Wydawnictwa Naukowo-Techniczne WNT, Warsaw (1984) (in Polish)
Norgard, M., Ravn, O., Poulsen, N., Hansen, L.: Networks for Modelling Control of Dynamic Systems. Springer, London (2000)
Olszewski, M.: Basics of Servopneumatics. VDI Verlag, Düsseldorf (2007)
Osowski, S.: Neural Networks for Processing Information. Warsaw University of Technology Press, Warsaw (2000) (in Polish)
O'Dwyer, A.: Hbook of PI and PID Controller Tuning Rules. Imperial College Press, London (2003)
Palm, R., Driankov, D.: Model Based Fuzzy Control. Springer, Heidelberg (1997)
Parlos, A.G., Chong, K.T., Atiya, A.F.: Application of the recurrent multilayer perceptron in modelling complex process dynamics. IEEE Transactions on Neural Networks 5(2), 255–266 (1994)
Patan, K.: Fault detection system for the sugar evaporator based on AI techniques. In: Proceedings of the 6th IEEE International Conference on Methods Models in Automation Robotics, MMAR 2000, Międzyzdroje, Poland, pp. 807–812 (2000)
Patan, K.: Training of the dynamic neural networks via constrained optimization. In: Proceedings of the IEEE International Joint Conference on Neural Networks, IJCNN 2004, Budapest, Hungary (2004) CD-ROM
Patan, K.: Robust fault diagnosis in catalytic cracking converter using artificial neural networks. In: Proceedings of the 16th IFAC World Congress, Prague, Czech Republic (2005) CD-ROM

Patan, K.: Robust faul diagnosis in a DC motor by means of artificial neural networks and model error modelling. In: Korbicz, J., Patan, K., Kowal, M. (eds.) Fault Diagnosis and Fault Tolerant Control, pp. 337–346. Akademicka Oficyna Wydawnicza EXIT, Warsaw (2007a)

Patan, K.: Stability analysis and the stabilization of a class of discrete-time dynamic neural networks. IEEE Transactions on Neural Networks 18(3), 660–673 (2007b)

Patan, K.: Aproximation of state-space trajectories by locally recurrent globally feed-forward neural networks. Neural Networks 21(1), 59–64 (2008a)

Patan, K.: Artificial Neural Networks for the Modelling and Fault Diagnosis of Technical Processes. Lecture Notes in Control Information Sciences, vol. 377. Springer, Berlin (2008b)

Patan, K.: Local stability conditions for discrete-time cascade locally recurrent neural networks. International Journal of Applied Mathematics and Computer Science 20(1), 23–34 (2010), doi:10.2478/v10006-010-0002-x

Patan, K., Korbicz, J.: Fault detection in catalytic cracking converter by means of probability density approximation. Engineering Applications of Artificial Intelligence 20(7), 912–923 (2007)

Patan, K., Korbicz, J., Głowacki, G.: DC motor fault diagnosis by means of artificial neural networks. In: Proceedings of the 4th International Conference on Informatics in Control, Automation and Robotics, ICINCO, Angers, France (2007) CD-ROM

Patan, K., Parisini, T.: Stochastic learning methods for dynamic neural networks: Simulated and real-data comparisons. In: Proceedings of the American Control Conference, ACC 2002, Anchorage, USA, pp. 2577–2582 (2002)

Patan, K., Parisini, T.: Identification of neural dynamic models for fault detection isolation: The case of a real sugar evaporation process. Journal of Process Control 15(1), 67–79 (2005)

Patan, K., Witczak, M., Korbicz, J.: Towards robustness in neural network based fault diagnosis. International Journal of Applied Mathematics and Computer Science 18(4), 443–454 (2008), doi:10.2478/v10006-008-0039-2

Patton, R., Lopez-Toribio, C., Uppal, F.: Artificial intelligence approaches to fault diagnosis for dynamic systems. International Journal of Applied Mathematics and Computer Science 9(3), 471–518 (1999)

Patton, R., Frank, P.: Issues of Fault Diagnosis for Dynamic Systems. Springer, Berlin (2000)

Pawlak, Z.: Rough Sets. Theoretical Aspects of Reasoning about Data. Kluwer Academic Publishers, Norwell (1991)

Pearl, J.: Probabilistic Reasoning in Intelligent Systems. Networks of Plausible Inference. Morgan Kaufmann, San Mateo (1988)

Pedrycz, W.: Fuzzy Control and Fuzzy Systems. Research Studies Press, Taunton (1993)

Pedrycz, W. (ed.): Granular Computing: An Emerging Paradigm. Physica-Verlag, Heidelberg (2001)

Pham, D.T., Liu, X.: Training of Elman networks and dynamic system modelling. International Journal of Systems Science 27(2), 221–226 (1996)

Piegat, A.: Fuzzy Modeling Control (Studies in Fuzziness Soft Computing). Physica-Verlag, Heidelberg (2001)

Platt, J.C.: Fast training of support vector machines using sequential minimal optimization. In: Schölkopf, B., Burges, C., Smola, A. (eds.) Advances in Kernel Methods—Support Vector Learning. MIT Press, Cambridge (1998)

Poddar, P., Unnikrishnan, K.P.: Memory neuron networks: A prolegomenon, Technical Report GMR-7493, General Motors Research Laboratories (1991)

Press, W.H., Teukolsky, S.A., Vetterling, W.T., Flannery, B.P.: Numerical Recipes in C. Cambridge University Press, Cambridge (1992)

Puig, V., Stancu, A., Escobet, T., Nejjari, F., Quevedo, J., Patton, R.J.: Passive robust fault detection using interval observers: Application to the DAMADICS benchmark problem. Control Engineering Practice 14, 621–633 (2006)

Puig, V., Witczak, M., Nejjari, F., Quevedo, J., Korbicz, J.: A GMDH neural network-based approach to passive robust fault detection using a constraint satisfaction backward test. Engineering Applications of Artificial Intelligence 20(7), 886–897 (2007)

Qin, S., Badgwell, T.: A survey of industrial model predictive control technology. Control Engineering Practice 11(7), 733–764 (2003)

Quinlan, J.R.: Induction of decision trees. Machine Learning 1(1), 81–106 (1986)

Quinn, S.L., Harris, T.J., Bacon, D.W.: Accounting for uncertainty in control-relevant statistics. Journal of Process Control 15(6), 675–690 (2005)

Rao, C.V., Rawlings, J.B.: Steady states and constraints in model predictive control. AIChE Journal 45(6), 1266–1278 (1999)

Reinelt, W., Garulli, A., Ljung, L.: Comparing different approaches to model error modeling in robust identification. Automatica 38(5), 787–803 (2002)

Rogala, T., Cholewa, W., Chrzanowski, P.: An example of an application of the diagnostic belief network based model. In: Zółtowski, B. (ed.) Elements of Machinery and Vehicle Diagnostics, pp. 249–256. Institute for Sustainable Technologies—Polish National Research Institute, Radom (2009)

Rosenwasser, E.N., Lampe, B.P.: Multivariable Computer-controlled Systems. A Transfer Function Approach. Springer, Berlin (2006)

Rossiter, J.: Model-Based Predictive Control. CRC Press, Boca Raton (2003)

Rutkowska, D., Piliñski, M., Rutkowski, L.: Neural Networks, Genetic Algorithms and Fuzzy Systems. Wydawnictwo Naukowe PWN, Warsaw (1997) (in Polish)

Rutkowski, L.: Flexible Neuro-Fuzzy Systems. Structures, Learning and Performance Evaluation. Kluwer Academic Publishers, Norwell (2004a)

Rutkowski, L.: New Soft Computing Techniques for System Modelling, Pattern Classification Image Processing. Springer, Berlin (2004b)

Sauter, D., Dubois, G., Levrat, E., Brémont, J.: Fault diagnosis in systems using fuzzy logic. In: Proceedings of the 1st European Congress on Fuzzy and Intelligent Technologies, EUFIT 1993, Aachen, Germany, pp. 781–788 (1993)

Schneider, H.: Implementation of a fuzzy concept for supervision and fault detection of robots. In: Proceedings of the 1st European Congress on Fuzzy and Intelligent Technologies, EUFIT 1993, Aachen, Germany, pp. 775–780 (1993)

Schölkopf, B., Smola, A.J.: Learning with Kernels: Support Vector Machines, Regularization, Optimization, and Beyond. MIT Press, Cambridge (2001)

Sędziak, D.: Method of Fault Localisation in Industrial Processes, PhD thesis, Faculty of Mechatronics, Warsaw University of Technology, Warsaw (2002) (in Polish)

Simani, S., Fantuzzi, C., Patton, R.J.: Model-based Fault Diagnosis in Dynamic Systems using Identification Techniques. Springer, Berlin (2003)

Shearer, J.: Fluid Power Control. John Wiley & Sons, New York (1960)

SimulationX: System documentation, `http://www.simulationx.com`

Sobhani-Tehrani, E., Khorasani, K.: Fault Diagnosis of Nonlinear Systems Using a Hybrid Approach. Springer, Berlin (2009)

Sobczyk, M.: Statistics. Wydawnictwo Naukowe PWN, Warsaw (2000) (in Polish)

Sontag, E.: Feedback stabilization using two-hidden-layer nets. IEEE Transactions on Neural Networks 3(6), 981–990 (1992)

Soderstrom, T., Stoica, P.: System Identification. Prentice Hall, Hemel Hampstead (1988)

Spriet, J., Vansteenkiste, G.: Computer Aided Modelling and Simulation. Academic Press, New York (1983)
Stornetta, W.S., Hogg, T., Hubermann, B.A.: A dynamic approach to temporal pattern processing. In: Anderson, D.Z. (ed.) Neural Information Processing Systems, pp. 750–759. American Institute of Physics, New York (1988)
Syfert, M., Kościelny, J.: Diagnostic reasoning based on symptom forming sequence. In: Proceedings of the 7th IFAC Symposium on Fault Detection, Supervision and Safety of Technical Processes, Barcelona, Spain (CD-ROM), pp. 89–94 (2009)
Szulim, R.: A Method of Knowledge Mining to Aid the Control of Complex Industrial Processes, PhD thesis, Uniwersity of Zielona Góra, Zielona Góra (2004) (in Polish)
Świątek, J.: Selected Problems of the Identification of Static Complex Systems. Warsaw University of Technology Press, Warsaw (2009) (in Polish)
Świder, Z., Trybus, L.: An alternative algorithm to EXACT self-tuning. Archives of Control Sciences 14(3), 287–298 (2004)
Świder, Z., Trybus, L.: Self-tuning adaptation in industrial controllers. Przegląd Elektrotechniczny 2009(9), 382–387 (2009) (in Polish)
Tadeusiewicz, R.: Neural Networks. Akademicka Oficyna Wydawnicza EXIT, Warsaw (1993) (in Polish)
Takahashi, Y., Rabins, M., Ausler, D.: Control and Dynamic Systems. Addison-Wesley, Reading (1972)
Talebi, H., Abdollahi, F., Patel, R., Khorasani, K.: Neural Network-Based State Estimation of Nonlinear Systems. Application to Fault Detection and Isolation. Springer, Berlin (2010)
Tatjewski, P.: Advanced Control of Industrial Processes Structures and Algorithms. Akademicka Oficyna Wydawnicza EXIT, Warszawa (2002) (in Polish)
Tatjewski, P.: Advanced Control of Industrial Processes. Springer, London (2007)
Tatjewski, P.: Advanced control and on-line process optimization in multilayer structures. Annual Reviews in Control 32(1), 71–85 (2008)
Tatjewski, P., Ławryńczuk, M.: Soft computing in model-based predictive control. International Journal of Applied Mathematics and Computer Science 16(1), 7–26 (2006)
Tenne, Y.: Computational Intelligence in Expensive Optimization Problems. Springer, Berlin (2010)
Tomasik, P.: Methods of Identification of Models of Slowly Changing Processes for Technical Diagnostics, PhD thesis, Silesian University of Technology, Department of Fundamentals of Machine Design (2006) (in Polish)
Trybus, L.: A set of PID tuning rules. Archives of Control Sciences 14(1), 5–17 (2005)
Tsoi, A.C., Back, A.D.: Locally recurrent globally feedforward networks: A critical review of architectures. IEEE Transactions on Neural Networks 5(2), 229–239 (1994)
Ying, H.: Fuzzy Control Modeling: Analytical Foundations and Applications. John Wiley & Sons, Hoboken (2000)
Vapnik, V.: The Nature of Statistical Learning Theory. Springer, New York (1995)
Wachla, D.: Identification of dynamic diagnostic models using methods of knowledge discovery in databases, PhD thesis, Silesian University of Technology, Department of Fundamentals of Machine Design (2006) (in Polish)
Walter, E., Pronzato, L.: Identification of Parametric Models from Experimental Data. Springer, London (1997)
Wang, H., Wang, D., Yang, S.: A memetic algorithm with adaptive hill climbing strategy for dynamic optimization problems. Soft Computing 13(8-9), 763–780 (2009)
Watanabe, K., Hou, L.: An optimal neural network for diagnosing multiple faults in chemical processes. In: Proceedings of the International Conference on Power Electronics Motion Control, San Diego, USA, vol. 2, pp. 1068–1073 (1992)

Williams, R.J.: Adaptive state representation and estimation using recurrent connectionist networks. In: Neural Networks for Control. MIT Press, London (1990)

Williams, R.J., Zipser, D.: Experimental analysis of the real-time recurrent learning algorithm. Connection Science 1(1), 87–111 (1989a)

Williams, R.J., Zipser, D.: A learning algorithm for continually running fully recurrent neural networks. Neural Computation 1(2), 270–289 (1989b)

Witczak, M.: Modelling and Estimation Strategies for Fault Diagnosis of Non-linear Systems. Lecture Notes in Control Information Sciences, vol. 354. Springer, Berlin (2007)

Witczak, M., Korbicz, J., Mrugalski, M., Patton, R.J.: A GMDH neural network-based approach to robust fault diagnosis: Application to the DAMADICS benchmark problem. Control Engineering Practice 14(6), 671–683 (2006)

Wnuk, P.: The use of evolutionary optimization in fuzzy tsk model identification. In: Proceedings of the IFAC Symposium on Fault Detection, Supervision and Safety of Technical Processes, Bejing, China (2006) CD-ROM

Wnuk, P., Kościelny, J., Leszczyński, M.: Diagnostic system decomposition with genetic optimization. In: Proceedings of the Symposium on Methods of Artificial Intelligence, AIMETH 2007, Gliwice, Poland, pp. 247–258 (2007)

Wright, S.: The method of path coefficients. Annals of Mathematical Statistics 5(3), 161–215 (1934)

Yager, R., Filev, D.: Essentials of Fuzzy Modelling Control. Wiley Interscience, New York (1994)

Zamarreno, J.M., Vega, P.: State space neural network. Properties application. Neural Networks 11(6), 1099–1112 (1998)

Zecevic, A., Siljak, D.: Structural Constraints and Uncertainty. Springer, Berlin (2010)

Żytkow, J.M., Zembowicz, R.: Database exploration in search of regularities. Journal of Intelligent Information Systems 2(1), 39–81 (1993)

Żytkow, J.M., Zembowicz, R.: Mining patterns at each scale in massive data. In: Raś, Z.W., Michalewicz, M. (eds.) ISMIS 1996. LNCS (LNAI), vol. 1079, pp. 139–148. Springer, Heidelberg (1996)

Index

GPSR Compliance

The European Union's (EU) General Product Safety Regulation (GPSR) is a set of rules that requires consumer products to be safe and our obligations to ensure this.

If you have any concerns about our products, you can contact us on ProductSafety@springernature.com

In case Publisher is established outside the EU, the EU authorized representative is:

Springer Nature Customer Service Center GmbH
Europaplatz 3
69115 Heidelberg, Germany

Batch number: 10755284

Printed by Printforce, the Netherlands